P9-83 $24.95

PRECALCULUS

PRECALCULUS
Second Edition

S.L. Salas
C.G. Salas

John Wiley & Sons

New York · Chichester · Brisbane · Toronto · Singapore

This book was printed and bound by Halliday Lithograph.
It was set in Times Roman by York Graphic Services.
The designer was Joseph Gillians.
The drawings were designed and executed by John Balbalis with the assistance
of the Wiley Illustration Department.
Lilly Kaufman supervised production.

Library of Congress Cataloging in Publication Data
Salas, Saturnino L
   Precalculus.

   Includes index.
    1. Mathematics—1961–   I. Salas, Charles G.,
joint author. II. Title.
QA39.2.S23 1979   512'.1   78–23236
ISBN 0-471-03124-0

Printed in the United States of America

10 9 8 7 6 5 4 3

To
Maria Gevaert Salas

PREFACE

Precalculus is a college text devoted entirely to the mathematical prerequisites for calculus.

It's our view that elaborate lengthy precalculus courses easily bog down in unnecessary details and tend to discourage students by delaying the calculus too long. In this book we've stayed with the material we consider most useful for calculus and tried to avoid all unnecessary complications.

Students with three or more years of high school mathematics can use this text to make better sense out of what they've already studied and focus on what's necessary for calculus. For students who have had less background in mathematics but are anxious to get to calculus, this book can provide all they need to know.

While trying to maintain the tone and the pace of the first edition, we've made many changes that render this edition more flexible in accommodating different programs. The material on lines, polynomials, and the trigonometric functions has been expanded, and new topics have been introduced, including exponential functions, the inverse trigonometric functions, and Cramer's rule. At the same time we've reduced the basic text by making some of the material optional.

There are other changes. The sections on absolute value, inequalities, and the conic sections have been rewritten. Selected exercises have been added throughout, and at the end of each of the first six chapters there is now a set of additional exercises with all answers given at the back of the book.

Extra care has been taken to avoid inaccuracies and to make sure that even the most elementary concepts are explained. All figures have been redrawn, many partly in color.

By moving quickly over the first three chapters and by omitting all optional material and much of the last chapter, an instructor can cover all the material absolutely necessary for calculus in a very short course. In its entirety *Precalculus* has more than enough material for a full course.

S.L. Salas and C.G. Salas
Haddam, Connecticut

ACKNOWLEDGMENTS

We are particularly grateful to Wiley's Mathematics Editor, Gary W. Ostedt who always seasoned his keen judgement with doses of warmth and good humor. Thanks also go to Lilly Kaufman, who supervised production.

Many of the changes we made were inspired by our reviewers. We are indebted to Professors Melvin F. Janowitz (University of Massachusetts), H.D. Perry (Texas A&M University), Arthur Riehl (University of Louisville), Doris S. Stockton (University of Massachusetts), and Ed R. Wheeler (Northern Kentucky State College).

Finally, special thanks go to Karen B. McWilliams (a student who analyzed the entire manuscript from a learner's point of view), to Danielle Vergnaud (who proofread all galleys and pages), and to Dagmar Noll (who checked the answers to the exercises).

S.L.S. and C.G.S.

CONTENTS

THE GREEK ALPHABET

A	α	alpha
B	β	beta
Γ	γ	gamma
Δ	δ	delta
E	ϵ	epsilon
Z	ζ	zeta
H	η	eta
Θ	θ	theta
I	ι	iota
K	κ	kappa
Λ	λ	lambda
M	μ	mu
N	ν	nu
Ξ	ξ	xi
O	o	omicron
Π	π	pi
P	ρ	rho
Σ	σ	sigma
T	τ	tau
Υ	υ	upsilon
Φ	ϕ	phi
X	χ	chi
Ψ	ψ	psi
Ω	ω	omega

NUMBERS AND ALGEBRAIC EXPRESSIONS

In the first two chapters we focus on those parts of high school algebra that are necessary for calculus.

1.1 Classification of Numbers

We begin with the set of *natural numbers*: $\{1, 2, 3, \cdots\}$. With these numbers we can count:

$$\begin{array}{ccccccc} \bullet & \bullet & \bullet & \bullet & \bullet & \bullet & \bullet \quad \cdots \\ 1 & 2 & 3 & 4 & 5 & 6 & 7 \end{array}$$

add and multiply:

$$2 + 3 = 5, \qquad 2 \cdot 3 = 6$$

but we cannot always subtract:

$$5 - 8 \quad \text{is not a natural number}$$

and we cannot always divide:

$$5 \div 8 \quad \text{is not a natural number.}$$

To be able to subtract arbitrarily, we enlarge the number system to include the set of *integers*: $\{0, \pm 1, \pm 2, \pm 3, \cdots\}$. Now, for instance,

$$5 - 8 = -3.$$

If we wish to divide, we must allow for fractions and use the set of *rational numbers*; a rational number, you will recall, is a number that can be written in the form

$$p/q \quad \text{with } p \text{ and } q \text{ integers, } q \neq 0.$$

The rational numbers include the integers ($n = n/1$) and the fractions, both *positive* and *negative.* They also include the *mixed numbers,* since these can be written as fractions:

$$2\tfrac{1}{3} = \tfrac{7}{3}, \qquad 4\tfrac{2}{5} = \tfrac{22}{5}, \text{ etc.}$$

With the rational numbers we can count, we can add and subtract:

$$\tfrac{5}{8} + \tfrac{1}{4} = \tfrac{5}{8} + \tfrac{2}{8} = \tfrac{7}{8}, \qquad \tfrac{5}{8} - \tfrac{1}{4} = \tfrac{5}{8} - \tfrac{2}{8} = \tfrac{3}{8}$$

multiply and divide:

$$\tfrac{5}{8} \cdot \tfrac{1}{4} = \tfrac{5}{32}, \qquad \tfrac{5}{8} \div \tfrac{1}{4} = \tfrac{5}{8} \cdot \tfrac{4}{1} = \tfrac{20}{8}$$

but we cannot always take roots and we cannot measure all distances. Consider, for example, a unit square. (Figure 1.1.1). By the Pythagorean theorem,† the distance between the opposite vertices of this square must be $\sqrt{2}$, the square root of 2. As we will show in Section 2.1, this is not a rational number. With only rational numbers at our disposal we could not measure this distance. Nor could we find $\sqrt{3}$, $\sqrt{5}$, $\sqrt[3]{13}$ (the cube root of 13), or $\sqrt[5]{5}$ (the fifth root of 5). The circle would also give us trouble.

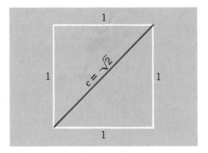

Figure 1.1.1

---

†The Pythagorean theorem states that in a right triangle, $a^2 + b^2 = c^2$, where a and b are the legs and c is the hypotenuse. (Figure 1.1.2) In the case of the unit square, $1^2 + 1^2 = c^2$, so that $c^2 = 2$ and $c = \sqrt{2}$.

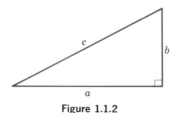

Figure 1.1.2

You are familiar with the formula

$$\text{circumference} = 2\pi r.$$

According to this formula, the distance around a unit circle ($r = 1$) would be 2π, but π is not rational and neither is 2π.† With only rational numbers, we could not even measure the distance around a wheel of radius 1.

    To be able to take roots of all positive numbers and to be able to measure all distances, we must enlarge the number system once again. We must accept not only the *rational* numbers but also the *irrational* numbers. Together these comprise the set of *real numbers*.‡

Real Numbers and the Number Line

You can visualize the real numbers on a horizontal line, as in Figure 1.1.3. Choose a point for 0 and a unit of length. Then display the number r that many units to the right of 0 if r is positive, and $-r$ units to the left of 0 if r is negative. The resulting pattern is the familiar *number line,* also called the *coordinate line.* Each point on it corresponds to a unique real number, and each real number corresponds to a unique point.

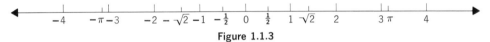

Figure 1.1.3

Decimal Expansions

Every decimal represents a real number:

$$n.a_1a_2a_3 \cdots = n + \frac{a_1}{10} + \frac{a_2}{100} + \frac{a_3}{1000} + \cdots$$

and every real number can be written as a decimal. Terminating decimals such as

$$0.5, \qquad 1.62, \qquad 5.0025$$

and repeating decimals such as

$$0.333 \cdots = 0.\overline{3}, \qquad 4.2121 \cdots = 4.\overline{21}, \qquad 0.1735735 \cdots = 0.1\overline{735}$$

represent rational numbers.¶ The nonrepeating infinite decimals represent the irrational numbers.

    To find a decimal expansion for a rational number p/q we divide the *denominator q* into the *numerator p.*

---

†You are asked to show this in Exercise 9.
‡To take even roots of negative numbers we must expand the number system once more, this time to include the *complex numbers.* We do this in Chapter 7.
¶We use a bar to indicate the repeating block.

Problem. Express $\frac{13}{5}$ as a decimal.

SOLUTION. We divide 5 into 13:

$$
\begin{array}{r}
2.6 \\
5\overline{\smash{)}13.0} \\
\underline{10} \\
30 \\
\underline{30} \\
0
\end{array}
$$

$\frac{13}{5} = 2.6$. ☐

Problem. Express $\frac{5}{11}$ as a decimal.

SOLUTION. We divide 11 into 5:

$$
\begin{array}{r}
0.45 \\
11\overline{\smash{)}5.00} \\
\underline{44} \\
60 \\
\underline{55} \\
50.
\end{array}
$$

From this point on the cycle repeats itself. (Once again we find ourselves dividing 11 into 50.) Thus

$$\tfrac{5}{11} = 0.454545\cdots = 0.\overline{45}. \quad \square$$

Problem. Express $0.555\cdots = 0.\overline{5}$ as the quotient of two integers.

SOLUTION

$$
\begin{array}{r}
10(0.\overline{5}) = 5.\overline{5} \\
-\quad 0.\overline{5} = 0.\overline{5} \\
\hline
9(0.\overline{5}) = 5
\end{array}
$$

Therefore $0.\overline{5} = \frac{5}{9}$. ☐

Problem. Express $0.141414\cdots = 0.\overline{14}$ as the quotient of two integers.

SOLUTION. In the previous problem the repeating block had length 1 and we multiplied by 10. Here the repeating block has length 2 and we multiply by 100. If the repeating block had length 3, we would multiply by 1000; and so on.

$$
\begin{array}{r}
100(0.\overline{14}) = 14.\overline{14} \\
-\quad 0.\overline{14} = 0.\overline{14} \\
\hline
99(0.\overline{14}) = 14
\end{array}
$$

Therefore $0.\overline{14} = \frac{14}{99}$.   □

Exercises†

1. Draw a number line and mark on it the points 0, $\frac{1}{2}$, -1, 1, 2, $\frac{7}{3}$, 3.3, -4.

2. Draw a number line and mark on it the points 0, 10, 50, -65, and -100.

3. Show that each of the following numbers is rational by expressing it as the quotient of two integers:
    *(a) $6\frac{4}{5}$.   (b) $-2\frac{1}{4}$.   *(c) $2 - \frac{1}{5}$.   (d) $-2 - \frac{1}{5}$.
    *(e) 3.1.   (f) 1.35.   *(g) 3.176.   (h) -2.115.

4. Write each fraction in *lowest terms* by cancelling all positive integer factors common to numerator and denominator:
    *(a) $\frac{16}{28}$.   (b) $\frac{120}{144}$.   *(c) $\frac{350}{490}$.   (d) $\frac{625}{1000}$.

5. Fractions are added, subtracted, multiplied, and divided according to the following rules:

    $$\frac{p}{q} + \frac{r}{s} = \frac{ps + qr}{qs}, \qquad \frac{p}{q} - \frac{r}{s} = \frac{ps - qr}{qs}$$

    $$\frac{p}{q} \cdot \frac{r}{s} = \frac{pr}{qs}, \qquad \frac{p}{q} \div \frac{r}{s} = \frac{p}{q} \cdot \frac{s}{r} = \frac{ps}{qr} \text{ provided } r \neq 0.$$

    Carry out the indicated operations expressing your answer as a fraction in lowest terms:
    *(a) $\frac{1}{2} + \frac{1}{3}$.   (b) $\frac{1}{2} - \frac{1}{3}$.   *(c) $\frac{5}{8} + \frac{3}{10}$.   (d) $\frac{7}{9} - \frac{4}{15}$.
    *(e) $\frac{1}{2} \cdot \frac{4}{9}$.   (f) $\frac{3}{25} \cdot \frac{5}{21}$.   *(g) $\frac{1}{2} \div \frac{3}{2}$.   (h) $\frac{6}{25} \div \frac{12}{5}$.

6. Express each fraction as a decimal:
    *(a) $\frac{14}{25}$.   (b) $\frac{1}{3}$.   *(c) $\frac{4}{11}$.   (d) $\frac{11}{4}$.
    *(e) $\frac{5}{9}$.   (f) $\frac{3}{8}$.   *(g) $\frac{2}{7}$.   (h) $\frac{1}{111}$.

7. Write each decimal as the quotient of two integers:
    *(a) $0.444 \cdots = 0.\overline{4}$.   (b) $0.212121 \cdots = 0.\overline{21}$.
    *(c) $0.373737 \cdots = 0.\overline{37}$.   (d) $4.5123123 \cdots = 4.5\overline{123}$.
    *(e) $0.a_1 a_1 a_1 \cdots = 0.\overline{a_1}$.   (f) $0.a_1 a_2 a_1 a_2 \cdots = 0.\overline{a_1 a_2}$.

*8. Given a unit length, use right triangles and the Pythagorean theorem to construct line segments of lengths $\sqrt{3}$ and $\sqrt{5}$. HINT: First construct a line segment of length $\sqrt{2}$.

---

†The starred exercises have answers at the back of the book.

Optional | 9. Show that, since π is irrational, 2π is irrational. HINT: Show that, if 2π were rational, then π would be rational.

1.2 Some Properties of the Real Number System

For easy reference and review, we briefly go over some of the basic arithmetic properties of real numbers. Undoubtedly you are already familiar with them.

I. Addition and multiplication are *associative*:

(1.2.1) $$a + (b + c) = (a + b) + c, \qquad a(bc) = (ab)c$$

and *commutative*:

(1.2.2) $$a + b = b + a, \qquad ab = ba.$$

II. Multiplication *distributes* over addition:

(1.2.3) $$a(b + c) = ab + ac \quad \text{and} \quad (b + c)a = ba + ca.$$

We will return to this in Section 1.6, for this is the basis of factoring.

III. The number 0 is an *additive identity*:

(1.2.4) $$a + 0 = 0 + a = a$$

and the number 1 is a *multiplicative identity*:

(1.2.5) $$a \cdot 1 = 1 \cdot a = a.$$

IV. The number 0 times any number is 0:

(1.2.6) $$0 \cdot a = a \cdot 0 = 0.$$

V. If the product of two numbers is 0, then at least one of the factors is 0:

(1.2.7) $$\text{if} \quad ab = 0, \qquad \text{then} \quad a = 0 \quad \text{or} \quad b = 0.$$

We use this property often when solving equations.

VI. Division by 0 is undefined:

(1.2.8) $\boxed{\text{there is no number } a/0.}$

In particular, $0/0$ is not defined.

VII. The product of positive factors is positive. The product of an even number of negative factors is positive:

$$(-2)(-3) = 6, \qquad (-1)(-4)(-2)(-3) = (4)(6) = 24;$$

the product of an odd number of negative factors is negative:

$$(-2)(3) = -6, \qquad (-1)(4)(-2)(-3) = (-4)(6) = -24.$$

1.3   Addition and Subtraction of Algebraic Expressions

In adding or subtracting algebraic expressions, we combine like terms. Thus,

$$(2x - 5y + 5) + (6x + 3y + 1) = 8x - 2y + 6,$$
$$(2x - 5y + 5) - (6x + 3y + 1) = -4x - 8y + 4, \text{ etc.}$$

Sometimes it is useful to first reorder the terms:

$$(x^2 + 1 + x) + (3x + 5x^2 + 1) = (x^2 + x + 1) + (5x^2 + 3x + 1)$$
$$= 6x^2 + 4x + 2.$$

Problem. Simplify

$$x^2 + 1 - [x^2 - x - (2x + 5)].$$

SOLUTION. Here there are two ways in which we can proceed. We can first remove the parentheses and write

$$x^2 + 1 - [x^2 - x - (2x + 5)] = x^2 + 1 - [x^2 - x - 2x - 5]$$
$$= x^2 + 1 - [x^2 - 3x - 5]$$
$$= x^2 + 1 - x^2 + 3x + 5$$
$$= 3x + 6.$$

Or, we can first remove the brackets:

$$x^2 + 1 - [x^2 - x - (2x + 5)] = x^2 + 1 - x^2 + x + (2x + 5)$$
$$= x^2 + 1 - x^2 + x + 2x + 5$$
$$= 3x + 6.$$

The result, of course, is the same. $\square$

Problem. Simplify

$$\tfrac{1}{2}(4x^2 + 6) - 5(1 - x^2).$$

SOLUTION.   By the distributive law,

$$\tfrac{1}{2}(4x^2 + 6) = 2x^2 + 3 \quad \text{and} \quad 5(1 - x^2) = 5 - 5x^2,$$

so that

$$\tfrac{1}{2}(4x^2 + 6) - 5(1 - x^2) = (2x^2 + 3) - (5 - 5x^2)$$
$$= 7x^2 - 2. \quad \square$$

Problem. Simplify

$$2x^2 + 3[4(x^2 + x + 1) - 2(x + 1)].$$

SOLUTION.   First we distribute the 3 and remove the brackets:

$$2x^2 + 12(x^2 + x + 1) - 6(x + 1).$$

Then we distribute the 12 and the 6:

$$2x^2 + 12x^2 + 12x + 12 - 6x - 6.$$

Finally we collect like terms:

$$14x^2 + 6x + 6. \quad \square$$

Problem. Evaluate

$$3(2x - y) + 2(4x + y - 1) \qquad \text{at } x = 1, \quad y = 3.$$

SOLUTION.   We substitute 1 for x and 3 for y:

$$3[(2)(1) - 3] + 2[(4)(1) + 3 - 1] = 3(2 - 3) + 2(4 + 3 - 1)$$
$$= (3)(-1) + (2)(6)$$
$$= -3 + 12$$
$$= 9. \quad \square$$

Exercises

Carry out the indicated operations and simplify:

*1. $4(x - 4)$.

2. $3a(x - 1)$.

*3. $14(x - y + 2)$.

*4. $\tfrac{1}{4}b(x - 4)$.

5. $2[x - (y + 1)]$.

*6. $2[3(y - a) + 4a]$.

*7. $\tfrac{2}{33}(33y - 11x)$.

8. $\tfrac{1}{19}(19y - 38x^2)$.

*9. $\tfrac{3}{5}(15a^2 - 20x^2)$.

*10. $x^2 - 1 - [x^2 - x - (2x + 5)]$.

11. $2(2x^2 - 1) + (x^2 + x + 3)$.

*12. $212[(y^2 + 1) - (x^4 + 1) - (y^2 - x^4)]$.

13. $\frac{1}{4}x + 4[x + (4 - 2x)]$.

*14. $\frac{1}{2}(4abc - 1) - \frac{1}{3}(15abc + 6)$.

15. $4(xy + ba) - \frac{1}{2}(yx - ab)$.

*16. $2[(y^2 - x^2) + 4(x^2 - y^2) - y(1 - x)]$.

17. $6[\frac{1}{2}(a - b) - \frac{2}{3}(a + b)]$.

*18. $\frac{3}{5}[10a^2x - \frac{5}{3}a^2(x - 1) + 10a(a - 1)]$.

19. $3[(a - b)x - 2(a + b)x]$.

Evaluate:

20. $6(x - y + 2)$ at $x = 3, y = 4$.

*21. $2(3x - y) - (4x + y)$ at $x = 0, y = 1$.

22. $4(xyz - 1) - 2(3 - xyz)$ at $x = 1, y = 2, z = -2$.

*23. $10(2x + 1) + 100(x - 5) - (3x - 2)$ at $x = 4$.

24. $a(ax^2 - y) - 2a(y + ax^2) - a(x^2 + y^2)$ at $x = -1, y = 1$.

1.4   Multiplication of Algebraic Expressions

The basic ideas here are the laws of exponents and the distributive law.

nth Powers

If a is a real number and n is a positive integer, then the *nth power* of a is defined by setting

(1.4.1)
$$a^n = \underbrace{a \cdot a \cdots a}_{n \text{ factors}}.$$

Here a is called the *base* and n the *exponent*.

By a to the power 1 we mean a itself:

$$a^1 = a.$$

As you probably remember,

(1.4.2) $\boxed{a^m \cdot a^n = a^{m+n}, \quad (a^m)^n = a^{mn}, \quad \text{and} \quad (ab)^n = a^n \cdot b^n.}$

PROOFS

$$a^m \cdot a^n = \underbrace{(a \cdot a \cdots a)}_{m \text{ factors}}\underbrace{(a \cdot a \cdots a)}_{n \text{ factors}} = \underbrace{a \cdot a \cdots a}_{m + n \text{ factors}} = a^{m+n}. \quad \square$$

$$(a^m)^n = \underbrace{a^m \cdot a^m \cdots a^m}_{n \text{ factors}} = \underbrace{a \cdot a \cdots a}_{mn \text{ factors}} = a^{mn}. \quad \square$$

$$(ab)^n = \underbrace{ab \cdot ab \cdots ab}_{n \text{ factors}} = \underbrace{(a \cdot a \cdots a)}_{n \text{ factors}}\underbrace{(b \cdot b \cdots b)}_{n \text{ factors}} = a^n b^n. \quad \square$$

Here are some sample calculations:

$$3^2 \cdot 3^5 = 3^7,$$
$$(3^2)^5 = 3^{10},$$
$$(2 \cdot 5)^3 = 2^3 \cdot 5^3,$$
$$(2ax^2)(-3a^2x^4) = -6a^3x^6,$$
$$(-2y)^3 = (-2)^3(y)^3 = -8y^3,$$
$$(a^2b^3)^2 = a^4b^6,$$
$$(6xy)^2(xy^2)^2 = (36x^2y^2)(x^2y^4) = 36x^4y^6.$$

In multiplying sums and differences of terms, the distributive law comes into play.

Problem. Multiply out
$$3ax(x^2 - y^2).$$

SOLUTION. The idea here is to distribute the $3ax$:
$$3ax(x^2 - y^2) = 3ax(x^2) - 3ax(y^2)$$
$$= 3ax^3 - 3axy^2. \quad \square$$

Problem. Compute

$$(a^2 + 2b)(a - b^2).$$

SOLUTION. Each term of each expression must be multiplied by each term of the other:

$$(a^2 + 2b)(a - b^2) = (a^2)(a) + (a^2)(-b^2) + (2b)(a) + (2b)(-b^2)$$
$$= a^3 - a^2b^2 + 2ab - 2b^3. \quad \square$$

Problem. Compute

$$(3x + 2x^2 + 4 + x^3)(2x - 3).$$

SOLUTION. We begin by reordering the terms of the first expression according to decreasing powers of x. This makes it easier to collect like terms at the end:

$$(3x + 2x^2 + 4 + x^3)(2x - 3) = (x^3 + 2x^2 + 3x + 4)(2x - 3).$$

Now we multiply each term of the first expression by each term of the second:

$$(x^3)(2x) + (2x^2)(2x) + (3x)(2x) + (4)(2x) + (x^3)(-3)$$
$$+ (2x^2)(-3) + (3x)(-3) + (4)(-3),$$

and thereby get

$$2x^4 + 4x^3 + 6x^2 + 8x - 3x^3 - 6x^2 - 9x - 12 = 2x^4 + x^3 - x - 12. \quad \square$$

Problem. Multiply out

$$(x - a)(x^2 - b)(x^3 - c).$$

SOLUTION. We multiply two of the factors first, and then the third. By associativity and commutativity it makes no difference which two factors we begin with. We may as well begin with the first two:

$$(x - a)(x^2 - b)(x^3 - c)$$
$$= [(x)(x^2) + (-a)(x^2) + (x)(-b) + (-a)(-b)](x^3 - c)$$
$$= (x^3 - ax^2 - bx + ab)(x^3 - c)$$
$$= (x^3)(x^3) + (-ax^2)(x^3) + (-bx)(x^3) + (ab)(x^3)$$
$$+ (x^3)(-c) + (-ax^2)(-c) + (-bx)(-c) + (ab)(-c)$$
$$= x^6 - ax^5 - bx^4 + abx^3 - cx^3 + acx^2 + bcx - abc$$
$$= x^6 - ax^5 - bx^4 + (ab - c)x^3 + acx^2 + bcx - abc. \quad \square$$

Problem. Carry out the indicated operations:

$$\tfrac{1}{3}x(x^2 + a^2) + \tfrac{2}{3}(x - a)(x^2 + 1).$$

SOLUTION. Multiplied out, we have in turn

$$(\tfrac{1}{3}x)(x^2) + (\tfrac{1}{3}x)(a^2) + \tfrac{2}{3}[(x)(x^2) + (x)(1) + (-a)(x^2) + (-a)(1)],$$

$$\tfrac{1}{3}x^3 + \tfrac{1}{3}a^2x + \tfrac{2}{3}(x^3 + x - ax^2 - a),$$

$$\tfrac{1}{3}x^3 + \tfrac{1}{3}a^2x + \tfrac{2}{3}x^3 + \tfrac{2}{3}x - \tfrac{2}{3}ax^2 - \tfrac{2}{3}a,$$

$$x^3 - \tfrac{2}{3}ax^2 + (\tfrac{2}{3} + \tfrac{1}{3}a^2)x - \tfrac{2}{3}a. \quad \square$$

Problem. Evaluate

$$(x + a)(x^2 + a^2)(x^3 + a^3) \quad \text{and} \quad (x^3 + x^2 + x + a)(x - a)$$

at $x = a$.

SOLUTION. At $x = a$, the first expression becomes

$$(2a)(2a^2)(2a^3) = 8a^6.$$

The second expression is 0 at $x = a$ since the factor $(x - a)$ is 0 there. $\square$

Exercises

Carry out the indicated operations:

*1. $x^3(5x^4)$.

2. $2a^2(ab^2)^2$.

*3. $(2x^2y)^3$.

*4. $(2a^2x)(-3a^5x^3)$.

5. $(x^2y)(y^2x)$.

*6. $(x - 3a)^2$.

*7. $(-3x^2y^5)^3$.

8. $(abc)^2(5a^2b^3)$.

*9. $u(u^2 + v)$.

*10. $u(u^2 - v) + 1$.

11. $(x + 1)(x - 2) + 2$.

*12. $4(x^2 + 1)(x^3 - 1)$.

13. $\tfrac{1}{2}x(x^2 + a^2) - \tfrac{1}{2}x^2(x - a^2)$.

*14. $(a + b + c)(a - b - c)$.

15. $2x(5x - y^2) - y^2(x^2 + x + 1)$.

*16. $2x^2y^3(6x - 2y)(3x + y)$.

17. $(x - y)[7x + 4(y - 2x) + 1]$.

*18. $25(x - 1)^3$.

19. $(x - 2)(x^2 - 4) - (x^2 + 1)(x + 3)$.

*20. $(4x^4 - y^3)(x - 4y) + (2x^5 - y^4)$.

21. $(4x^4 - y)(x - y)^2$.

*22. $(ax^2 + b)(bx - a^2) + a^2b$.

23. $(x + 1)(x + 2)(x + 3) - (x^3 + 4x - 1)$.

*24. $(ax^2 + bx + c)(2x - 1)$.

25. $(ax^2 + bx + c)^2 + 2x^4$.

*26. $s[r - 2sr + 2s(r + 1) - 2s][p + p(v - 1)]$.

Evaluate:

27. $2x(x - 1)(x - 2)$ (a) at $x = \frac{1}{2}$,   (b) at $x = 1$.

*28. $\frac{1}{4}n^2(n + 1)$ (a) at $n = 2$,   (b) at $n = 4$.

29. $(2x - c)(x^2 - c^2)$ (a) at $x = c$,   (b) at $x = 2c$.

*30. $3x^3 + 2x^2 + 5x - 7$ (a) at $x = 1$,   (b) at $x = 0$.

31. $(x^2 - c^2)(x^2 + 2c^2) - (x + 4c)(x + c)$ (a) at $x = -c$,   (b) at $x = -4c$.

1.5   Some Special Products

The following special products occur so frequently that you should memorize them.

(i)	$(a + b)(a - b) = a^2 - b^2$.	(product of a sum and difference)	
(ii)	$(a + b)^2 = a^2 + 2ab + b^2$.	(square of a sum)	
(iii)	$(a - b)^2 = a^2 - 2ab + b^2$.	(square of a difference)	
(iv)	$(a + b)^3 = a^3 + 3a^2b + 3ab^2 + b^3$.	(cube of a sum)	
(v)	$(a - b)^3 = a^3 - 3a^2b + 3ab^2 - b^3$.	(cube of a difference)	

(1.5.1)

Each of these products is easily verifiable by direct computation. What we want to emphasize here is the use of these formulas.

Problem. Compute

$$(3x + 5)(3x - 5) \quad \text{and} \quad (ct - d)(ct + d).$$

SOLUTION. Each is the product of a sum and difference:

$$(3x + 5)(3x - 5) = (3x)^2 - (5)^2 = 9x^2 - 25,$$

$$(ct - d)(ct + d) = (ct)^2 - (d)^2 = c^2t^2 - d^2. \quad \square$$

Problem. Compute

$$(2x + 3y)^2 \quad \text{and} \quad (2x - 3y)^2.$$

SOLUTION. The first is the square of a sum:

$$(2x + 3y)^2 = (2x)^2 + 2(2x)(3y) + (3y)^2$$
$$= 4x^2 + 12xy + 9y^2.$$

The second is the square of a difference:

$$(2x - 3y)^2 = (2x)^2 - 2(2x)(3y) + (3y)^2$$
$$= 4x^2 - 12xy + 9y^2. \quad \square$$

Problem. Compute

$$(2x + 5)^3 \quad \text{and} \quad (2x - 5)^3.$$

SOLUTION. The first is the cube of a sum:

$$(2x + 5)^3 = (2x)^3 + 3(2x)^2(5) + 3(2x)(5)^2 + (5)^3$$
$$= 8x^3 + 60x^2 + 150x + 125.$$

The second is the cube of a difference:

$$(2x - 5)^3 = (2x)^3 - 3(2x)^2(5) + 3(2x)(5)^2 - (5)^3$$
$$= 8x^3 - 60x^2 + 150x - 125. \quad \square$$

Exercises

Compute by means of the appropriate formula. Do the problem in your head if you can.

*1. $(x + 1)(x - 1)$.   2. $(x + 1)^2$.   *3. $(x - 3)^2$.

*4. $(2x + 1)^2$.   5. $(2x + 3)(2x - 3)$.   *6. $(5x - 1)^2$.

*7. $(4t + 5)^2$.   8. $(4t - 5)^2$.   *9. $(4t - 5)(4t + 5)$.

*10. $(x + 2)^2$.   11. $(10x + 4y)^2$.   *12. $(2y - 3x)^2$

*13. $(2x + 1)^3$.   14. $(x + 2y)(x - 2y)$.   *15. $(2r + 3s)^2$.

*16. $(2r - 3s)^2$.   17. $(x - 2a)^3$.   *18. $(x + 2a)^3$.

*19. $(x + \frac{1}{2})(x - \frac{1}{2})$.   20. $(x + \frac{1}{2})^2$.   *21. $(x + \frac{1}{2})^3$.

*22. $(x - \frac{1}{2})^3$.   23. $(\frac{1}{2}x + 1)^2$.   *24. $(\frac{1}{2}x - 1)^2$.

*25. $(\frac{1}{2}x + 1)^3$.   26. $(\frac{1}{2}x - 1)^3$.   *27. $(\frac{1}{2}x + \frac{1}{3})^2$.

*28. $(x^2 - a^2)^3$.   29. $(x^2 - 4b^2)(x^2 + 4b^2)$.   *30. $(x^4 - 1)^3$.

*31. $(x^2 + a^2)^3$.   32. $[1 + (x - 1)^2]^2$.   *33. $[1 - (x - 1)^2]^2$.

1.6   The Rudiments of Factoring

The basic factoring formulas are the multiplication formulas written in reverse order. The distributive law,

$$a(b + c) = ab + ac,$$

written as

(1.6.1)
$$\boxed{ab + ac = a(b + c),}$$

tells us that from the sum $ab + ac$ we can factor out an a.

Examples

$$a^2 t - a = a(at - 1), \qquad\qquad (a \text{ is a common factor})$$
$$2x + 4y = 2(x + 2y), \qquad\qquad (2 \text{ is a common factor})$$
$$b^2 x^3 + 2b^3 y - b^4 = b^2(x^3 + 2by - b^2), \qquad (b^2 \text{ is a common factor})$$
$$x^5 - x^4 + 2x = x(x^4 - x^3 + 2). \qquad\qquad (x \text{ is a common factor}) \quad \square$$

The product formula

$$(a + b)(a - b) = a^2 - b^2,$$

written as

(1.6.2)
$$\boxed{a^2 - b^2 = (a + b)(a - b),}$$

tells us how to factor the difference of two squares.

Examples

$$a^2 - 1 = (a + 1)(a - 1),$$
$$4x^2 - 9 = (2x)^2 - 3^2 = (2x + 3)(2x - 3),$$
$$(a + 1)^2 - x^2 = (a + 1 + x)(a + 1 - x). \quad \square$$

Formulas

(1.6.3)
$$\boxed{a^2 + 2ab + b^2 = (a + b)^2, \qquad a^2 - 2ab + b^2 = (a - b)^2}$$

tell us how to recognize perfect squares.

Examples

$$x^2 + 2x + 1 = (x + 1)^2,$$
$$x^2 - 4x + 4 = x^2 - 2(x)(2) + 2^2 = (x - 2)^2,$$

$$x^4 + 2x^2y^2 + y^4 = (x^2)^2 + 2x^2y^2 + (y^2)^2 = (x^2 + y^2)^2,$$
$$4t^2 - 12t + 9 = (2t)^2 - 2(2t)(3) + 3^2 = (2t - 3)^2. \quad \square$$

From the sum of two cubes, $a^3 + b^3$, we can factor $a + b$:

(1.6.4)
$$a^3 + b^3 = (a + b)(a^2 - ab + b^2),$$

and from the difference of two cubes, $a^3 - b^3$, we can factor $a - b$:

(1.6.5)
$$a^3 - b^3 = (a - b)(a^2 + ab + b^2).$$

You can check these formulas by multiplying out the right-hand sides. In the first case,

$$(a + b)(a^2 - ab + b^2) = (a)(a^2) + (a)(-ab) + (a)(b^2)$$
$$+ (b)(a^2) + (b)(-ab) + (b)(b^2)$$
$$= a^3 - a^2b + ab^2 + a^2b - ab^2 + b^3$$
$$= a^3 + b^3;$$

and in the second case,

$$(a - b)(a^2 + ab + b^2) = (a)(a^2) + (a)(ab) + (a)(b^2)$$
$$+ (-b)(a^2) + (-b)(ab) + (-b)(b^2)$$
$$= a^3 + a^2b + ab^2 - a^2b - ab^2 - b^3$$
$$= a^3 - b^3. \quad \square$$

Examples

$$x^3 + 8 = x^3 + 2^3 = (x + 2)[x^2 - (x)(2) + 2^2] = (x + 2)(x^2 - 2x + 4),$$
$$x^3 - 8 = x^3 - 2^3 = (x - 2)[x^2 + (x)(2) + 2^2] = (x - 2)(x^2 + 2x + 4),$$
$$27x^3 - 1 = (3x)^3 - (1)^3 = (3x - 1)[(3x)^2 + (3x)(1) + 1^2]$$
$$= (3x - 1)(9x^2 + 3x + 1),$$
$$(x + 1)^3 + 27 = (x + 1)^3 + 3^3 = [(x + 1) + 3][(x + 1)^2 - (x + 1)(3) + 3^2]$$
$$= (x + 4)(x^2 + 2x + 1 - 3x - 3 + 9)$$
$$= (x + 4)(x^2 - x + 7),$$
$$4x^3 - 32b^3 = 4(x^3 - 8b^3) = 4[x^3 - (2b)^3]$$
$$= 4(x - 2b)[x^2 + (x)(2b) + (2b)^2]$$
$$= 4(x - 2b)(x^2 + 2bx + 4b^2). \quad \square$$

Exercises

Factor:

*1. $3x + 6$.

2. $5x - 100$.

*3. $ab + a$.

*4. $ab + bc$.

5. $abc + 3ab$.

*6. $a^2b + ab^2$.

*7. $xy + 2x$.

8. $3x + 12y - 9$.

*9. $2x^2 - 4x$.

*10. $x^2 - c^2$.

11. $2x^2 - 8c^2$.

*12. $36 - 4x^2$.

*13. $16x^2 - a^2$.

14. $x^2 + 4x + 4$.

*15. $x^2 - 4x + 4$.

*16. $x^2 + 6x + 9$.

17. $x^2 + 16x + 64$.

*18. $x^2 - 16x + 64$.

*19. $x^3 - 8$.

20. $27x^3 - 8$.

*21. $8x^3 + 27$.

*22. $x^2y^3 - xy^2 + xy$.

23. $81x^2 + 72xy + 16y^2$.

*24. $25x^2 - (3x + y)^2$.

*25. $(4x - 3)^2 - (x - 3)^2$.

26. $(x - b)^2 - (x - d)^2$.

*27. $(ax + b)^2 - (cx + d)^2$.

*28. $3x^2 + 66x + 363$.

29. $4x^2 - 40x + 100$.

*30. $5x^2 - 30x + 45$.

*31. $2a^3x^3 - 16$.

32. $2x^3 - 16a^3$.

*33. $16x^3 + 2a^3$.

34. What is wrong with the following argument?

Set $a = b \neq 0$. Then

$$a^2 = ab$$
$$a^2 - b^2 = ab - b^2$$
$$(a + b)(a - b) = b(a - b)$$
$$a + b = b$$
$$2b = b$$
$$2 = 1. \quad \square$$

1.7   The Product of Two Linear Expressions

We begin with a definition.

Definition   1.7.1   Linear Expression

A *linear expression in x* is an expression of the form

$$ax + b \quad \text{with } a \text{ and } b \text{ real, } a \neq 0.$$

The numbers a and b are called the *coefficients*.†

It is useful to be able to multiply linear expressions with some facility. The basic result can be written

(1.7.2) $$(ax + b)(cx + d) = acx^2 + (bc + ad)x + bd.$$

This is illustrated in Figure 1.7.1. Each term of the first expression is multiplied by each term of the second expression. If a and c are both 1, we have simply

(1.7.3) $$(x + b)(x + d) = x^2 + (b + d)x + bd.$$

To verify these formulas, just carry out the multiplication on the left.

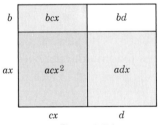

Figure 1.7.1

Examples.   We begin with the second formula:

$$(x + 2)(x + 3) = x^2 + (2 + 3)x + (2)(3) = x^2 + 5x + 6,$$

$$(x + 2)(x - 3) = x^2 + [2 + (-3)]x + (2)(-3) = x^2 - x - 6,$$

$$(x - 2)(x + 3) = x^2 + [(-2) + 3]x + (-2)(3) = x^2 + x - 6,$$

$$(x - 2)(x - 3) = x^2 + [(-2) + (-3)]x + (-2)(-3) = x^2 - 5x + 6.$$

---

†One can also form linear expressions with complex coefficients, but these are outside the scope of this text. When we say "linear expression" in this text, we always mean "linear expression with real coefficients."

Now we illustrate the first formula, the more general one:

$$(2x + 1)(3x + 5) = (2)(3)x^2 + [(1)(3) + (2)(5)]x + (1)(5)$$
$$= 6x^2 + 13x + 5,$$
$$(2x + 1)(3x - 5) = (2)(3)x^2 + [(1)(3) + (2)(-5)]x + (1)(-5)$$
$$= 6x^2 - 7x - 5.$$

Similarly,

$$(2x - 1)(3x + 5) = 6x^2 + 7x - 5$$

and

$$(2x - 1)(3x - 5) = 6x^2 - 13x + 5. \quad \square$$

Exercises

Carry out the following multiplications in your head:

*1. $(x + 1)(x + 4)$.	2. $(x + 1)(x - 4)$.	*3. $(x + 4)(x - 1)$.
*4. $(x - 1)(x - 4)$.	5. $(x + 3)(x + 5)$.	*6. $(x - 3)(x - 5)$.
*7. $(x + 3)(x - 5)$.	8. $(2x + 3)(3x + 5)$.	*9. $(2x + 3)(3x - 5)$.
*10. $(2x - 3)(3x + 5)$.	11. $(2x - 3)(3x - 5)$.	*12. $(4x - 1)(3x + 2)$.
*13. $(4x + 1)(3x - 2)$.	14. $(4x + 1)(3x + 2)$.	*15. $(4x - 1)(3x - 2)$.
*16. $(3x - 2)^2$.	17. $(3x - 2)(2x - 3)$.	
*18. $(3x + 2)(2x - 3)$.	*19. $(5x - 2)(x - 1)$.	
20. $(5x - 2)(x + 1)$.	*21. $(2x + 7)(5x - 1)$.	
*22. $(5x^2 + 2)(x^2 + 1)$.	23. $(2x^2 - 5)(7x^2 + 1)$.	

1.8 Factoring Quadratic Expressions

Definition 1.8.1   Quadratic Expression

A *quadratic in x* is an expression of the form

$$ax^2 + bx + c \qquad \text{with } a, b, c \text{ real and } a \neq 0.$$

The numbers a, b, c are called the *coefficients*.†

---

† One can also form quadratics with complex coefficients, but these are outside the scope of this text. When we say "quadratic" in this text, we always mean "quadratic with real coefficients."

In the last section you saw that multiplication of two linear expressions yields a quadratic expression. In particular, you verified that

$$(x + b)(x + d) = x^2 + (b + d)x + bd.$$

Rewriting this as

(1.8.2)
$$\boxed{x^2 + (b + d)x + bd = (x + b)(x + d),}$$

we have a factoring formula.

Problem. Factor

$$x^2 + 5x + 6.$$

SOLUTION.   To apply our formula, we set

$$bd = 6, \quad b + d = 5.$$

We seek numbers b and d whose product is 6 and whose sum is 5. Obviously we can take $b = 2$, $d = 3$. Doing this, we have

$$x^2 + 5x + 6 = (x + 2)(x + 3).$$

Or we can take $b = 3$, $d = 2$, which gives

$$x^2 + 5x + 6 = (x + 3)(x + 2).$$

Since multiplication is commutative, order doesn't matter here. Both factorizations are the same. □

Problem. Factor

$$x^2 + x - 6.$$

SOLUTION.   Here

$$bd = -6, \quad b + d = 1.$$

The product of the numbers is -6 and the sum is 1. The numbers are 3 and -2:

$$x^2 + x - 6 = (x + 3)(x + (-2)) = (x + 3)(x - 2). \quad □$$

Problem. Factor

$$x^2 - 10x + 21.$$

SOLUTION.   Here

$$bd = 21, \quad b + d = -10.$$

The product of the numbers is 21 and the sum is -10. The numbers are -3 and -7:

$$x^2 - 10x + 21 = (x - 3)(x - 7). \quad \square$$

Until now the coefficient of x^2 has always been 1. To factor quadratics such as

$$3x^2 + 7x + 4 \quad \text{and} \quad 10x^2 - x - 3,$$

we have to turn to the more general formula

(1.8.3) $\boxed{acx^2 + (bc + ad)x + bd = (ax + b)(cx + d).}$

Problem. Factor

$$3x^2 + 7x + 4.$$

SOLUTION. Here

$$ac = 3, \qquad bd = 4, \qquad bc + ad = 7.$$

We need numbers a and c whose product is 3. We try 3 and 1 and write

$$(3x \quad)(x \quad).$$

We now need numbers b and d whose product is 4. We try 2 and 2 and write

$$(3x + 2)(x + 2).$$

But this doesn't work because the product fails to meet the last condition:

$$(3x + 2)(x + 2) = 3x^2 + 8x + 4$$

and we wanted

$$3x^2 + 7x + 4.$$

Instead of 2 and 2, we try 4 and 1:

$$(3x + 4)(x + 1).$$

This works:

$$(3x + 4)(x + 1) = 3x^2 + 7x + 4. \quad \square$$

Problem. Factor

$$10x^2 - x - 3.$$

SOLUTION. The guessing game begins again. As factors of 10, we try 10 and 1:

$$(10x \quad)(x \quad).$$

As factors of -3, we try combinations of 3 and 1 with opposite signs:

$$(10x + 3)(x - 1), \qquad (10x - 3)(x + 1),$$
$$(10x + 1)(x - 3), \qquad (10x - 1)(x + 3).$$

Unfortunately none of these combinations work. We therefore go back to the 10 and factor it as 5 times 2:

$$(5x \qquad)(2x \qquad).$$

The combination

$$(5x - 3)(2x + 1)$$

works:

$$(5x - 3)(2x + 1) = 10x^2 - x - 3. \quad \square$$

A more systematic way of dealing with such quadratics is discussed in Chapter 2.

CAUTION: Some quadratics ($x^2 + 1$ for example) cannot be factored without invoking complex numbers. Complex numbers are discussed in Chapter 7.

Exercises

Factor:

*1. $x^2 + 4x + 3$.

2. $x^2 - 4x + 3$.

*3. $x^2 + 2x - 3$.

*4. $x^2 + 6x + 8$.

5. $x^2 + 8x + 15$.

*6. $x^2 - 8x + 15$.

*7. $x^2 - x - 2$.

8. $x^2 - 6x + 5$.

*9. $x^2 - 12x + 27$.

*10. $x^2 - 12x + 35$.

11. $x^2 + 50x + 400$.

*12. $x^2 - 12x + 20$.

*13. $x^2 + 3x - 28$.

14. $x^2 + x - 6$.

*15. $x^2 - 10x - 24$.

*16. $x^2 + (C - A)x - AC$.

17. $x^2 + (1 - C)x - C$.

*18. $x^2 - (r + 1)x + r$.

*19. $2x^2 + 5x + 2$.

20. $2x^2 - 3x - 2$.

*21. $2x^2 + x - 3$.

*22. $6x^2 + 19x + 10$.

23. $6x^2 - 11x - 10$.

*24. $2x^2 + 23x + 30$.

*25. $25x^2 + 53x + 6$.

26. $12x^2 - 16x - 3$.

*27. $2x^2 - 25x + 50$.

*28. $12x^2 - 11x + 2$.

29. $4x^2 + 20x + 25$.

Optional 30. Show that the quadratic $x^2 + 1$ is not the product of two linear expressions with real coefficients. HINT: If

$$x^2 + 1 = (ax + b)(cx + d),$$

then $x^2 + 1$ has to be zero at $x = -b/a$ and at $x = -d/c$.

1.9 More on Factoring

Often we can factor more than once.

Problem. Factor

$$x^4 - 16.$$

SOLUTION. We can begin by writing

$$x^4 - 16 = (x^2)^2 - 4^2$$
$$= (x^2 + 4)(x^2 - 4),$$

but we can still factor some more. Namely,

$$x^2 - 4 = (x + 2)(x - 2).$$

Thus

$$x^4 - 16 = (x^2 + 4)(x + 2)(x - 2). \quad \square$$

Problem. Factor

$$x^3 - 12x^2 - 45x.$$

SOLUTION. First we factor out an x:

$$x^3 - 12x^2 - 45x = x(x^2 - 12x - 45).$$

We can now factor the quadratic:

$$x^2 - 12x - 45 = (x + 3)(x - 15).$$

All together we have

$$x^3 - 12x^2 - 45x = x(x + 3)(x - 15). \quad \square$$

Problem. Factor

$$x^4 - 2x^2 + 1.$$

SOLUTION. First we note that this is a perfect square:

$$x^4 - 2x^2 + 1 = (x^2)^2 - 2(x^2) + 1 = (x^2 - 1)^2.$$

Since

$$x^2 - 1 = (x + 1)(x - 1),$$

we have

$$x^4 - 2x^2 + 1 = [(x + 1)(x - 1)]^2 = (x + 1)^2(x - 1)^2. \quad \square$$

Problem. Factor

$$9(x - y)^2 - (x + y)^2.$$

SOLUTION. The expression has the form $A^2 - B^2$ with $A = 3(x - y)$, $B = x + y$. In general

$$A^2 - B^2 = (A + B)(A - B),$$

so that here

$$
\begin{aligned}
9(x - y)^2 - (x + y)^2 &= [3(x - y) + (x + y)][3(x - y) - (x + y)] \\
&= (3x - 3y + x + y)(3x - 3y - x - y) \\
&= (4x - 2y)(2x - 4y) \\
&= (2)(2x - y)(2)(x - 2y) \\
&= 4(2x - y)(x - 2y). \quad \square
\end{aligned}
$$

Problem. Factor

$$x^3 + 3x^2 - x - 3.$$

SOLUTION. Once you recognize that $(x + 3)$ is a factor, the rest is straightforward:

$$
\begin{aligned}
x^3 + 3x^2 - x - 3 &= x^2(x + 3) - (x + 3) \\
&= (x + 3)(x^2 - 1) \\
&= (x + 3)(x + 1)(x - 1). \quad \square
\end{aligned}
$$

Exercises

Factor these expressions as far as you can:

*1. $x^3 - x.$ 2. $50x - 2x^3.$

*3. $x^4 - 6x^3 + 5x^2.$ 4. $x^8 - 1.$

*5. $3x^5 - 36x^4 + 105x^3.$ 6. $16x^4 + 16y^4 - 32x^2y^2.$

*7. $x^6 - a^6$.

8. $3x^4 - 18x^3 + 27x^2$.

*9. $2x^3 + 3x^2 - 2x$.

10. $(x - 1)^2 - (x^2 - 1)$.

*11. $x^3 - x^2 - x + 1$.

12. $r^2 - a^2 - 2ac - c^2$.

*13. $3x^4 - 24x^2 + 48$.

14. $(x + c)^3 - (x^3 + c^3)$.

*15. $x^4 - (x - 1)^4$.

16. $6x^5 - 4x^4 - 10x^3$.

*17. $(x + 2)^2 - (x^2 - 4)^2$.

18. $64x^2 - (x^2 + 16)^2$.

*19. $x^3 + x^2 + (12x - 45)(x + 1)$.

20. $6(x - y) - 10(x^2 - y^2)$.

Optional | *21. $x^2 + 9y^2 + 6xy - 1$.

22. $x^3 - c^2 + 2cx - c^3 - x^2$.

*23. $12 - 5(2x - 1) - 3(2x - 1)^2$.

24. $(x^2 + y^2)^2 + x^2y^2 + 3x^4 - 2y^4$.

1.10 Quotients of Algebraic Expressions

You have seen how to add, subtract, and multiply algebraic expressions. Here we take up division.

A common way to indicate a quotient is to write it as a fraction: for $B \neq 0$

$$A \div B \quad \text{can be written} \quad \frac{A}{B}.$$

As in arithmetic you can test whether two quotients are equal by cross-multiplying: for $B \neq 0$, $D \neq 0$

(1.10.1)
$$\boxed{\frac{A}{B} = \frac{C}{D} \quad \text{if and only if} \quad AD = BC.}$$

If both numerator and denominator are multiplied by the same nonzero factor, then the value of the quotient is unchanged: if $B \neq 0$,

(1.10.2)
$$\boxed{\frac{CA}{CB} = \frac{A}{B} \quad \text{provided that } C \neq 0.}^{\dagger}$$

In other words, nonzero factors common to both numerator and denominator can be cancelled.

---

†This follows directly from (1.10.1).

Thus

$$\frac{14x}{16x} = \frac{(2x)7}{(2x)8} = \frac{7}{8} \qquad \text{for } x \neq 0;$$

$$\frac{3x + 15}{9x^2 + 3} = \frac{3(x + 5)}{3(3x^2 + 1)} = \frac{x + 5}{3x^2 + 1} \qquad \text{for all real } x;$$

$$\frac{r^3 s^2}{r^2 s^2 + r^2} = \frac{r^2(rs^2)}{r^2(s^2 + 1)} = \frac{rs^2}{s^2 + 1} \qquad \text{for } r \neq 0, \text{ all real } s;$$

$$\frac{x^2 + x - 2}{x^2 + 2x - 3} = \frac{(x - 1)(x + 2)}{(x - 1)(x + 3)} = \frac{x + 2}{x + 3} \qquad \text{for } x \neq 1, -3;$$

$$\frac{(x^2 + 4)^2}{(x^4 + 8x^2 + 16)^2(x - 1)} = \frac{(x^2 + 4)^2}{(x^2 + 4)^4(x - 1)} = \frac{1}{(x^2 + 4)^2(x - 1)} \qquad \text{for } x \neq 1. \quad \square$$

To add or subtract algebraic fractions, we first give them a common denominator, one that contains each of the initial denominators as a factor:

$$\frac{6x}{7} + \frac{2x}{3} = \frac{18x}{21} + \frac{14x}{21} = \frac{32x}{21}, \qquad \frac{6x}{7} - \frac{2x}{3} = \frac{18x}{21} - \frac{14x}{21} = \frac{4x}{21}, \qquad \text{(all real } x)$$

$$\frac{a}{b} + \frac{c}{d} = \frac{ad}{bd} + \frac{bc}{bd} = \frac{ad + bc}{bd}, \qquad \frac{a}{b} - \frac{c}{d} = \frac{ad}{bd} - \frac{bc}{bd} = \frac{ad - bc}{bd}. \qquad (b, d \neq 0)$$

For $x \neq -1, -2$

$$\frac{1}{x + 1} + \frac{1}{x + 2} = \frac{x + 2}{(x + 1)(x + 2)} + \frac{x + 1}{(x + 1)(x + 2)} = \frac{2x + 3}{(x + 1)(x + 2)}$$

and

$$\frac{2x}{(x + 1)^2} - \frac{3}{(x + 1)(x + 2)} = \frac{2x(x + 2)}{(x + 1)^2(x + 2)} - \frac{3(x + 1)}{(x + 1)^2(x + 2)}$$

$$= \frac{2x(x + 2) - 3(x + 1)}{(x + 1)^2(x + 2)}$$

$$= \frac{2x^2 + 4x - 3x - 3}{(x + 1)^2(x + 2)}$$

$$= \frac{2x^2 + x - 3}{(x + 1)^2(x + 2)}. \quad \square$$

Exercises

Reduce these fractions by cancelling the factors common to numerator and denominator.

*1. $\dfrac{2 + 4x}{2x^2 - 2}$.

2. $\dfrac{4x^2 + 20x}{8x^4}$.

*3. $\dfrac{100x^2}{20x^3 + 10x}$.

4. $\dfrac{x^2 + 1}{x^4 - 1}$.

*5. $\dfrac{x^2 - 5x + 6}{2x - 6}$.

6. $\dfrac{4x^3}{x^3 + 2x}$.

*7. $\dfrac{(x^2 + 4)^2(x - 1)}{(x - 1)^2(x^2 + 4)}$.

8. $\dfrac{-2^{2n}x^n}{2^n x^{2n}}$.

*9. $\dfrac{x^2 + 4x + 3}{x^2 - 4x - 5}$.

10. $\dfrac{x^2 + 9x + 14}{2x^2 + 13x - 7}$.

Complete the equation.

*11. $\dfrac{1}{2x} + \dfrac{1}{3x} = \dfrac{}{6x}$, $x \neq 0$.

12. $\dfrac{1}{x + 1} - \dfrac{1}{x + 2} = \dfrac{}{(x + 1)(x + 2)}$, $x \neq -1, -2$.

*13. $\dfrac{2}{x} + \dfrac{3}{x^2 + 1} = \dfrac{}{x(x^2 + 1)}$, $x \neq 0$

14. $\dfrac{2}{x + 4} - \dfrac{1}{(x + 4)(x + 2)} = \dfrac{}{(x + 4)(x + 2)}$, $x \neq -2, -4$.

*15. $\dfrac{2x - 1}{x^2} - \dfrac{3x + 4}{x^2(x + 1)} = \dfrac{}{x^2(x + 1)}$, $x \neq 0, -1$.

16. $x + \dfrac{1}{x} + \dfrac{2}{x^2} = \dfrac{}{x^2}$, $x \neq 0$.

*17. $x - \dfrac{2}{x} + \dfrac{3}{x^2} = \dfrac{}{x^2}$, $x \neq 0$.

18. $\dfrac{1}{x - 1} - \dfrac{2}{x - 2} + \dfrac{3}{x - 3} = \dfrac{}{(x - 1)(x - 2)(x - 3)}$, $x \neq 1, 2, 3$.

*19. $\dfrac{4x}{(x + 2)^2} - \dfrac{5}{(x + 2)(x + 5)} = \dfrac{}{(x + 2)^2(x + 5)}$, $x \neq -2, -5$.

20. $\dfrac{3}{x^2 + 1} - \dfrac{2}{x^2 + 2} + \dfrac{1}{x^2 + 3} = \dfrac{}{(x^2 + 1)(x^2 + 2)(x^2 + 3)}$,   all real x.

1.11   Long Division of Polynomials

You are familiar with arithmetical long division:

$$
\begin{array}{r}
284 \\
17\,\overline{)\,4836} \\
34 \\
\hline
1436 \\
136 \\
\hline
76 \\
68 \\
\hline
8
\end{array}
$$

so that

$$\tfrac{4836}{17} = 284\tfrac{8}{17}.$$

You can check this by verifying that $(284)(17) + 8 = 4836$.

A similar technique works with algebraic expressions of the form

$$a_0 x^n + a_1 x^{n-1} + \cdots + a_{n-1}x + a_n,$$

called *polynomials*.† For example, to carry out the division

$$3x^4 - x^3 + x^2 - x + 2 \quad \div \quad x^2 + 2x + 4$$

we write

$$x^2 + 2x + 4\,\overline{)\,3x^4 - x^3 + x^2 - x + 2}.$$

Since x^2 goes into $3x^4$ exactly $3x^2$ times, we write

$$
\begin{array}{r}
3x^2 \\
x^2 + 2x + 4\,\overline{)\,3x^4 - x^3 + x^2 - x + 2}.
\end{array}
$$

We multiply $3x^2$ by $x^2 + 2x + 4$, record the product, and subtract:

$$
\begin{array}{r}
3x^2 \\
x^2 + 2x + 4\,\overline{)\,3x^4 - x^3 + x^2 - x + 2} \\
3x^4 + 6x^3 + 12x^2 \\
\hline
- 7x^3 - 11x^2 - x + 2.
\end{array}
$$

---

† Here $a_0, a_1, \cdots, a_{n-1}, a_n$ are real numbers and n is a positive integer. Polynomials are discussed at length in Chapter 4.

Now we continue, first dividing x^2 into $-7x^3 - 11x^2 - x + 2$:

$$
\begin{array}{r}
3x^2 - 7x + 3 \\
x^2 + 2x + 4\,\overline{\smash{)}\,3x^4 - x^3 + x^2 - x + 2} \\
\underline{3x^4 + 6x^3 + 12x^2} \\
-7x^3 - 11x^2 - x + 2 \\
\underline{-7x^3 - 14x^2 - 28x} \\
3x^2 + 27x + 2 \\
\underline{3x^2 + 6x + 12} \\
21x - 10.
\end{array}
$$

We can go no further:

$$\frac{3x^4 - x^3 + x^2 - x + 2}{x^2 + 2x + 4} = 3x^2 - 7x + 3 + \frac{21x - 10}{x^2 + 2x + 4}.$$

You can check the result by verifying that

$$(3x^2 - 7x + 3)(x^2 + 2x + 4) + 21x - 10 = 3x^4 - x^3 + x^2 - x + 2. \quad \square$$

Problem. Carry out the indicated division:

$$x^3 + 2 \;\div\; x + 1.$$

SOLUTION. We could begin by writing

$$
\begin{array}{r}
x^2 \\
x + 1\,\overline{\smash{)}\,x^3 + 2} \\
\underline{x^3 + x^2}
\end{array}
$$

but since there are no x^2's and no x's in the dividend, it works out more simply to move the 2 out to the proper column:

$$
\begin{array}{r}
x^2 - x + 1 \\
x + 1\,\overline{\smash{)}\,x^3 \qquad\quad + 2} \\
\underline{x^3 + x^2} \\
-x^2 \quad + 2 \\
\underline{-x^2 - x} \\
x + 2 \\
\underline{x + 1} \\
1
\end{array}
$$

so that

$$\frac{x^3 + 2}{x + 1} = x^2 - x + 1 + \frac{1}{x + 1}. \quad \square$$

Problem.   Calculate by long division:

$$\frac{x^2 - x^3 + 5x - 125}{5 + x}.$$

SOLUTION.   First we rewrite the numerator and denominator according to descending powers of x:

$$\frac{-x^3 + x^2 + 5x - 125}{x + 5}.$$

The rest is straightforward:

$$
\begin{array}{r}
-x^2 + 6x\ -\ 25 \\
x + 5\ \overline{\smash{\big)}\ -x^3 +\ x^2 +\ 5x\ -\ 125} \\
\underline{-x^3 -\ 5x^2} \\
6x^2 +\ 5x\ -\ 125 \\
\underline{6x^2 + 30x} \\
-25x\ -\ 125 \\
\underline{-25x\ -\ 125} \\
0.
\end{array}
$$

The quotient is

$$-x^2 + 6x - 25. \quad \square$$

Problem.   Factor

$$x^3 + 3x^2 - 4x - 12$$

completely, given that $x - 2$ is a factor.

SOLUTION.   We begin by dividing $x - 2$ into $x^3 + 3x^2 - 4x - 12$:

$$
\begin{array}{r}
x^2 + 5x + 6 \\
x - 2\ \overline{\smash{\big)}\ x^3 + 3x^2 -\ 4x - 12} \\
\underline{x^3 - 2x^2} \\
5x^2 -\ 4x - 12 \\
\underline{5x^2 - 10x} \\
6x - 12 \\
\underline{6x - 12} \\
0.
\end{array}
$$

We know now that

$$x^3 + 3x^2 - 4x - 12 = (x - 2)(x^2 + 5x + 6),$$

but we still have not factored completely; namely,

$$x^2 + 5x + 6 = (x + 3)(x + 2),$$

so that

$$x^3 + 3x^2 - 4x - 12 = (x - 2)(x + 3)(x + 2). \quad \square$$

Problem. Factor

$$x^4 + x^3 - 7x^2 - 13x - 6$$

completely, given that $x^2 - x - 6$ is a factor.

SOLUTION. We begin by dividing:

$$
\begin{array}{r}
x^2 + 2x + 1 \\
x^2 - x - 6\overline{\smash{\big)}\ x^4 + x^3 - 7x^2 - 13x - 6} \\
\underline{x^4 - x^3 - 6x^2} \\
2x^3 - x^2 - 13x - 6 \\
\underline{2x^3 - 2x^2 - 12x} \\
x^2 - x - 6 \\
\underline{x^2 - x - 6} \\
0.
\end{array}
$$

From this, we get

$$x^4 + x^3 - 7x^2 - 13x - 6 = (x^2 - x - 6)(x^2 + 2x + 1),$$

but we can still factor some more:

$$x^2 - x - 6 = (x - 3)(x + 2)$$

and

$$x^2 + 2x + 1 = (x + 1)^2.$$

Factored completely,

$$x^4 + x^3 - 7x^2 - 13x - 6 = (x - 3)(x + 2)(x + 1)^2. \quad \square$$

Exercises

Calculate by long division:

*1. $\dfrac{x^2 - 10}{x + 5}$.

2. $\dfrac{x^5 + 4x^3 + 3x^2 + 12}{x^2 + 4}$.

*3. $\dfrac{2x^3 + 150}{x - 5}$.

4. $\dfrac{x^4 + 12x^3 + 33x^2 + 36x + 9}{x^2 + 9x + 3}$.

*5. $\dfrac{x^4 + 2x^2 + 1}{x^2 - 2}$.

6. $\dfrac{x^4 - 2x + 1}{x^2 - 2x}$.

*7. $\dfrac{2x^5 - 7x^3 + 2x + 1}{2x^3 + x + 1}$.

8. $\dfrac{6x^4 + 2x + 1}{2x^2 + x + 1}$.

Factor as far as you can:

*9. $x^4 + 2x^3 - 13x^2 - 14x + 24$ given that $x^2 + x - 12$ is a factor.

10. $20x^4 - 60x^3 - 60x^2 + 220x - 120$ given that $(x - 1)^2$ is a factor.

*11. $x^3 - (2a + b)x^2 + (a^2 + 2ab)x - a^2b$ given that $x - b$ is a factor.

12. $2x^4 - 6x^3 + 5x^2 - 3x + 2$ given that $2x^2 + 1$ is a factor.

1.12 Additional Exercises

*1. Which of the following numbers are rational?
     (a) $1 + \sqrt{2}$,   (b) 3.1416,   (c) $(1 + \sqrt{2})(1 - \sqrt{2})$,   (d) $(1 + \sqrt{2})^2$.

*2. Show that the sum and product of irrational numbers need not be irrational.

Multiply out:

*3. $(x + y)(x - y)(x^2 + y^2)$.   *4. $(x - a)(x^2 - 3x + 1)$.

*5. $(x^2 + 2x - 1)(2x^2 - 3x + 4)$.   *6. $(x + y)^2(x - y)^2$.

*7. $(x - y)(x^3 + y^3)$.   *8. $(x + 2a)(x - a)^3$.

Factor as far as you can:

*9. $9x^2 - 100$.   *10. $x^2 - 5$.   *11. $27x^3 + 8$.

*12. $x^2 + 8x + 7$.   *13. $5x^2 + 12x + 7$.   *14. $2x^2 + 3x - 9$.

*15. $6ax^2 + 8a$.   *16. $a^4x^4 - b^4$.   *17. $16x^3 + 12x^2 - 10x$.

*18. $a^4b^6x^6 - a^4$.   *19. $4x^2 - (x - 2)^2$. *20. $2x^3 - x + 2a^3 - a$.

Complete the equation:

*21. $\dfrac{x^2 + 6x + 9}{(x + 3)(x - 4)} = \dfrac{x + 3}{}$, $x \neq -3, 4$.

*22. $\dfrac{}{(x - a)(x - b)} = \dfrac{x - a}{x - b}$, $x \neq a, b$.

*23. $\dfrac{1}{x^2 + 6x + 8} = \dfrac{1}{x + 4}$, $x \neq -4, -2$.

*24. $\dfrac{x^3 + a^3}{} = \dfrac{x + a}{2}$, all real x.

*25. $\dfrac{1}{2x} - \dfrac{3}{x + 1} = \dfrac{1}{2x(x + 1)}$, $x \neq 0, -1$.

*26. $\dfrac{1}{ax + b} - \dfrac{1}{cx + d} = \dfrac{1}{(ax + b)(cx + d)}$, $x \neq -b/a, -d/c$.

*27. $\dfrac{3x + 2}{2x^2 + 1} - \dfrac{2x + 1}{3x^2 + 1} = \dfrac{1}{(3x^2 + 1)(2x^2 + 1)}$, all real x.

*28. $\dfrac{1}{(x + 1)(x + 2)} - \dfrac{1}{(x + 2)(x + 3)} = \dfrac{1}{(x + 1)(x + 2)(x + 3)}$, $x \neq -1, -2, -3$.

Calculate by long division:

*29. $\dfrac{x^3 - x^2 - x + 10}{x + 2}$.

*30. $\dfrac{4x^3 + 5x + 3}{2x^2 - x + 3}$.

*31. $\dfrac{x^5}{x^2 + 1}$.

*32. Factor $3x^5 + x^3 + 24x^2 + 8$ as far as you can given that $3x^2 + 1$ is a factor.

*33. Factor $x^4 + 14x^3 + 65x^2 + 112x + 60$ as far as you can given that $(x + 2)$ and $(x + 5)$ are factors.

*34. Show that if p is an odd integer, then p^2 is odd. HINT: If p is odd, then p is of the form $2n + 1$ with n an integer.

*35. Show that if an integer p is not divisible by 3, then p^2 is not divisible by 3.

QUADRATIC EQUATIONS, INEQUALITIES, ABSOLUTE VALUE

2

2.1 Square Roots

We begin by discussing square roots in some detail; nth roots appear in Section 4.7.

If a is a nonnegative number, then $\sqrt{a}$, called *the square root of a*, is the unique nonnegative number whose square is a:

(2.1.1)
$$\boxed{\sqrt{a} \geq 0 \quad \text{and} \quad (\sqrt{a})^2 = a.}$$

For example,

$$\sqrt{4} = 2, \qquad \sqrt{6.25} = 2.5, \qquad \sqrt{\tfrac{1}{100}} = \tfrac{1}{10}.\dagger$$

These examples are somewhat artificial because, in general, the square root of a rational number is not rational and therefore cannot be expressed as a terminating decimal, nor as a repeating decimal. Usually, the best we can do is give an approximating decimal. The following are all approximations, in each case to the nearest thousandth‡:

$$\sqrt{2} \cong 1.414, \qquad \sqrt{3} \cong 1.732, \qquad \sqrt{9.5} \cong 3.082.$$

---

†A word on negative square roots seems in order. If a is positive, then there are *two* real numbers whose square is a:

$$\sqrt{a} \quad \text{and} \quad -\sqrt{a}.$$

Only the first of these is called *the square root of a*. The second one, $-\sqrt{a}$, is called *the negative square root of a*.

‡The symbol $\cong$ will be used throughout this book to indicate approximate equality. By "$a \cong b$" we mean "a is approximately equal to b."

In dealing with square roots, keep in mind that, for a and b positive,

(2.1.2)
$$\sqrt{ab} = \sqrt{a}\sqrt{b} \quad \text{and} \quad \sqrt{\frac{a}{b}} = \frac{\sqrt{a}}{\sqrt{b}}.$$

You can verify these relations by squaring both sides. Their usefulness is illustrated by the following computations:

$$\sqrt{18} = \sqrt{(9)(2)} = \sqrt{9}\sqrt{2} = 3\sqrt{2},$$

$$\sqrt{14400} = \sqrt{(144)(100)} = \sqrt{144}\sqrt{100} = (12)(10) = 120,$$

$$\sqrt{\frac{25}{64}} = \frac{\sqrt{25}}{\sqrt{64}} = \frac{5}{8},$$

$$\sqrt{\frac{1}{100}} = \frac{\sqrt{1}}{\sqrt{100}} = \frac{1}{10},$$

$$\sqrt{0.0121} = \sqrt{(121)(0.0001)} = \sqrt{121}\sqrt{0.0001} = (11)(0.01) = 0.11.$$

Table 2.1.1 gives the square roots of the integers 2 through 19 rounded off to the nearest thousandth.†

Table 2.1.1

n	$\sqrt{n}$	n	$\sqrt{n}$
2	1.414	11	3.317
3	1.732	12	3.464
4	2.000	13	3.606
5	2.236	14	3.742
6	2.449	15	3.873
7	2.646	16	4.000
8	2.828	17	4.123
9	3.000	18	4.243
10	3.162	19	4.359

Problem. Use Table 2.1.1 to estimate the following square roots:

(a) $\sqrt{40}$.   (b) $\sqrt{72}$.   (c) $\sqrt{0.63}$.   (d) $\sqrt{\frac{5}{16}}$.

(Use a hand calculator, if available, to check your answers.)

---

†A more extended table appears at the end of the book, but throughout this section all numerical calculations will be based on Table 2.1.1.

SOLUTION

(a) $\sqrt{40} = \sqrt{(4)(10)} = \sqrt{4}\sqrt{10} = 2\sqrt{10} \cong 2(3.162) = 6.324.$

(b) $\sqrt{72} = \sqrt{(36)(2)} = \sqrt{36}\sqrt{2} = 6\sqrt{2} \cong 6(1.414) = 8.484.$

(c) $\sqrt{0.63} = \sqrt{\dfrac{63}{100}} = \dfrac{\sqrt{9}\sqrt{7}}{\sqrt{100}} = \dfrac{3}{10}\sqrt{7} \cong \dfrac{3}{10}(2.646) = 0.7938.$

(d) $\sqrt{\dfrac{5}{16}} = \dfrac{\sqrt{5}}{\sqrt{16}} = \dfrac{\sqrt{5}}{4} \cong \dfrac{2.236}{4} = 0.559.$ $\square$

Rationalizing the Denominator

Radicals, $\sqrt{}$, left in the denominator tend to be troublesome. To estimate $1/\sqrt{3}$ we could write

$$\frac{1}{\sqrt{3}} \cong \frac{1}{1.732},$$

but then we would have to divide 1.732 into 1. The computations become easier if we first clear the radical from the denominator. We call this "rationalizing the denominator," and in the case of $1/\sqrt{3}$, we do it by multiplying both numerator and denominator by $\sqrt{3}$:

$$\frac{1}{\sqrt{3}} = \frac{1}{\sqrt{3}} \cdot \frac{\sqrt{3}}{\sqrt{3}} = \frac{\sqrt{3}}{3} \cong \frac{1.732}{3} \cong 0.577.$$

In general, we rationalize the denominator of $1/\sqrt{n}$ by multiplying numerator and denominator by $\sqrt{n}$:

$$\frac{1}{\sqrt{n}} = \frac{1}{\sqrt{n}} \cdot \frac{\sqrt{n}}{\sqrt{n}} = \frac{\sqrt{n}}{n}.$$

To rationalize a denominator of the form $a\sqrt{m} - b\sqrt{n}$ we multiply numerator and denominator by $a\sqrt{m} + b\sqrt{n}$, and vice versa:

$$\frac{1}{5 - \sqrt{2}} = \frac{1}{5 - \sqrt{2}} \cdot \frac{5 + \sqrt{2}}{5 + \sqrt{2}} = \frac{5 + \sqrt{2}}{25 - (\sqrt{2})^2} = \frac{5 + \sqrt{2}}{25 - 2} = \frac{5 + \sqrt{2}}{23},$$

$$\frac{1}{5\sqrt{3} + \sqrt{2}} = \frac{1}{5\sqrt{3} + \sqrt{2}} \cdot \frac{5\sqrt{3} - \sqrt{2}}{5\sqrt{3} - \sqrt{2}}$$

$$= \frac{5\sqrt{3} - \sqrt{2}}{(5\sqrt{3})^2 - (\sqrt{2})^2} = \frac{5\sqrt{3} - \sqrt{2}}{(25)(3) - 2} = \frac{5\sqrt{3} - \sqrt{2}}{73}.$$

$$\frac{\sqrt{2}}{\sqrt{5} - 2\sqrt{2}} = \frac{\sqrt{2}}{\sqrt{5} - 2\sqrt{2}} \cdot \frac{\sqrt{5} + 2\sqrt{2}}{\sqrt{5} + 2\sqrt{2}}$$

$$= \frac{(\sqrt{2})(\sqrt{5}) + (\sqrt{2})(2\sqrt{2})}{(\sqrt{5})^2 - (2\sqrt{2})^2} = \frac{\sqrt{10} + 4}{5 - (4)(2)} = -\frac{4 + \sqrt{10}}{3}.$$

Problem. Use Table 2.1.1 to estimate

$$\frac{4}{\sqrt{15} + 2\sqrt{3}}.$$

SOLUTION. First we rationalize the denominator:

$$\frac{4}{\sqrt{15} + 2\sqrt{3}} = \frac{4}{\sqrt{15} + 2\sqrt{3}} \cdot \frac{\sqrt{15} - 2\sqrt{3}}{\sqrt{15} - 2\sqrt{3}}$$

$$= \frac{4\sqrt{15} - 8\sqrt{3}}{(\sqrt{15})^2 - (2\sqrt{3})^2} = \frac{4\sqrt{15} - 8\sqrt{3}}{15 - 12} = \frac{4\sqrt{15} - 8\sqrt{3}}{3}.$$

From Table 2.1.1,

$$\frac{4\sqrt{15} - 8\sqrt{3}}{3} \cong \frac{4(3.873) - 8(1.732)}{3}$$

$$= \frac{15.492 - 13.856}{3} = \frac{1.636}{3} \cong 0.545. \quad \square$$

Optional | **Problem.** Show that $\sqrt{2}$ is not rational.

SOLUTION. If on the contrary $\sqrt{2}$ is rational, then we can write it as a fraction. By cancelling all factors common to numerator and denominator we reduce the fraction to lowest terms. Suppose then that

(*) $\sqrt{2} = p/q$ with p/q in lowest terms.

Squaring both sides, we have

$$2 = p^2/q^2,$$

and therefore

(**) $2q^2 = p^2.$

This is an equation between positive integers. It tells us that p^2 is a multiple of 2, and thus that p^2 is even, which means that p itself must be

Optional | even.† Therefore we can write

$$p = 2r.$$

Substituting $2r$ for p in (∗∗), we get

$$2q^2 = (2r)^2 = 4r^2$$

and, dividing by 2,

$$q^2 = 2r^2.$$

This last equation tells us that q^2 is even, and therefore that q is even.
 With both p and q even, the fraction p/q *cannot be* in lowest terms;
but by (∗) it *is* in lowest terms. Thus our assumption that $\sqrt{2}$ is rational
has led to a contradiction. It follows that $\sqrt{2}$ is not rational.   □

Exercises

Use Table 2.1.1 to estimate the following square roots:‡

*1. $\sqrt{50}$.   2. $\sqrt{44}$.   *3. $\sqrt{180}$.   4. $\sqrt{7500}$.

*5. $\sqrt{0.5}$.   6. $\sqrt{0.76}$.   *7. $\sqrt{0.0019}$.   8. $\sqrt{1.12}$.

*9. $\sqrt{\frac{3}{4}}$.   10. $\sqrt{\frac{5}{36}}$.   *11. $\sqrt{\frac{24}{49}}$.   12. $\sqrt{\sqrt{171} - \sqrt{76}}$.

Rationalize the denominator:

*13. $\dfrac{1}{\sqrt{2}}$ 14. $\dfrac{1}{\sqrt{5}}$.   *15. $\dfrac{1}{1 + \sqrt{6}}$.

*16. $\dfrac{1}{\sqrt{11} - \sqrt{10}}$.   17. $\dfrac{\sqrt{3}}{\sqrt{5} + \sqrt{3}}$.   *18. $\dfrac{\sqrt{5}}{\sqrt{5} - \sqrt{3}}$.

*19. $\dfrac{\sqrt{5} + \sqrt{2}}{\sqrt{5} - \sqrt{2}}$.   20. $\dfrac{\sqrt{x} + \sqrt{y}}{\sqrt{x} - \sqrt{y}}$.   *21. $\dfrac{\sqrt{3}}{\sqrt{3} + \sqrt{x}}$.

Rationalize the denominator and, if possible, simplify:

*22. $\dfrac{3\sqrt{2}}{\sqrt{6}}$.   23. $\dfrac{\sqrt{3}}{\sqrt{15}}$.   *24. $\dfrac{12\sqrt{7}}{\sqrt{14} - \sqrt{2}}$.

*25. $\dfrac{x - y}{\sqrt{x} - \sqrt{y}}$.   26. $\dfrac{1}{\sqrt{x + 1} + \sqrt{x}}$.   *27. $\dfrac{1}{\sqrt{x - h} - \sqrt{x + h}}$.

---

†If p were odd, p^2 would be odd. (Exercise 34, Section 1.12)
‡Use a hand calculator, if available, to check your answers.

*28.   Express $\sqrt{\frac{3}{5}} + \sqrt{\frac{5}{3}}$ as a rational multiple of $\sqrt{15}$.

29.   Express $\sqrt{\frac{9}{5}} - \sqrt{\frac{5}{4}}$ as a rational multiple of $\sqrt{5}$.

30.   Show that $\sqrt{a^2}$ is not necessarily a.

*31.   Give a decimal estimate for the hypotenuse given that the legs of the right triangle have lengths $4\sqrt{3}$ and $5\sqrt{2}$.

*32.   Estimate

$$100 - 20\sqrt{\tfrac{1}{5}p}$$

(a) at $p = 1$.   (b) at $p = 2$.

33.   Estimate

$$\frac{\sqrt{x + h} - \sqrt{x}}{h}$$

(a) at $x = 1$, $h = 0.6$.   (b) at $x = 1$, $h = 0.1$.

Optional | 34.   Show that $\sqrt{3}$ is not rational.

2.2   Quadratic Equations

A *quadratic* equation in x is an equation of the form

$$ax^2 + bx + c = 0 \qquad \text{with } a, b, c \text{ real and } a \neq 0.$$

To *solve* such an equation is to find the numbers that satisfy it. These numbers are called the *roots* of the equation.

Some quadratics can be readily solved by factoring.

Problem.   Solve

$$x^2 - x - 2 = 0.$$

SOLUTION

$$x^2 - x - 2 = (x + 1)(x - 2) = 0.$$

Since a product is 0 iff† one of the factors is 0, we know that either

$$x + 1 = 0 \quad \text{or} \quad x - 2 = 0.$$

---

† By "iff" we mean "if and only if." This expression is used so often in mathematics that it's convenient to have an abbreviation for it.

Thus either

$$x = -1 \quad \text{or} \quad x = 2.$$

The roots are -1 and 2.

You can check this result by substituting -1 and 2 into the original equation:

$$(-1)^2 - (-1) - 2 = 1 + 1 - 2 \overset{\checkmark}{=} 0,$$
$$(2)^2 - 2 - 2 = 4 - 2 - 2 \overset{\checkmark}{=} 0. \quad \square$$

Problem. Solve

$$3x^2 + 12x + 12 = 0.$$

SOLUTION

$$3x^2 + 12x + 12 = 3(x^2 + 4x + 4) = 3(x + 2)^2 = 0.$$

Obviously,

$$x + 2 = 0$$

and

$$x = -2.$$

The number -2 is the only root.

Checking:

$$3(-2)^2 + 12(-2) + 12 = 3(4) - 24 + 12 \overset{\checkmark}{=} 0. \quad \square$$

Completing the Square

The equations in the last two problems were easy to handle because in each instance the quadratic was easy to factor. When this is not the case, we use a technique called "completing the square."

The technique is based on the observation that we can turn an expression of the form

$$x^2 + Ax$$

into a perfect square simply by adding $(\frac{1}{2}A)^2$:

$$x^2 + Ax + (\tfrac{1}{2}A)^2 = (x + \tfrac{1}{2}A)^2. \qquad \text{(check this)}$$

The next two problems will serve to illustrate the idea.

Problem. Solve

$$2x^2 + 6x - 1 = 0.$$

SOLUTION. Once we have satisfied ourselves that we cannot factor the quadratic by inspection, we proceed as follows. First we divide by 2, the *leading* coefficient. This gives

$$x^2 + 3x - \tfrac{1}{2} = 0.$$

We then clear the constant term from the left side by adding $\tfrac{1}{2}$ to both sides:

$$x^2 + 3x = \tfrac{1}{2}.$$

The left side is now of the form

$$x^2 + Ax \qquad \text{with} \quad A = 3.$$

To make it a perfect square, we add $(\tfrac{1}{2}A)^2 = (\tfrac{3}{2})^2$ to both sides:

$$x^2 + 3x + (\tfrac{3}{2})^2 = \tfrac{1}{2} + (\tfrac{3}{2})^2.$$

Rewriting the left-hand side as $(x + \tfrac{3}{2})^2$ and carrying out the arithmetic on the right, we have

$$(x + \tfrac{3}{2})^2 = \tfrac{11}{4},$$

so that

$$x + \tfrac{3}{2} = \pm\sqrt{\tfrac{11}{4}} = \pm\tfrac{1}{2}\sqrt{11}$$

and

$$x = -\tfrac{3}{2} \pm \tfrac{1}{2}\sqrt{11}.$$

The roots are

$$-\tfrac{3}{2} + \tfrac{1}{2}\sqrt{11} \quad \text{and} \quad -\tfrac{3}{2} - \tfrac{1}{2}\sqrt{11}. \quad \square$$

Problem. Solve

$$20x^2 + 31x + 12 = 0.$$

SOLUTION

$$20x^2 + 31x + 12 = 0,$$

$$x^2 + \tfrac{31}{20}x + \tfrac{12}{20} = 0, \qquad \text{(we divide by 20)}$$

$$x^2 + \tfrac{31}{20}x = -\tfrac{12}{20}, \qquad \text{(transfer the constant term)}$$

$$x^2 + \tfrac{31}{20}x + (\tfrac{31}{40})^2 = -\tfrac{12}{20} + (\tfrac{31}{40})^2, \qquad \text{(complete the square on the left)}$$

$$(x + \tfrac{31}{40})^2 = \tfrac{1}{1600},$$

$$x + \tfrac{31}{40} = \pm\sqrt{\tfrac{1}{1600}} = \pm\tfrac{1}{40},$$

$$x = -\tfrac{31}{40} \pm \tfrac{1}{40},$$

so that

$$x = -\tfrac{31}{40} + \tfrac{1}{40} = -\tfrac{30}{40} = -\tfrac{3}{4} \quad \text{or} \quad x = -\tfrac{32}{40} = -\tfrac{4}{5}.$$

The roots are

$$-\tfrac{3}{4} \quad \text{and} \quad -\tfrac{4}{5}. \quad \square$$

The equation we just solved by completing the square could have been solved directly by factoring. Had we seen from the beginning that

$$20x^2 + 31x + 12 = (5x + 4)(4x + 3),$$

we could have set

$$(5x + 4)(4x + 3) = 0$$

and quickly concluded that

$$x = -\tfrac{4}{5} \quad \text{or} \quad x = -\tfrac{3}{4}.$$

But this is all hindsight.

Problem. Solve

$$x^2 - 2x + 7 = 0.$$

SOLUTION. Since the leading coefficient is already 1, we begin by transferring the constant term:

$$x^2 - 2x = -7.$$

To complete the square on the left, we add 1 to both sides. This gives

$$x^2 - 2x + 1 = -7 + 1$$

and therefore

$$(x + 1)^2 = -6.$$

There is no point going further. With x real, $(x + 1)^2$ is nonnegative and cannot possibly equal -6.

The equation has no real roots. $\square$

Summary

To solve a quadratic equation, first try factoring the quadratic. If you cannot do this by inspection, then divide by the leading coefficient, transfer the constant term, and complete the square. There are three possibilities: two real roots, one real root, no real roots.

Exercises

Solve by factoring directly:

*1. $x^2 + 5x + 6 = 0.$ 2. $x^2 + x - 6 = 0.$

*3. $x^2 + 3x - 4 = 0.$ 4. $9x^2 - 12x + 4 = 0.$

*5. $x^2 - 5x - 14 = 0.$ 6. $5x^2 - 30x + 45 = 0.$

*7. $3x^2 + 7x + 4 = 0.$ 8. $8x^2 + 31x - 4 = 0.$

*9. $12x^2 - 11x + 2 = 0.$ 10. $3x^2 + 14x - 5 = 0.$

Solve by completing the square:

*11. $x^2 - 12x + 35 = 0.$ 12. $2x^2 + 8x - 6 = 0.$

*13. $x^2 + 6x + 6 = 0.$ 14. $2x^2 + 6x + 3 = 0.$

*15. $5x^2 + 10x + 3 = 0.$ 16. $2x^2 + 4x + 6 = 0.$

*17. $4x^2 + 4x + 1 = 0.$ 18. $x^2 + 2x + 2 = 0.$

*19. $3x^2 + x - 6 = 0.$ 20. $x^2 + 4x + 5 = 0.$

*21. $16x^2 + 5x - 1 = 0.$ 22. $3x^2 + 2\sqrt{6}x + 2 = 0.$

2.3 The General Quadratic Formula

In the last section you saw how to solve quadratic equations by completing the square. Here we apply that method to the general quadratic equation

$$ax^2 + bx + c = 0, \qquad a \neq 0.$$

This will give us *the general quadratic formula.*
First we divide by the leading coefficient a:

$$x^2 + \frac{b}{a}x + \frac{c}{a} = 0.$$

Then we transfer the constant term c/a to the right-hand side:

$$x^2 + \frac{b}{a}x = -\frac{c}{a}.$$

To complete the square on the left, we add $(b/2a)^2$ to both sides of the equation. This gives

$$x^2 + \frac{b}{a}x + \left(\frac{b}{2a}\right)^2 = -\frac{c}{a} + \left(\frac{b}{2a}\right)^2$$

which, as you can check, we can rewrite as

$$(*) \qquad \left(x + \frac{b}{2a}\right)^2 = \frac{b^2 - 4ac}{4a^2}.$$

The rest depends on the sign of $b^2 - 4ac$. This expression is called the *discriminant*.

CASE 1. $b^2 - 4ac \geq 0$. With $b^2 - 4ac$ nonnegative, we can take the square root of both sides of ($*$) and conclude that

$$x + \frac{b}{2a} = \pm \sqrt{\frac{b^2 - 4ac}{4a^2}} = \frac{\pm \sqrt{b^2 - 4ac}}{2a}$$

and therefore

(2.3.1)
$$\boxed{x = -\frac{b \pm \sqrt{b^2 - 4ac}}{2a}.}$$

This is the *general quadratic formula*. It gives two real roots when $b^2 - 4ac > 0$; but only one root,

$$x = -\frac{b}{2a},$$

when $b^2 - 4ac = 0$.

CASE 2. $b^2 - 4ac < 0$. With $b^2 - 4ac$ negative, the right-hand side of equation ($*$) is negative, and since the left-hand side is obviously nonnegative, the equation cannot be satisfied by real numbers. In this case the quadratic has no real roots. $\square$

Summary 2.3.2

A quadratic equation

$$ax^2 + bx + c = 0, \qquad a \neq 0$$

has:

two real roots	$x = \dfrac{-b \pm \sqrt{b^2 - 4ac}}{2a}$	if $b^2 - 4ac > 0$;
one real root	$x = -\dfrac{b}{2a}$	if $b^2 - 4ac = 0$;
no real roots		if $b^2 - 4ac < 0$.

Problem. Solve

$$3x^2 + 5x - 1 = 0.$$

SOLUTION.   Here

$$a = 3, \qquad b = 5, \qquad c = -1,$$

so that

$$b^2 - 4ac = (5)^2 - 4(3)(-1) = 25 + 12 = 37$$

and

$$x = \frac{-b \pm \sqrt{b^2 - 4ac}}{2a} = \frac{-5 \pm \sqrt{37}}{6}. \quad \square$$

Problem. Solve

$$9x^2 - 30x + 25 = 0.$$

SOLUTION.   Here

$$a = 9, \qquad b = -30, \qquad c = 25,$$

so that

$$b^2 - 4ac = (-30)^2 - 4(9)(25) = 900 - 900 = 0$$

and

$$x = -\frac{b}{2a} = -\frac{(-30)}{18} = \frac{15}{9} = \frac{5}{3}$$

is the only root. $\square$

Problem. Solve

$$3(x^2 + 1) = x^2 + 2x - 7.$$

SOLUTION.   To handle this equation, we first simplify it and put it in the form $ax^2 + bx + c = 0$:

$$3(x^2 + 1) = x^2 + 2x - 7,$$
$$3x^2 + 3 = x^2 + 2x - 7,$$
$$2x^2 - 2x + 10 = 0,$$
$$x^2 - x + 5 = 0.$$

Now we can use the formula:

$$a = 1, \qquad b = -1, \qquad c = 5,$$

so that
$$b^2 - 4ac = (-1)^2 - 4(1)(5) = 1 - 20 = -19.$$
Since the discriminant is negative, the equation has no real roots. $\square$

Exercises

Solve the following quadratic equations:

*1. $x^2 + 12x + 5 = 0.$

2. $2x^2 - 3x + 2 = 0.$

*3. $6x^2 + 66x - 60 = 0.$

4. $3x^2 + 10x + 8 = 0.$

*5. $2x^2 + 7x + 6 = 0.$

6. $2x^2 - 2x - 3 = 0.$

*7. $2x^2 - 6x - 3 = 0.$

8. $3x^2 + 13x + 4 = 0.$

*9. $2x^2 + 4x - 20 = 0.$

10. $7x^2 + 9x + 3 = 0.$

*11. $x^2 - 5x - 20 = 0.$

12. $6x^2 - 15x + 2 = 0.$

*13. $5x^2 - 4x + 1 = 0.$

14. $4x^2 + 11x + 8 = 0.$

*15. $3x^2 + 4x - 5 = 0.$

16. $2x^2 - 4x + 1 = 0.$

17. In a right triangle, the hypotenuse is 6 and the perimeter is 14. How long are the legs?

Optional | *18. Solve the equation
$$x^2y^2 - 2x + y = 0$$
(a) for x in terms of y.   (b) for y in terms of x.

19. Solve the equation
$$2x^2y^2 + xy + 2x - 1 = 0$$
(a) for x in terms of y.   (b) for y in terms of x.

2.4 Order

In this section we review some of the *order* properties of the real number system. First, if a and b are real numbers, then

(2.4.1)   either $a < b$,   or $a > b$,   or $a = b$.

There are no other possibilities.

Second, the order is *transitive*; namely,

(2.4.2)

$$\text{if } a < b \text{ and } b < c, \quad \text{then } a < c.$$

We use this frequently.

The remaining properties relate the order to the operations of arithmetic. If you start with an inequality and add the same number to both sides, then the order is preserved:

(2.4.3)

$$\text{if } a < b, \quad \text{then } a + c < b + c \text{ for all real numbers } c.$$

You can also subtract the same number from both sides:

(2.4.4)

$$\text{if } a < b, \quad \text{then } a - c < b - c \text{ for all real numbers } c.$$

Thus, for example, since $2 < 5$,

$$2 + \tfrac{1}{2} < 5 + \tfrac{1}{2}, \quad 2 + 7 < 5 + 7, \quad 2 + \sqrt{a^2 + b^2} < 5 + \sqrt{a^2 + b^2}$$

and

$$2 - \tfrac{1}{2} < 5 - \tfrac{1}{2}, \quad 2 - 7 < 5 - 7, \quad 2 - \sqrt{a^2 + b^2} < 5 - \sqrt{a^2 + b^2}. \quad \square$$

If you multiply an inequality by a positive number, then the order is preserved:

(2.4.5)

$$\text{if } a < b \text{ and } c > 0, \quad \text{then } ac < bc;$$

but if you multiply an inequality by a negative number, then the order is reversed:

(2.4.6)

$$\text{if } a < b \text{ and } c < 0, \quad \text{then } ac > bc.$$

Since $2 < 5$,

$$(2)(3) < (5)(3) \quad \text{but} \quad (2)(-3) > (5)(-3),$$
$$(2)(15) < (5)(15) \quad \text{but} \quad (2)(-15) > (5)(-15), \text{ etc.} \quad \square$$

Problem. Compare $-3a$ and $-7b$ given that $0 < a < b$.

SOLUTION. With $a < b$ we know that

$$3a < 3b. \qquad \text{(we multiplied by 3)}$$

Since $3 < 7$ and $b > 0$,

$$3b < 7b \qquad \text{(we multiplied by } b\text{)}$$

so that by transitivity

$$3a < 7b.$$

Multiplication by -1 reverses the inequality:

$$-3a > -7b. \quad \square$$

Problem.   Show that

$$\text{if} \quad 0 < a < b, \qquad \text{then} \quad a^2 < b^2.$$

SOLUTION.   With $0 < a < b$ we have

$$a < b \quad \text{with } a \text{ and } b \text{ both positive.}$$

Multiplication by a gives

$$a^2 < ab,$$

and multiplication by b gives

$$ab < b^2.$$

With $a^2 < ab$ and $ab < b^2$, we know by transitivity that

$$a^2 < b^2. \quad \square$$

Exercises

Arrange the following numbers in order:

*1. $0, 1, \sqrt{2}, \frac{1}{3}, \dfrac{1}{\sqrt{3}}, 0.333, 0.3334, 1.41, 1.142.$

2. $\sqrt{2}, \sqrt{3}, \dfrac{1}{1 - \sqrt{3}}, \dfrac{1}{1 - \sqrt{2}}, -\dfrac{1}{\sqrt{3}}, -\dfrac{1}{\sqrt{2}}, 1 - \sqrt{3}, 1 - \sqrt{2}.$

Compare, given that $0 < a < b$:

*3. $2a$ and $2b$.	4. $2b$ and $3b$.	*5. $-2a$ and $-2b$.
*6. $-b$ and $-2b$.	7. $-2a$ and $-3b$.	*8. ab and a^2.
*9. $-2a^2$ and $-2b^2$.	10. $-2a^2$ and $-3b^2$.	*11. $1 - ab$ and $1 - a^2$.

Compare, given that $a < b < 0$:

*12. $2a$ and $2b$.	13. $2b$ and $3b$.	*14. $-2a$ and $3b$.
*15. $-2a$ and $-2b$.	16. a^2 and b^2.	*17. ab and b^2.

Compare, given that $0 < a < b$:

*18. $\sqrt{a}$ and $\sqrt{b}$.   19. $1/\sqrt{a}$ and $1/\sqrt{b}$.   *20. $\sqrt{a} + \sqrt{b}$ and $\sqrt{a + b}$.

Optional 21.   Arrange the following in order:

$$1, x, \sqrt{x}, \frac{1}{x}, \frac{1}{\sqrt{x}}$$

given that   *(a) $1 < x$.   (b) $0 < x < 1$.

22.   Show that, if $0 \leq a \leq b$, then

$$\frac{a}{1 + a} \leq \frac{b}{1 + b}.$$

23.   Show that, if a, b, c are nonnegative and $a \leq b + c$, then

$$\frac{a}{1 + a} \leq \frac{b}{1 + b} + \frac{c}{1 + c}.$$

2.5   Sets

A *set* is a collection of objects. The objects in a set are called the *elements* (or *members*) of the set.

We might, for example, consider the set of capital letters appearing on this page, or the set of motorcycles licensed in Idaho, or the set of rational numbers. Suppose, however, that we wanted to find the set of rational people. Everybody might have a different collection. Which would be the right one? To avoid such problems we insist that sets be unambiguously defined. Collections based on highly subjective judgments—such as "all good football players" or "all likeable children"—are not sets.

Notation

To indicate that an object x is in the set A, we write

$$x \in A.$$

To indicate that x is not in A, we write

$$x \notin A.$$

Thus

$$\sqrt{2} \in \text{the set of real numbers}$$

but

$$\sqrt{2} \notin \text{the set of rational numbers.}$$

Sets are often denoted by braces. The set consisting of a alone is written $\{a\}$; that consisting of a, b is written $\{a, b\}$; that consisting of $a, b, c, \{a, b, c\}$; and so on. Thus

$$0 \in \{0, 1, 2\}, \quad 1 \in \{0, 1, 2\}, \quad 2 \in \{0, 1, 2\}, \quad \text{but} \quad 3 \notin \{0, 1, 2\}.$$

We can also use braces for infinite sets:

$\{1, 2, 3, \cdots\}$ is the set of positive integers,

$\{-1, -2, -3, \cdots\}$ is the set of negative integers,

$\{1, 3, 5, 7, \cdots\}$ is the set of odd positive integers.

Sets are often defined by a property. To indicate the set of all x *such that*——, we write

$$\{x: —\}.$$

Thus,

$\{x: x > 2\}$ is the set of all numbers greater than 2;

$\{x: x^2 > 9\}$ is the set of all numbers with squares greater than 9;

$\{p/q: p, q, \text{ integers}, q \neq 0\}$ is the set of all rational numbers.

If A is a set, then

$$\{x: x \in A\} \text{ is } A \text{ itself.}$$

Containment and Equality

If A and B are sets, then A is said to be *contained* in B, in symbols $A \subseteq B$, iff every element of A is also an element of B. For example,

the set of equilateral triangles $\subseteq$ the set of all triangles,

the set of all college freshmen $\subseteq$ the set of all college students,

the set of rational numbers $\subseteq$ the set of real numbers.

If A is contained in B, then A is called a *subset* of B. Thus

the set of equilateral triangles is a subset of the set of all triangles,

the set of college freshmen is a subset of the set of all college students,

the set of rational numbers is a subset of the set of real numbers.

Two sets are said to be *equal* iff they have exactly the same membership. In symbols,

(2.5.1) $\qquad\qquad \boxed{A = B \quad \text{iff} \quad A \subseteq B \quad \text{and} \quad B \subseteq A.}$

Examples

$$\text{The set of even integers} = \{2n: n \text{ an integer}\}.$$
$$\{x: x^2 = 1\} = \{-1, 1\},$$
$$\{x: x^2 < 4\} = \{x: -2 < x < 2\}. \quad \square$$

The Intersection of Two Sets

The set of elements common to two sets A and B is called the *intersection* of A and B and is denoted by $A \cap B$. The idea is illustrated in Figure 2.5.1. In symbols,

(2.5.2)

$$x \in A \cap B \quad \text{iff} \quad x \in A \quad \text{and} \quad x \in B.$$

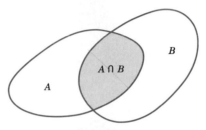

Figure 2.5.1

Examples

1. If A is the set of nonnegative numbers and B is the set of nonpositive numbers, then

    $$A \cap B = \{0\}.$$

2. If A is the set of all multiples of 3 and B is the set of all multiples of 4, then

    $$A \cap B = \text{the set of all multiples of 12}.$$

3. If $A = \{a, b, c, d, e\}$ and $B = \{c, d, e, f\}$, then

    $$A \cap B = \{c, d, e\}.$$

4. If $A = \{x: x > 1\}$ and $B = \{x: x < 4\}$, then

    $$A \cap B = \{x: 1 < x < 4\}. \quad \square$$

The Union of Two Sets

The *union* of two sets A and B, written $A \cup B$, is the set of elements which are either in A or in B. This does not exclude objects which are elements of both A and B. (See Figure 2.5.2) In symbols,

(2.5.3)

$$x \in A \cup B \quad \text{iff} \quad x \in A \quad \text{or} \quad x \in B.$$

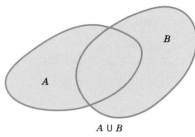

$A \cup B$

Figure 2.5.2

Examples

1. If A is the set of nonnegative numbers and B is the set of nonpositive numbers, then

$$A \cup B = \text{the set of real numbers.}$$

2. If $A = \{a, b, c, d, e\}$ and $B = \{c, d, e, f\}$, then

$$A \cup B = \{a, b, c, d, e, f\}.$$

3. If $A = \{x: 0 < x < 1\}$ and $B = \{0, 1\}$, then

$$A \cup B = \{x: 0 \leq x \leq 1\}.$$

4. If $A = \{x: x > 1\}$ and $B = \{x: x > 2\}$, then

$$A \cup B = \{x: x > 1\}. \quad \square$$

The Empty Set

If the sets A and B have no elements in common, we say that A and B are *disjoint* and write $A \cap B = \emptyset$. We regard $\emptyset$ as a set with no elements and refer to it as *the empty set*.

Examples

1. If A is the set of positive numbers and B is the set of negative numbers, then

$$A \cap B = \emptyset.$$

2. If $A = \{0, 1, 2, 3\}$ and $B = \{4, 5, 6, 7, 8\}$, then

$$A \cap B = \emptyset.$$

3. The set of all irrational rational numbers is empty; so is the set of all even odd integers; so is the set of real numbers whose squares are negative. □

The empty set $\emptyset$ plays a role in the theory of sets which is strikingly similar to the role played by 0 in the arithmetic of numbers. Without pursuing the matter very far, note that for numbers,

$$a + 0 = 0 + a = a, \qquad a \cdot 0 = 0 \cdot a = 0,$$

and for sets,

$$A \cup \emptyset = \emptyset \cup A = A, \qquad A \cap \emptyset = \emptyset \cap A = \emptyset.$$

Exercises

For Exercises 1–9 take

$$A = \{0, 2\}, \qquad B = \{-1, 0, 1\}, \qquad C = \{1, 2, 3, 4\},$$
$$D = \{2, 4, 6, 8, \cdots\}, \qquad E = \{-2, -4, -6, -8, \cdots\},$$

and determine the following sets:

*1. $A \cup B$.	2. $A \cap B$.	*3. $B \cup C$.
*4. $B \cap C$.	5. $A \cup D$.	*6. $C \cap D$.
*7. $B \cap D$.	8. $D \cup E$.	*9. $D \cap E$.

For Exercises 10–15 take

$$A = \{x : x > 2\}, \qquad B = \{x : x \le 4\}, \qquad C = \{x : x > 3\},$$

and determine the following sets:

*10. $A \cup C$.	11. $A \cap C$.	*12. $B \cup C$.
*13. $B \cap C$.	14. $A \cap (B \cap C)$.	*15. $A \cap (B \cup C)$.

16. Given that $A \subseteq B$, find (a) $A \cup B$.   (b) $A \cap B$.

*17. Find all the nonempty subsets of $\{0, 1, 2\}$.

18. What can you conclude about A and B given that
(a) $A \cup B = A$?   (b) $A \cap B = A$?
(c) $A \cup B = A$ and $A \cap B = A$?

2.6   Intervals; Boundedness

First we introduce some subsets of the number line to which we shall refer time and again. The notation, perhaps new to some of you, is standard in calculus.

Take two real numbers a and b with $a < b$, and picture them on the number line. (Figure 2.6.1)

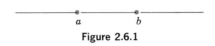

Figure 2.6.1

The *open interval* (a, b) is the set of all numbers between a and b:

$(a, b) = \{x : a < x < b\}.$

The *closed interval* $[a, b]$ is the open interval (a, b) together with the endpoints a and b:

$[a, b] = \{x : a \leq x \leq b\}.$

There are seven other types of intervals:

$(a, b] = \{x : a < x \leq b\}.$

$[a, b) = \{x : a \leq x < b\}.$

$(a, \infty) = \{x : a < x\}.$

$[a, \infty) = \{x : a \leq x\}.$

$(-\infty, b) = \{x : x < b\}.$

$(-\infty, b] = \{x : x \leq b\}.$

$(-\infty, \infty) =$ set of real numbers.

This interval notation is easy to remember: we use a bracket to include an endpoint; otherwise a parenthesis. The symbol ∞, read "infinity," has no meaning by itself. Just as in ordinary language we use syllables to construct words and need assign no meaning to the syllables themselves, so in mathematics we can use symbols to construct mathematical expressions, assigning no separate meaning to the individual symbols. While ∞ has no meaning, (a, ∞), $[a, \infty)$, $(-\infty, b)$, $(-\infty, b]$, $(-\infty, \infty)$ do have meaning, and that meaning is given above.

Figure 2.6.2 depicts the open interval $(0, 1)$ and below it the closed interval $[\frac{1}{2}, 3]$. The intersection of these two intervals consists of the points that lie in both intervals:

$$(0, 1) \cap [\tfrac{1}{2}, 3] = [\tfrac{1}{2}, 1).$$

The union of the two intervals consists of the points that lie in at least one of the intervals:

$$(0, 1) \cup [\tfrac{1}{2}, 3] = (0, 3].$$

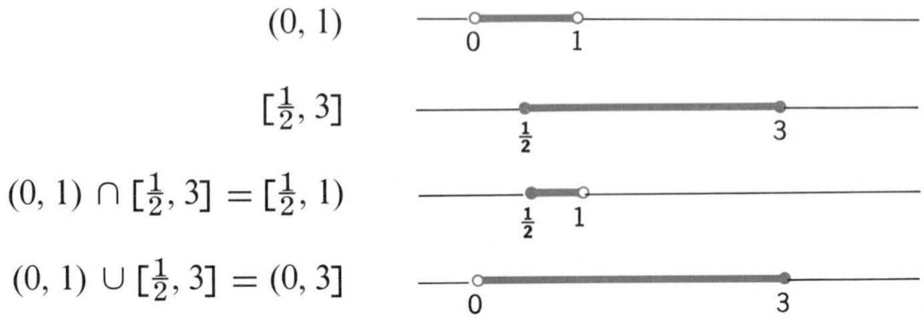

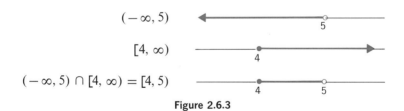

$(0, 1)$	
$[\tfrac{1}{2}, 3]$	
$(0, 1) \cap [\tfrac{1}{2}, 3] = [\tfrac{1}{2}, 1)$	
$(0, 1) \cup [\tfrac{1}{2}, 3] = (0, 3]$	

Figure 2.6.2

Problem.   Find $(-\infty, 5) \cap [4, \infty)$.

SOLUTION.   The first interval is the set of all numbers less than 5. The second interval is the set of all numbers greater than or equal to 4. The intersection of these two intervals is the set of all numbers that are simultaneously less than 5 and greater than or equal to 4:

$$(-\infty, 5) \cap [4, \infty) = [4, 5).$$

See Figure 2.6.3.   ☐

$(-\infty, 5)$

$[4, \infty)$

$(-\infty, 5) \cap [4, \infty) = [4, 5)$

Figure 2.6.3

Boundedness

Definition   2.6.1   Upper and Lower Bounds

Let S be a set of real numbers. A number M greater than or equal to all numbers in S is called an *upper bound for S*. A number m less than or equal to all numbers in S is called a *lower bound for S*.

Examples

1. As an upper bound for the interval $[0, 1)$ you can take 1 (or any number greater than 1) and as a lower bound 0 (or any number less than 0).

2. As an upper bound for $(-\infty, 5)$ you can take 5 (or any number greater than 5). The set has no lower bounds: there is no number less than or equal to all numbers in the set.

3. The interval $[-1, \infty)$ has no upper bounds: there is no number greater than or equal to all numbers in the set. As a lower bound for $[-1, \infty)$ you can take -1 (or any number less than -1). $\square$

You have seen that some sets have upper bounds, and some do not; some sets have lower bounds, and some do not. Sets that have upper bounds are said to be *bounded above*; sets that have lower bounds are said to be *bounded below*; sets that have both upper and lower bounds are said to be *bounded*. Thus, for example, $(-\infty, 1)$ and $(-\infty, 1]$ are bounded above, $(1, \infty)$ and $[1, \infty)$ are bounded below, $(-1, 1)$ and $[-1, 1]$ are bounded. The set of positive integers is bounded below but not above; the set of negative integers is bounded above but not below.

Exercises

Find the following:

*1. $(0, 1) \cup (\frac{1}{2}, 2)$.

2. $(0, 1) \cap (\frac{1}{2}, 2)$.

*3. $(0, 1) \cup \{1\}$.

4. $(0, 1) \cap \{1\}$.

*5. $[0, 1] \cup (\frac{1}{2}, 2]$.

6. $[0, 1] \cap (\frac{1}{2}, 2]$.

*7. $[0, 2] \cup [0, 1]$.

8. $[0, 2] \cap [0, 1]$.

*9. $(-3, \infty) \cup [-2, 0)$.

10. $(-3, \infty) \cap [-2, 0)$.

*11. $(-\infty, 3) \cup [-2, \infty)$.

12. $(-\infty, 3) \cap [-2, \infty)$.

*13. $(0, 1) \cup \{0\} \cup \{1\}$.

14. $[-1, 0] \cap [0, 1]$.

*15. Express the set of all real numbers, excluding $-3, 0, 3$, as the union of intervals.

Determine whether the given set is (a) bounded above, (b) bounded below, (c) bounded:

*16. The set of negative rationals.

17. $(1, \infty)$.

*18. $(-100, 100)$.

19. $(-\infty, 4) \cap (3, \infty)$.

*20. $(-\infty, 3) \cup (5, \infty)$.

21. $[0, 1) \cup (2, \infty)$.

*22. $\{1, \frac{1}{2}, 2, \frac{1}{3}, 3, \frac{1}{4}, 4, \cdots\}$.

23. $\{\frac{1}{2}, \frac{2}{3}, \frac{3}{4}, \frac{4}{5}, \cdots\}$.

*24. $\{2\frac{1}{2}, -3\frac{1}{3}, 4\frac{1}{4}, -5\frac{1}{5}, \cdots\}$.

25. $\{\frac{1}{2}, 4, \frac{1}{8}, 16, \cdots\}$.

*26. $\{x : x^2 < 1\}$.

27. $\{x : x^2 < 4\}$.

*28. $\{x^2 : x < 1\}$.

29. $\{x^3 : x < 1\}$.

2.7 Some Inequalities

We come now to inequalities that involve a variable. To solve such an inequality is to find the set of numbers that satisfy it.

Problem. Solve the inequality

$$\tfrac{1}{2}(1 + x) \le 6.$$

Solution. The idea is to isolate x:

$$\tfrac{1}{2}(1 + x) \le 6,$$

$$1 + x \le 12, \qquad \text{(we multiplied by 2)}$$

$$x \le 11. \qquad \text{(subtracted 1)}$$

The solution is the interval $(-\infty, 11]$. □

Problem. Solve the inequality

$$-3(4 - x) < 12.$$

Solution. Dividing both sides by -3 (or, equivalently, multiplying by $-\frac{1}{3}$), we have

$$4 - x > -4. \qquad \text{(the inequality has been reversed)}$$

The subtraction of 4 gives

$$-x > -8.$$

To isolate x, we multiply by -1. This gives

$$x < 8. \qquad \text{(the inequality has been reversed again)}$$

The solution is the interval $(-\infty, 8)$. □

There are generally several ways to solve a given inequality. For example, the last inequality could have been solved this way:

$$-3(4 - x) < 12,$$
$$-12 + 3x < 12,$$
$$3x < 24, \qquad \text{(we added 12)}$$
$$x < 8. \qquad \text{(divided by 8)}$$

The solution, of course, is once again $(-\infty, 8)$. $\square$

Problem. Solve the inequality

$$\tfrac{1}{5}(x^2 - 4x + 3) < 0.$$

SOLUTION

$$\tfrac{1}{5}(x^2 - 4x + 3) < 0,$$
$$x^2 - 4x + 3 < 0, \qquad \text{(we multiplied by 5)}$$
$$(x - 1)(x - 3) < 0. \qquad \text{(factored)}$$

On a number line we mark the points 1 and 3 where the product $(x - 1)(x - 3)$ is zero. These points separate three intervals

$$(-\infty, 1), \qquad (1, 3), \qquad (3, \infty)$$

on which each factor keeps a constant sign:

for $x \in (-\infty, 1)$

$$\text{the sign of } (x - 1)(x - 3) = (-)(-) = +;$$

for $x \in (1, 3)$

$$\text{the sign of } (x - 1)(x - 3) = (+)(-) = -;$$

for $x \in (3, \infty)$

$$\text{the sign of } (x - 1)(x - 3) = (+)(+) = +.$$

All this is illustrated in Figure 2.7.1.

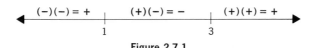

Figure 2.7.1

The product $(x - 1)(x - 3)$ is negative only for $x \in (1, 3)$. The solution is thus the open interval $(1, 3)$. $\square$

Problem. Solve the inequality

$$2x^2 - 2x > 4.$$

SOLUTION

$$2x^2 - 2x > 4,$$

$$2x^2 - 2x - 4 > 0, \qquad\qquad \text{(we subtracted 4)}$$

$$x^2 - x - 2 > 0, \qquad\qquad \text{(divided by 2)}$$

$$(x + 1)(x - 2) > 0. \qquad\qquad \text{(factored)}$$

On the number line we mark the points -1 and 2 where the product $(x + 1)(x - 2)$ is zero. See Figure 2.7.2. These points separate the intervals

$$(-\infty, -1), \qquad (-1, 2), \qquad (2, \infty)$$

on which each factor keeps a constant sign:

for $x \in (-\infty, -1)$

$$\text{the sign of } (x + 1)(x - 2) = (-)(-) = +;$$

for $x \in (-1, 2)$

$$\text{the sign of } (x + 1)(x - 2) = (+)(-) = -;$$

for $x \in (2, \infty)$

$$\text{the sign of } (x + 1)(x - 2) = (+)(+) = +.$$

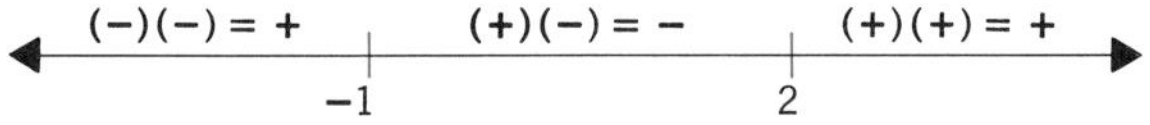

Figure 2.7.2

The product $(x + 1)(x - 2)$ is positive for $x \in (-\infty, -1)$ and also for $x \in (2, \infty)$. The solution is the union $(-\infty, -1) \cup (2, \infty)$. $\square$

Problem. Solve the inequality

$$x^2 + 4x - 2 \leq 0.$$

SOLUTION.   Here we must complete the square:

$$x^2 + 4x - 2 \leq 0,$$

$$x^2 + 4x \leq 2,$$

$$x^2 + 4x + 4 \le 2 + 4,$$

$$(x + 2)^2 \le 6,$$

$$(x + 2)^2 - 6 \le 0,$$

$$(x + 2 + \sqrt{6})(x + 2 - \sqrt{6}) \le 0. \qquad \text{(we factored the difference of two squares)}$$

This time we mark the points $-2 - \sqrt{6}$ and $-2 + \sqrt{6}$ (where the product is 0) and proceed as before. See Figure 2.7.3.

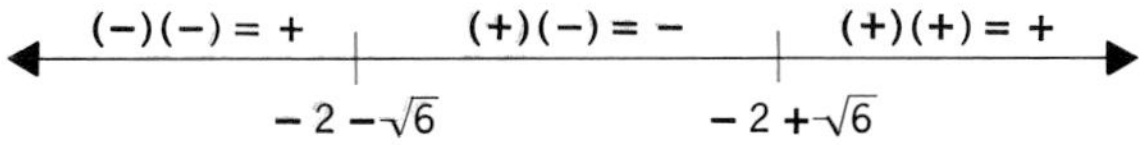

Figure 2.7.3

The product is negative only for $x \in (-2 - \sqrt{6}, -2 + \sqrt{6})$ and it is zero only at $x = -2 - \sqrt{6}$ and $x = -2 + \sqrt{6}$. The solution is thus the closed interval $[-2 - \sqrt{6}, -2 + \sqrt{6}]$. $\square$

Problem. Solve the inequality

$$(x + 1)(x - 1)(x - 2) > 0.$$

SOLUTION. The product is zero at $-1, 1, 2$. These points separate four intervals

$$(-\infty, -1), \qquad (-1, 1), \qquad (1, 2), \qquad (2, \infty)$$

on which each factor keeps a constant sign. The sign of the product on each interval is recorded in Figure 2.7.4.

$(-)(-)(-) = -$	$(+)(-)(-) = +$	$(+)(+)(-) = - \qquad (+)(+)(+) = +$

$-1 \qquad\qquad\qquad\qquad 1 \qquad\qquad 2$

Figure 2.7.4

The product is positive for $x \in (-1, 1)$ and for $x \in (2, \infty)$. The solution is the union $(-1, 1) \cup (2, \infty)$. $\square$

Exercises

Solve the following inequalities:

*1. $2 + 3x < 5.$

2. $\frac{1}{2}(2x + 3) < 6.$

*3. $16x + 64 \le 16.$

4. $3x + 5 > \frac{1}{4}(x - 2).$

*5. $3x - 2 \le 1 + 6x.$

6. $(x + 1)(x - 2) \le 0.$

*7. $x^2 - 1 < 0$.

8. $x^2 + x - 2 \le 0$.

*9. $4(x^2 - 3x + 2) > 0$.

10. $x^2 + 9x + 20 < 0$.

*11. $x^2 - 2x + 1 \ge 0$.

12. $x^2 - 4x + 4 \le 0$.

*13. $\frac{1}{2}(1 + x) < \frac{1}{3}(1 - x)$.

14. $2 - x^2 \ge -4x$.

*15. $\frac{1}{2}(1 + x)^2 < \frac{1}{3}(1 - x)^2$.

16. $2x^2 + 9x + 6 \ge x + 2$.

*17. $1 - 3x^2 < \frac{1}{2}(2 - x^2)$.

18. $6x^2 + 4x + 4 \le 4(x - 1)^2$.

*19. $x(x - 1)(x - 2) > 0$.

20. $x(x - 1)(x - 2) < 0$.

*21. $(x - 1)(x - 2)(x - 3) < 0$.

22. $x(2x - 1)(3x - 5) \le 0$.

2.8 More on Inequalities

Here we come to inequalities which involve quotients. The central idea is that

(2.8.1)
$$\frac{a}{b} > 0 \quad \text{iff} \quad ab > 0 \qquad \text{and} \qquad \frac{a}{b} < 0 \quad \text{iff} \quad ab < 0.$$

Problem. Solve the inequality

$$\frac{x + 2}{x - 1} > 0.$$

SOLUTION. By (2.8.1)

$$\frac{x + 2}{x - 1} > 0 \qquad \text{iff} \qquad (x + 2)(x - 1) > 0.$$

The product $(x + 2)(x - 1)$ is zero at -2 and 1. These points separate three intervals

$$(-\infty, -2), \qquad (-2, 1), \qquad (1, \infty)$$

on which each factor keeps a constant sign. The sign of the product on each interval is recorded in Figure 2.8.1.

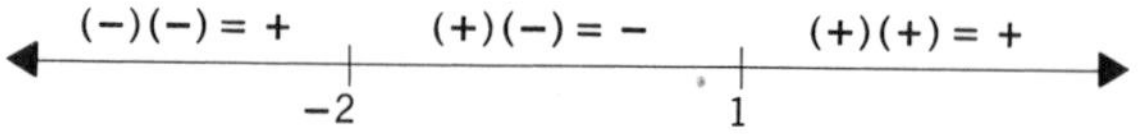

Figure 2.8.1

The product is positive for $x \in (-\infty, -2)$ and for $x \in (1, \infty)$. The solution is the union $(-\infty, -2) \cup (1, \infty)$. □

Problem. Solve the inequality

$$\frac{x + 2}{1 - x} > 1.$$

SOLUTION

$$\frac{x + 2}{1 - x} > 1,$$

$$\frac{x + 2}{1 - x} - 1 > 0,$$

$$\frac{x + 2 - (1 - x)}{1 - x} > 0,$$

$$\frac{2x + 1}{1 - x} > 0,$$

$$(2x + 1)(1 - x) > 0, \qquad\qquad\qquad \text{[by (2.8.1)]}$$

$$(2x + 1)(x - 1) < 0, \qquad\qquad \text{(we multiplied by } -1)$$

$$(x + \tfrac{1}{2})(x - 1) < 0. \qquad\qquad \text{(divided by 2)}$$

The product $(x + \tfrac{1}{2})(x - 1)$ is zero at $-\tfrac{1}{2}$ and 1. These points separate three intervals

$$(-\infty, -\tfrac{1}{2}), \qquad (-\tfrac{1}{2}, 1), \qquad (1, \infty)$$

on which each factor keeps a constant sign. The sign of the product on each interval is recorded in Figure 2.8.2.

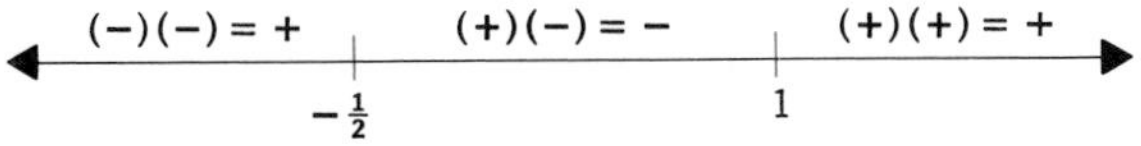

Figure 2.8.2

The product is negative only for $x \in (-\tfrac{1}{2}, 1)$. The solution is the open interval $(-\tfrac{1}{2}, 1)$. □

Problem. Solve the inequality

$$\frac{x^2 - 2}{1 - 2x} > 1.$$

SOLUTION

$$\frac{x^2 - 2}{1 - 2x} - 1 > 0,$$

$$\frac{x^2 - 2 - (1 - 2x)}{1 - 2x} > 0,$$

$$\frac{x^2 + 2x - 3}{1 - 2x} > 0,$$

$$\frac{(x + 3)(x - 1)}{1 - 2x} > 0,$$

$$(x + 3)(x - 1)(1 - 2x) > 0, \qquad\qquad \text{[by (2.8.1)]}$$

$$(x + 3)(2x - 1)(x - 1) < 0, \qquad \text{(we multiplied by } -1)$$

$$(x + 3)(x - \tfrac{1}{2})(x - 1) < 0. \qquad \text{(divided by 2)}$$

We leave it to you to verify that the product

$$(x + 3)(x - \tfrac{1}{2})(x - 1)$$

is negative for $x \in (-\infty, -3)$ and for $x \in (\tfrac{1}{2}, 1)$. The solution is thus the union $(-\infty, -3) \cup (\tfrac{1}{2}, 1)$. ☐

Problem. Solve

$$x + \frac{1}{x} > 1.$$

SOLUTION. There is no point doing a lot of work here. Suppose that $x > 0$. If $x > 1$, the inequality obviously holds; if $x = 1$, then $x + 1/x = 2$, and the inequality still holds. If $0 < x < 1$, then $1/x > 1$, and once again the inequality holds.

Suppose now that $x \leq 0$. If $x < 0$, then $x + 1/x < 0$ and the inequality fails. If $x = 0$, then $1/x$ is not even defined and the inequality again fails.

The solution is the interval $(0, \infty)$. ☐

Exercises

Solve the inequality.

*1. $\dfrac{x - 2}{x - 5} < 0.$ 2. $\dfrac{x - 2}{x - 5} > 1.$ *3. $\dfrac{x - 2}{x - 5} < -2.$

*4. $\dfrac{x}{x - 5} > \dfrac{1}{4}.$ 5. $\dfrac{1 - x}{x^2 - 4} > 0.$ *6. $\dfrac{1 - x}{x^2 + 4} < 0.$

*7. $\dfrac{1}{3x - 5} < 2.$

8. $\dfrac{1}{2x - 3} > 2.$

*9. $\dfrac{x - 1}{3x - 2} < 1.$

*10. $\dfrac{x}{x - 1} > 2.$

11. $\dfrac{x^2}{x - 2} \le 1.$

*12. $\dfrac{1}{x} < x + 2.$

*13. $\dfrac{1}{3x - 5} \le \dfrac{1}{x}.$

14. $\dfrac{1}{x} < \dfrac{2}{x - 1}.$

*15. $\dfrac{1}{3x + 5} > \dfrac{1}{2x^2}.$

2.9 Absolute Value

Think of the real numbers as being strung out along the number line, and pick out a number a. What is the distance between a and 0? To answer this question, we have to take into account the sign of a. If a is nonnegative, (see Figure 2.9.1), the distance between a and 0 is a itself;

Figure 2.9.1

but, if a is negative, the distance between a and 0 is not a, but rather $-a$ (see Figure 2.9.2).†

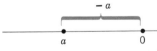

Figure 2.9.2

This leads directly to the notion of *absolute value*.

Definition 2.9.1   Absolute Value

The *absolute value* of a real number a is a if a is nonnegative and $-a$ if a is negative. In symbols,

$$|a| = \left\{ \begin{array}{ll} a, & \text{if } a \ge 0 \\ -a, & \text{if } a < 0 \end{array} \right].$$

---

† Take, for example, $a = -2$. The distance between -2 and 0 is not -2 but rather $-(-2) = 2$. You can convince yourself that the labeling in Figure 2.9.2 is correct by taking $a = -2$. The figure then reads

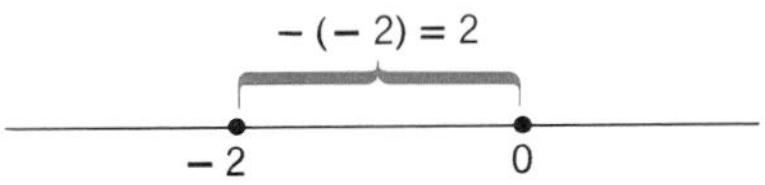

$$-(-2) = 2$$

Figure 2.9.3

Examples

1. Since $3 > 0$,

$$|3| = 3,$$

and, since $-3 < 0$,

$$|-3| = -(-3) = 3.$$

2. Since $\sqrt{2} > 0$,

$$|\sqrt{2}| = \sqrt{2},$$

and, since $-\sqrt{2} < 0$,

$$|-\sqrt{2}| = -(-\sqrt{2}) = \sqrt{2}.$$

3. If $b \geq 1$, then

$$|b - 1| = b - 1,$$

but, if $b < 1$, then

$$|b - 1| = -(b - 1) = 1 - b. \quad \square$$

Here are two other characterizations of absolute value:

(2.9.2)
$$\boxed{|a| = \sqrt{a^2}, \qquad |a| = \max\{-a, a\}.}$$

The first equation says that $|a|$ is simply the square root of a^2. The second equation says that $|a|$ is the maximum of $-a$ and a. We leave it to you to verify that these statements are correct.

We come now to the basic properties of absolute value — properties that we'll be using again and again:

(2.9.3)

> 1. $|a| \geq 0$; $|a| = 0$ iff $a = 0$.
>
> 2. $|-a| = |a|$.
>
> 3. $-|a| \leq a \leq |a|$.
>
> 4. $|ab| = |a||b|$.
>
> 5. $|a + b| \leq |a| + |b|$.   (the triangle inequality)†
>
> 6. $||a| - |b|| \leq |a - b|$.   (another form of the triangle inequality)

---

† The absolute value of the sum of two numbers cannot exceed the sum of their absolute values just as, in a triangle, the length of a side cannot exceed the sum of the lengths of the other two sides.

Properties 1 and 2 are obvious, so let's go directly to the others.

A PROOF OF PROPERTY 3.   Remember that

$$|a| = \max\{-a, a\}.$$

From this it follows that

(∗) $a \le |a|$

and also that

$$-a \le |a|.$$

Multiplying the second inequality by -1, we have

(∗∗) $-|a| \le a.$

Combining (∗) and (∗∗), we have

$$-|a| \le a \le |a|. \quad \square$$

A PROOF OF PROPERTY 4.   If we use the square root characterization of absolute value, the proof is simple:

$$|ab| = \sqrt{(ab)^2} = \sqrt{a^2 b^2} = \sqrt{a^2}\sqrt{b^2} = |a||b|. \quad \square$$

A PROOF OF PROPERTY 5. $|a + b|^2 = (a + b)^2$

$$= a^2 + 2ab + b^2$$
$$\le |a|^2 + 2|a||b| + |b|^2$$
$$= (|a| + |b|)^2.$$

We have shown that

$$|a + b|^2 \le (|a| + |b|)^2.$$

Taking square roots, we have the triangle inequality

$$|a + b| \le |a| + |b|. \quad \square$$

The proof of Property 6 is left to you as an optional exercise. $\square$

From the introduction to this section you can see that $|a|$ is just the distance between a and 0 along the number line. Now take two numbers a and b and mark them on the number line. If $a - b \ge 0$, then $b \le a$, and, as you can see in Figure 2.9.4, the distance between a and b is simply $a - b$.

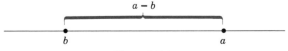

Figure 2.9.4

If, on the other hand, $a - b < 0$, then $a < b$, and, as you can see in Figure 2.9.5, the distance between a and b is $b - a$.

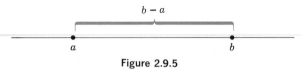

Figure 2.9.5

Thus no matter how a and b are related,

(2.9.4)

$$\text{the distance between } a \text{ and } b = |a - b|.$$

We come now to an equation and two related inequalities that are important in calculus. Picture a number c on the number line and choose a positive number d. A number x will satisfy the equation $|x - c| = d$ iff its distance from c is d, which will happen iff $x = c - d$ or $x = c + d$:

(2.9.5)

$$|x - c| = d \qquad \text{iff} \qquad x = c - d \quad \text{or} \quad x = c + d.$$

A number x will satisfy the inequality $|x - c| < d$ iff its distance from c is less than d, which will happen iff x lies between $c - d$ and $c + d$:

(2.9.6)

$$|x - c| < d \qquad \text{iff} \qquad c - d < x < c + d.$$

Finally, a number x will satisfy the inequality $|x - c| > d$ iff its distance from c is greater than d, which will happen iff $x < c - d$ or $x > c + d$:

(2.9.7)

$$|x - c| > d \qquad \text{iff} \qquad x < c - d \quad \text{or} \quad x > c + d.$$

These statements are illustrated in Figure 2.9.6.

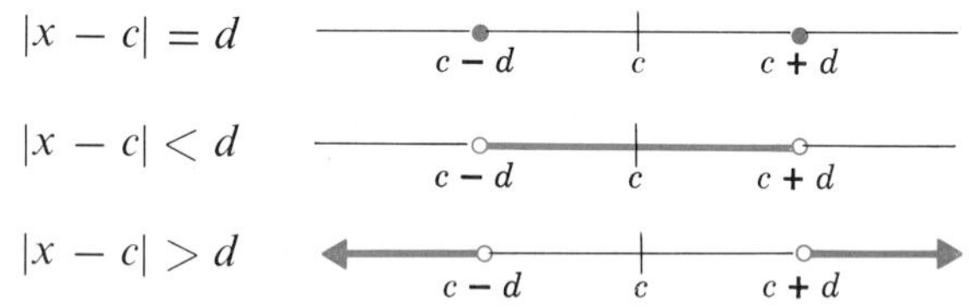

$$|x - c| = d$$
$$|x - c| < d$$
$$|x - c| > d$$

Figure 2.9.6

Problem.   Solve the equation

$$|3x - 1| = 2.$$

SOLUTION.   By (2.9.5) we see that

$$3x = 1 - 2 = -1 \qquad \text{or} \qquad 3x = 1 + 2 = 3.$$

This forces

$$x = -\tfrac{1}{3} \quad \text{or} \quad x = 1. \quad \square$$

Problem.   Solve the inequality

$$|2x - 5| < 1.$$

SOLUTION.   By (2.9.6) we have

$$5 - 1 < 2x < 5 + 1,$$

so that

$$4 < 2x < 6$$

and

$$2 < x < 3.$$

The solution is the open interval $(2, 3)$. $\quad \square$

Problem.   Solve the inequality

$$|5x - 3| > 4.$$

SOLUTION.   By (2.9.7) we have

$$5x < 3 - 4 \quad \text{or} \quad 5x > 3 + 4,$$

so that

$$5x < -1 \quad \text{or} \quad 5x > 7$$

and therefore

$$x < -\tfrac{1}{5} \quad \text{or} \quad x > \tfrac{7}{5}.$$

The solution is the union $(-\infty, -\tfrac{1}{5}) \cup (\tfrac{7}{5}, \infty)$. $\quad \square$

Problem.   Solve the inequality

$$|4x + 1| \leq 2.$$

SOLUTION.   We begin by rewriting the inequality as

$$|4x - (-1)| \leq 2.$$

By (2.9.5) and (2.9.6) we have

$$-1 - 2 \leq 4x \leq -1 + 2$$

and thus

$$-3 \leq 4x \leq 1.$$

Division by 4 gives

$$-\tfrac{3}{4} \leq x \leq \tfrac{1}{4}.$$

The solution is the closed interval $[-\tfrac{3}{4}, \tfrac{1}{4}]$. $\quad \square$

Exercises

Solve:

*1. $|x - 4| = 0$.   2. $|2x - 3| = 0$.

*3. $|x - 4| = 1$.   4. $|2x - 3| = 1$.

*5. $|x^2 - 1| = 0$.   6. $|x(x - 1)(x - 2)| = 0$.

*7. $|x - 4| < 1$.   8. $|2x - 3| < 1$.

*9. $|x - 3| \leq 2$.   10. $|x - 3| > 2$.

*11. $|4x - 3| \geq 4$.   12. $|4x - 3| < 4$.

*13. $|2x + 1| \leq 6$.   14. $|6 - 2x| \geq 4$.

*15. $|5x - 4| < \frac{1}{2}$.   16. $|4x + 5| \geq \frac{1}{2}$.

*17. $|x| < x$.   18. $|x| \leq x$.

*19. $x \leq |x|$.   20. $x < |x|$.

Optional | Solve:

*21. $|x - 1| < |x - 2|$.   22. $|x + 3| \leq |x - 1|$.

*23. $|x - 2| < |x - 1|$.   24. $|2x - 4| < |2x - 6|$.

25. Show that
$$|a| = \sqrt{a^2}.$$

26. Show that
$$|a| = \max\{-a, a\}.$$

27. Show that
$$||a| - |b|| \leq |a - b|.$$
HINT: $||a| - |b||^2 = (|a| - |b|)^2$ etc.

2.10 Additional Exercises

Rationalize the denominator:

*1. $\dfrac{1}{3 - 2\sqrt{5}}$.   *2. $\dfrac{1}{2\sqrt{2} + \sqrt{5}}$.   *3. $\dfrac{1}{(1 + \sqrt{2})(1 - \sqrt{5})}$.

*4. Express $\sqrt{\frac{3}{5}} - \sqrt{\frac{5}{3}}$ as a rational multiple of $\sqrt{15}$.

*5. Express $\sqrt{\frac{9}{5}} + \sqrt{\frac{5}{4}}$ as a rational multiple of $\sqrt{5}$.

Solve the following quadratic equations:

*6. $12x^2 - 19x + 4 = 0$.   *7. $x^2 + 2\sqrt{5}x + 5 = 0$.

*8. $x^2 + 16x - 3 = 0$.   *9. $x^2 + 3x + 16 = 0$.

*10. $16x^2 + 24x + 9 = 0$.   *11. $-2x^2 + x + 6 = 0$.

*12. $3x^2 + 3x + 2 = 0$.   *13. $2x^2 + (\sqrt{5} - 2\sqrt{2})x - \sqrt{10} = 0$.

Solve the following inequalities:

*14. $4x - 3 < 2x + 7$.   *15. $3x \geq 4x + 9$.

*16. $x^2 - x - 20 < 0$.   *17. $x^2 - x - 20 \geq 0$.

*18. $4x^2 - 9 \leq 0$.   *19. $x^2 - 4x + 4 < 0$.

*20. $|5x - 4| < 2$.   *21. $|4x + 5| \geq 3$.

*22. $|2x - 1| < x$.   *23. $|x^2 - 4| < 1$.

*24. $(x + 2)(x + 1)(x - 1) > 0$.   *25. $(2x + 1)(x - 1)(x - 3) < 0$.

*26. $\dfrac{x^2}{(x - 1)(x - 2)} < 0$.   *27. $\dfrac{x}{x - 1} > \dfrac{1}{x - 2}$.

*28. Let A be a set of real numbers. Determine whether the statement is true or false.
   (a) If A is bounded above, then $\{|x|: x \in A\}$ is bounded above.
   (b) Whether A is bounded or not, $\{|x|: x \in A\}$ is bounded below.
   (c) If A is bounded above, then $\{-x: x \in A\}$ is bounded below.
   (d) If A is bounded and $0 \notin A$, then $\left\{\dfrac{1}{x}: x \in A\right\}$ is bounded.

Optional | *29. Given that $c < 0$, simplify
$$\frac{1 + \sqrt{c^2}}{c - 1}.$$

*30. Given that $a < 0$ and $b > 0$, simplify
$$a^2|b^2| - 2ab|ab|.$$

*31. Show that
$$|a - b| \leq |a| + |b|$$
for all real numbers a and b.

Optional ***32.** Show that

$$ab \leq \tfrac{1}{2}(a^2 + b^2)$$

for all real numbers a and b.

*33. Given that $r > 0$, show that

$$\text{if} \quad a < b, \qquad \text{then} \quad a < \frac{a + rb}{1 + r} < b.$$

ANALYTIC GEOMETRY

3

The geometric foundation for calculus is provided by analytic geometry. Here we review the fundamentals. Further topics in analytic geometry are discussed in Chapter 6.

3.1 Plane Rectangular Coordinates

We start with a pair of coordinate lines (number lines) that intersect at right angles as in Figure 3.1.1. The horizontal line is called the *x-axis* and the vertical line is called the *y-axis*. The point of intersection is called the *origin* and is labeled O. It corresponds to the number 0 on both axes.

Each point of the plane can now be specified by a pair of coordinates. As in

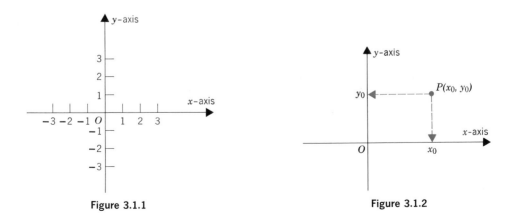

Figure 3.1.1   Figure 3.1.2

Figure 3.1.2, a point P has coordinates (x_0, y_0) iff its projection onto the x-axis has linear coordinate x_0 and its projection onto the y-axis has linear coordinate y_0.

To indicate that P has coordinates (x_0, y_0), we write $P(x_0, y_0)$. Figure 3.1.3 gives some examples.

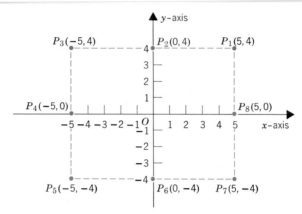

Figure 3.1.3

Points of the form $P(a, 0)$ lie on the x-axis, and those of the form $P(0, b)$ lie on the y-axis. The remaining points, those with two nonzero coordinates, lie in one of four quarter-planes, called *quadrants*. These are indicated in Figure 3.1.4. In the first quadrant, both coordinates are positive; in the second, the x-coordinate is negative but the y-coordinate is positive; and so on, as you can read from the figure.

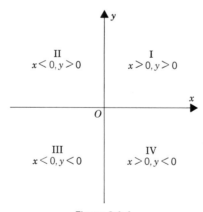

Figure 3.1.4

Returning to Figure 3.1.3, note that the points $P_1(5, 4)$ and $P_7(5, -4)$ are symmetric about the x-axis; in general,

(3.1.1) $P(a, b)$ and $Q(a, -b)$ are symmetric about the x-axis.

In the same figure, the points $P_1(5, 4)$ and $P_3(-5, 4)$ are symmetric about the y-axis; in general,

(3.1.2) $P(a, b)$ and $Q(-a, b)$ are symmetric about the y-axis.

Finally, note that $P_1(5, 4)$ and $P_5(-5, -4)$ are symmetric about the origin; in general,

(3.1.3) $P(a, b)$ and $Q(-a, -b)$ are symmetric about the origin.

These three types of symmetry are illustrated in Figures 3.1.5, 3.1.6, and 3.1.7.

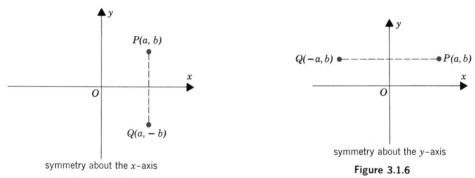

symmetry about the x-axis

Figure 3.1.5

symmetry about the y-axis

Figure 3.1.6

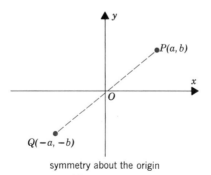

symmetry about the origin

Figure 3.1.7

You have probably run across the terms "abscissa" and "ordinate." The *abscissa* of a point is its first coordinate (the x-coordinate), and the *ordinate* is the second coordinate (the y-coordinate). Thus $P(a, b)$ has abscissa a and ordinate b. We will use these terms only sparingly.

With rectangular coordinates defined, we are in a position to use them. We start with a simple proposition about line segments.

Theorem   3.1.4   The Midpoint of a Line Segment

If P_1 has coordinates (x_1, y_1) and P_2 has coordinates (x_2, y_2), then the midpoint of the line segment $\overline{P_1 P_2}$ has coordinates

$$(\tfrac{1}{2}[x_1 + x_2], \tfrac{1}{2}[y_1 + y_2]).$$

PROOF.   We begin by supposing that

$$x_1 < x_2, \qquad y_1 < y_2$$

and take P_1 and P_2 as in Figure 3.1.8.

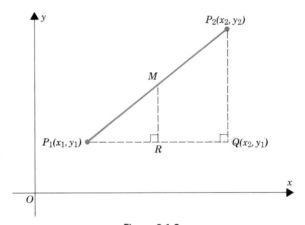

Figure 3.1.8

We take M as the midpoint of $\overline{P_1 P_2}$ and the points Q and R as marked in the figure. It's not hard to see that

$$\text{the } x\text{-coordinate of } M = x_1 + \text{length of } \overline{P_1 R},$$

$$\text{the } y\text{-coordinate of } M = y_1 + \text{length of } \overline{RM}.$$

Since triangles $P_1 R M$ and $P_1 Q P_2$ have equal angles, they are similar and their corresponding sides are proportional. Therefore

$$\text{length of } \overline{P_1 R} = \tfrac{1}{2}(\text{length of } \overline{P_1 Q}) = \tfrac{1}{2}(x_2 - x_1),$$

$$\text{length of } \overline{RM} = \tfrac{1}{2}(\text{length of } \overline{QP_2}) = \tfrac{1}{2}(y_2 - y_1).$$

(Figure 3.1.9)

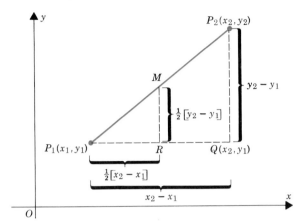

Figure 3.1.9

It follows that

the x-coordinate of $M = x_1 + \frac{1}{2}(x_2 - x_1) = x_1 + \frac{1}{2}x_2 - \frac{1}{2}x_1 = \frac{1}{2}(x_1 + x_2)$,

the y-coordinate of $M = y_1 + \frac{1}{2}(y_2 - y_1) = y_1 + \frac{1}{2}y_2 - \frac{1}{2}y_1 = \frac{1}{2}(y_1 + y_2)$.

Similar arguments work if

$$x_1 \geq x_2 \quad \text{or} \quad y_1 \geq y_2. \quad \square$$

Problem. Find the midpoint of the line segment that joins $P(-1, 5)$ to $Q(5, 3)$.

SOLUTION. By the theorem, the x-coordinate of the midpoint is

$$\tfrac{1}{2}(-1 + 5) = \tfrac{1}{2}(4) = 2$$

and the y-coordinate is

$$\tfrac{1}{2}(5 + 3) = \tfrac{1}{2}(8) = 4.$$

The midpoint is $M(2, 4)$. $\quad \square$

Optional   ***Problem.*** Given that P_1 has coordinates $(1, 2)$ and P_2 has coordinates $(7, -1)$, find the point T on the line segment $\overline{P_1 P_2}$ which lies one-third of the way from P_1 to P_2.

SOLUTION. We refer to Figure 3.1.10. First of all,

$$\text{the } x\text{-coordinate of } T = 1 + \text{length of } \overline{RT},$$
$$\text{the } y\text{-coordinate of } T = 2 - \text{length of } \overline{P_1 R}. \qquad \text{(explain)}$$

Optional

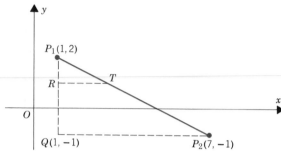

Figure 3.1.10

By similar triangles,

$$\text{length of } \overline{RT} = \tfrac{1}{3}(\text{length of } \overline{QP_2}) = \tfrac{1}{3}(6) = 2,$$

$$\text{length of } \overline{P_1R} = \tfrac{1}{3}(\text{length of } \overline{P_1Q}) = \tfrac{1}{3}(3) = 1.$$

It follows that

$$\text{the } x\text{-coordinate of } T = 1 + 2 = 3,$$

$$\text{the } y\text{-coordinate of } T = 2 - 1 = 1. \quad \square$$

Exercises

1. Specify the quadrant that contains the given point:
   *(a) $P(-3, 2)$.   (b) $Q(2, 4)$.   *(c) $R(2, -1)$.   (d) $S(-\tfrac{1}{3}, -\pi)$.

2. Find the coordinates of the point that is symmetric to $P(2, -3)$ about
   *(a) the x-axis.   (b) the y-axis.   *(c) the origin.

3. Find the coordinates of the point that is symmetric to $P(-\tfrac{1}{2}, \pi)$ about
   (a) the x-axis.   *(b) the y-axis.   (c) the origin.

*4. Sketch the line segment with endpoints $P_1(1, -2)$ and $P_2(2, -6)$ and determine the coordinates of the midpoint.

5. Find the midpoint of the line segment that joins the following points:
   *(a) $P_1(1, 0)$, $P_2(3, 2)$. (b) $P_3(0, -4)$, $P_4(4, 0)$. *(c) $P_5(1, -\sqrt{2})$, $P_6(\sqrt{2}, -1)$.

*6. Sketch the rectangle with vertices $V_1(-1, -2)$, $V_2(5, -2)$, $V_3(5, 6)$. Where is the fourth vertex? Where do the diagonals intersect?

7. The point M bisects the line segment $\overline{P_1P_2}$. Find the coordinates of P_1 if P_2 is at $(4, 8)$ and M is at $(1, -5)$.

*8. Find the third vertex of an equilateral triangle given that O and $P(0, b)$ are two of the vertices.

9. Sketch the triangle with vertices $P(-1, 0)$, $Q(0, 2)$, $R(3, 0)$.
    *(a) Find the midpoint of each of the sides.
    (b) Find the midpoint of each of the medians. (A median is a line segment that joins a vertex to the midpoint of the opposite side.)

10. Find the numbers r and s given that the points $P(r, s)$ and $Q(2r + 1, 2s - 1)$ are symmetric about *(a) the x-axis,   (b) the y-axis,   *(c) the origin.

Optional | 11. Find the coordinates of S and T in Figure 3.1.11.

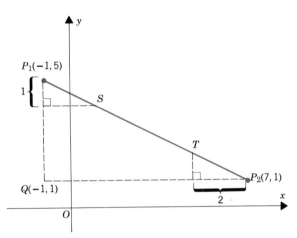

Figure 3.1.11

3.2   The Distance Between Two Points; The Equation of a Circle

The distance between two points $P_1(x_1, y_1)$ and $P_2(x_2, y_2)$, denoted by $d(P_1, P_2)$, can be found by the Pythagorean theorem.

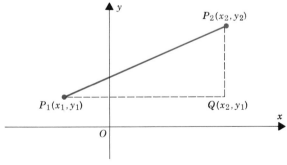

Figure 3.2.1

With Q as in Figure 3.2.1, $\triangle P_1 Q P_2$ is a right triangle. Consequently

$$[d(P_1, P_2)]^2 = [d(P_1, Q)]^2 + [d(Q, P_2)]^2 \quad \text{(the Pythagorean theorem)}$$
$$= (x_2 - x_1)^2 + (y_2 - y_1)^2.$$

Taking square roots, we have the *distance formula*

(3.2.1)
$$\boxed{d(P_1, P_2) = \sqrt{(x_2 - x_1)^2 + (y_2 - y_1)^2}.}$$

Problem. Find the distance between

$$P_1(1, 2) \quad \text{and} \quad P_2(-1, 5).$$

SOLUTION

$$d(P_1, P_2) = \sqrt{(-1 - 1)^2 + (5 - 2)^2} = \sqrt{4 + 9} = \sqrt{13}. \quad \square$$

Problem. Show that the points $P_1(4, 8)$ and $P_2(5, 1)$ are equidistant from $Q(1, 4)$.

SOLUTION

$$d(P_1, Q) = \sqrt{(1 - 4)^2 + (4 - 8)^2} = \sqrt{9 + 16} = \sqrt{25} = 5.$$
$$d(P_2, Q) = \sqrt{(1 - 5)^2 + (4 - 1)^2} = \sqrt{16 + 9} = \sqrt{25} = 5. \quad \square$$

Problem. Show that the points

$$P_1(3, 5), \quad P_2(-3, 7), \quad P_3(-6, -2)$$

are the vertices of a right triangle. Find the area of the triangle.

SOLUTION. The lengths of the sides are

$$d(P_1, P_2) = \sqrt{(-3 - 3)^2 + (7 - 5)^2} = \sqrt{36 + 4} = \sqrt{40},$$
$$d(P_2, P_3) = \sqrt{(-6 + 3)^2 + (-2 - 7)^2} = \sqrt{9 + 81} = \sqrt{90},$$
$$d(P_1, P_3) = \sqrt{(-6 - 3)^2 + (-2 - 5)^2} = \sqrt{81 + 49} = \sqrt{130}.$$

Since

$$[d(P_1, P_2)]^2 + [d(P_2, P_3)]^2 = 40 + 90 = 130 = [d(P_1, P_3)]^2,$$

we know from the Pythagorean theorem that the triangle is a right triangle.

The legs of the right triangle are the line segments $\overline{P_1P_2}$ and $\overline{P_2P_3}$, and the area is

$$\tfrac{1}{2}d(P_1, P_2)\, d(P_2, P_3) = \tfrac{1}{2}\sqrt{40}\sqrt{90} = \tfrac{1}{2}\sqrt{3600} = \tfrac{1}{2}(60) = 30.\dagger \quad \square$$

The Equation of a Circle

The *circle* with radius r centered at the point Q_0 is the set of all points P such that

$$d(P, Q_0) = r.$$

If the coordinates of Q_0 are (x_0, y_0), then a point $P(x, y)$ will lie on the given circle iff

$$\sqrt{(x - x_0)^2 + (y - y_0)^2} = r$$

or, equivalently, iff

(3.2.2)
$$\boxed{(x - x_0)^2 + (y - y_0)^2 = r^2.}$$

This is the usual equation for the circle of radius r centered at (x_0, y_0). If the center is placed at the origin, then x_0 and y_0 are both 0, and the equation reduces to

(3.2.3)
$$\boxed{x^2 + y^2 = r^2.}$$

Examples

1. The equation

    $$(x - 2)^2 + (y - 1)^2 = 4$$

    represents the circle of radius 2 centered at $(2, 1)$.

2. The equation

    $$(x - 2)^2 + (y + 1)^2 = 5$$

    represents the circle of radius $\sqrt{5}$ centered at $(2, -1)$. (Here $y_0 = -1$.)

3. The equation

    $$x^2 + y^2 = 1$$

    represents the circle of radius 1 centered at the origin. This is generally called the *unit circle*. $\square$

$\dagger$ In general, the area of a triangle is given by the formula $A = \tfrac{1}{2}bh$. In a right triangle, we can take either leg as the base and the other as the height.

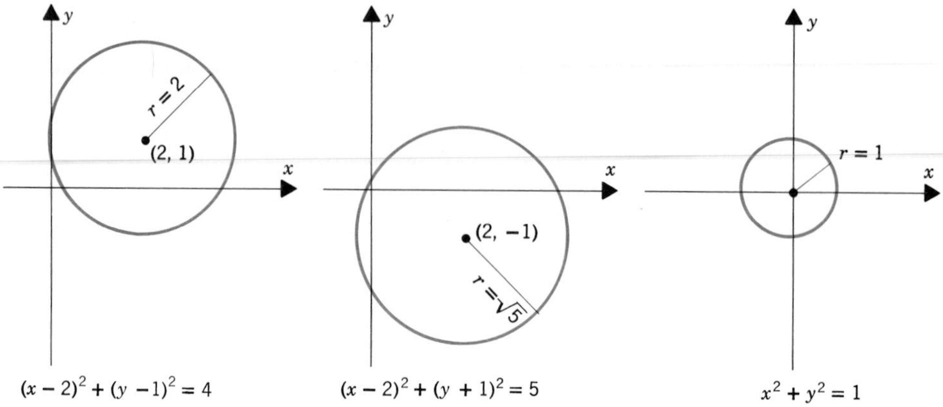

$(x - 2)^2 + (y - 1)^2 = 4$ $(x - 2)^2 + (y + 1)^2 = 5$ $x^2 + y^2 = 1$

Figure 3.2.2

Problem. Show that the equation

$$x^2 + y^2 - 2x - 6y - 15 = 0$$

represents a circle. Find the radius and the center.

SOLUTION. The idea is to rewrite the equation in the form

$$(x - x_0)^2 + (y - y_0)^2 = r^2.$$

We begin by writing

$$(x^2 - 2x \quad) + (y^2 - 6y \quad) = 15.$$

Completing the squares in the parentheses, we get

$$(x^2 - 2x + 1) + (y^2 - 6y + 9) = 15 + 1 + 9$$

and thus

$$(x - 1)^2 + (y - 3)^2 = 25.$$

This is the equation of a circle. The radius is 5 and the center is at $(1, 3)$.   ☐

Exercises

1. Find the distance between the given points:
   *(a) $P_1(3, -1)$, $P_2(2, 4)$.   (b) $P_1(-1, -2)$, $P_2(-4, -2)$.
   *(c) $P_1(x_0, 0)$, $P_2(0, y_0)$.   (d) O and $P(x_0, y_0)$.

2. *(a) Find a given that $P(a, 0)$ is equidistant from $Q(-2, -4)$ and $R(4, 3)$.
   (b) Find b given that $P(0, b)$ is equidistant from these same points.

3. Determine whether the following points are the vertices of a right triangle:

*(a) $P_1(-1, 0)$, $P_2(1, 0)$, $P_3(\frac{1}{2}\sqrt{2}, -\frac{1}{2}\sqrt{2})$.
(b) $P_1(-1, -4)$, $P_2(4, -2)$, $P_3(-3, 1)$.
*(c) $P_1(4, 1)$, $P_2(2, 4)$, $P_3(1, 3)$.

4. Write an equation for the circle
    *(a) of radius 2 centered at $Q(1, 2)$.
    (b) of radius 2 centered at $Q(-1, -2)$.
    *(c) of radius 7 centered at the origin.
    (d) of radius 7 centered at $Q(-4, 5)$.

5. Which of the following points lie on the circle of radius 13 centered at $Q(1, 2)$?

$$P_1(6, 14), \quad P_2(1, 15), \quad P_3(-4, -10), \quad P_4(-1, 12).$$

6. Write an equation for the circle that passes through the point
    *(a) $P(1, -2)$ and is centered at $Q(2, -1)$.
    (b) $P(2, -1)$ and is centered at $Q(1, -2)$.

7. Find the radius and the center of the following circles:
    *(a) $x^2 + y^2 - 8x - 2y + 13 = 0$.
    (b) $x^2 + y^2 + 6x + 4y + 12 = 0$.
    *(c) $x^2 + y^2 + 2\sqrt{2}x + 2\sqrt{3}y - 95 = 0$.
    (d) $x^2 + y^2 - x + y - \frac{7}{2} = 0$.

8. Find all possible values of x_0 given that O, $P(-3, 1)$, and $R(x_0, 0)$ are the vertices of a right triangle.

*9. Find the area of the rectangle ($A = bh$) if the vertices are O, $P(2, 1)$, $Q(-4, 8)$, $R(-2, 9)$.

10. Sketch the triangle with the indicated vertices and find its area. ($A = \frac{1}{2}bh$)
    *(a) O, $P(5, 0)$, $Q(6, 4)$.
    (b) O, $P(0, 5)$, $Q(6, 4)$.
    *(c) $P_1(-4, 0)$, $P_2(1, 0)$, $P_3(-\frac{11}{5}, \frac{12}{5})$.

11. *(a) Determine the area of the triangle with vertices

$$P(a, 0), \quad Q(0, b), \quad R(c, 0).$$

    (b) What is the area of the triangle with vertices at the midpoints of the sides of $\triangle PQR$?

3.3 The Slope of a Line

In Figure 3.3.1 on page 84 we have drawn some nonvertical lines through a common point. These lines differ in their rate of climb:

On l_1, as the x-coordinate increases by 3, the y-coordinate increases by 5 and the rate of climb is $\frac{5}{3}$. (5 divided by 3)

On l_2, as the x-coordinate increases by 3, the y-coordinate increases by 1; the rate of climb is therefore $\frac{1}{3}$. (1 divided by 3)

On l_3, no matter how much x increases, y always stays the same; the rate of climb is 0.

On l_4, as x increases by 3, y *decreases* by 4 (or, equivalently, increases by -4); the rate of climb is therefore $-\frac{4}{3}$. (-4 divided by 3)

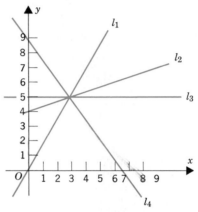

Figure 3.3.1

In general, to find the rate of climb of a nonvertical line, it is enough to know the coordinates of two points. For the line that passes through $P_0(x_0, y_0)$ and $P_1(x_1, y_1)$ the rate of climb is

$$\frac{y_1 - y_0}{x_1 - x_0}.$$ (see Figure 3.3.2)

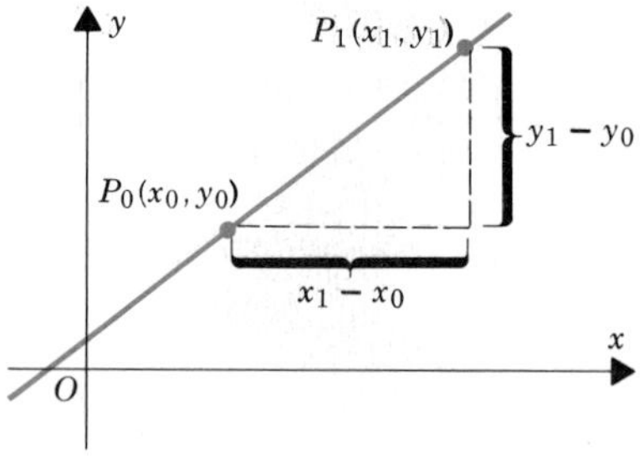

Figure 3.3.2

This ratio is called the *slope* of the line and is usually denoted by the letter m. This notion is important enough to be cast formally as a definition.

Definition   3.3.1   The Slope of a Line

The *slope m* of the nonvertical line that passes through the points $P_0(x_0, y_0)$ and $P_1(x_1, y_1)$ is the ratio

$$\frac{y_1 - y_0}{x_1 - x_0}.$$

Since

$$\frac{y_1 - y_0}{x_1 - x_0} = \frac{y_0 - y_1}{x_0 - x_1},$$

the points can be taken in either order. Make sure, however, to follow the same order in both numerator and denominator. The slope of the line through $P_0(1, 4)$ and $P_1(3, 7)$ can be written as

$$\frac{7 - 4}{3 - 1} = \frac{3}{2} \quad \text{or as} \quad \frac{4 - 7}{1 - 3} = \frac{-3}{-2} = \frac{3}{2}$$

but not as

$$\frac{7 - 4}{1 - 3} \quad \text{or} \quad \frac{4 - 7}{3 - 1}.$$

Problem.   Find the point at which the line through $P_0(1, 3)$ and $P_1(4, -2)$ intersects the *x*-axis.

SOLUTION.   The line intersects the *x*-axis at a point of the form $Q(a, 0)$. The slope of the line as measured from P_0 and P_1 is

$$\frac{-2 - 3}{4 - 1} = -\frac{5}{3}.$$

As measured from Q and P_1, the slope is

$$\frac{-2 - 0}{4 - a} = \frac{2}{a - 4}.$$

Therefore we must have

$$-\frac{5}{3} = \frac{2}{a - 4}.$$

We now solve the equation for a:

$$-5(a - 4) = 6,$$
$$-5a + 20 = 6,$$
$$-5a = -14,$$
$$a = \tfrac{14}{5}.$$

The point is $Q(\tfrac{14}{5}, 0)$. $\square$

The Sign of the Slope (Figure 3.3.3)

(i)   A positive slope indicates a positive rate of climb: the line rises from left to right.

(ii)   Zero slope indicates a zero rate of climb: the line is horizontal.

(iii)   A negative slope indicates a negative rate of climb: the line slants down from left to right.

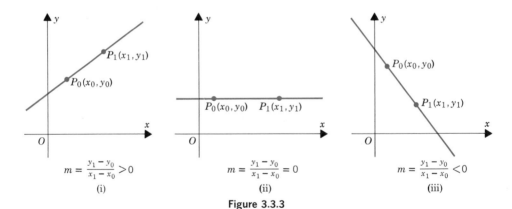

Figure 3.3.3

Vertical Lines

The notion of slope does not apply to vertical lines. On a vertical line (Figure 3.3.4) all points have the same x-coordinate. With $x_1 = x_0$, the difference $x_1 - x_0$ is 0 and the quotient

$$\frac{y_1 - y_0}{x_1 - x_0}$$

is undefined.

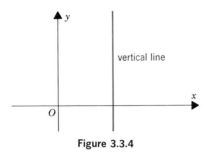

Figure 3.3.4

Parallel Lines and Perpendicular Lines

It is easy to see that

(3.3.2)   *two nonvertical lines are parallel iff they have the same slope:*
$$l_1 \parallel l_2 \quad \text{iff} \quad m_1 = m_2.$$

Not so obvious is that

(3.3.3)   *two nonvertical lines are perpendicular iff the product of their slopes is* -1:
$$l_1 \perp l_2 \quad \text{iff} \quad m_1 m_2 = -1.$$

You can find a proof of these assertions in the supplement at the end of this section. ☐

Problem.   Given the lines

l_1 through $A(2, 1)$ and $B(3, 5)$, l_2 through $C(-1, 3)$ and $D(4, y_0)$,

find y_0 if   (a) $l_1 \parallel l_2$.   (b) $l_1 \perp l_2$.

SOLUTION.   First we find the slope of each line:

$$m_1 = \frac{5 - 1}{3 - 2} = 4 \quad \text{and} \quad m_2 = \frac{y_0 - 3}{4 - (-1)} = \frac{y_0 - 3}{5}.$$

(a) If $l_1 \parallel l_2$, then

$$4 = \frac{y_0 - 3}{5}$$

so that

$$20 = y_0 - 3 \quad \text{and} \quad y_0 = 23.$$

(b) If $l_1 \perp l_2$, then

$$4\left(\frac{y_0 - 3}{5}\right) = -1,$$

so that

$$4y_0 - 12 = -5$$
$$4y_0 = 7$$
$$y_0 = \tfrac{7}{4}. \quad \square$$

Exercises

1. Find the slope of the line through the given points:
   *(a) $P_0(-1, 2)$, $P_1(3, 4)$.   (b) $P_0(4, 5)$, $P_1(6, -3)$.
   *(c) $P_0(3, 5)$, $P_1(6, 5)$.   (d) O, $P(x_0, y_0)$.
   *(e) $P(x_0, 0)$, $Q(0, y_0)$.   (f) $P(a, b)$, $Q(b, a)$.
   *(g) $P(a, a)$, $Q(-a, -a)$.   (h) $P(a, b)$, $Q(-a, b)$.

2. The following lines pass through the given points:
   l_1: $A(4, 2)$, $B(7, 1)$. l_2: $C(2, 5)$, $D(2, -1)$.
   l_3: $E(0, 0)$, $F(1, 3)$. l_4: $G(2, 4)$, $H(1, 0)$.
   l_5: $I(-1, 2)$, $J(5, 2)$. l_6: $K(3, 2)$, $L(2, -1)$.
   *(a) Which lines are horizontal?
   (b) Which lines are vertical?
   *(c) Which pairs of lines are parallel?
   (d) Which pairs of lines are perpendicular?

3. Where does the line through $P(-2, 5)$ and $Q(2, 3)$ intersect
   *(a) the x-axis?   (b) the y-axis?

4. Where does the line through $P(5, 2)$ and $Q(1, 1)$ intersect
   (a) the x-axis?   *(b) the y-axis?

*5. Find a and b given that the following points all lie on the same line:

$$P_1(-3, 5), \quad P_2(4, 1), \quad P_3(a, 8), \quad P_4(2, b).$$

6. Given the lines

   l_1 through $A(2, -3)$ and $B(7, 7)$, l_2 through $C(5, -1)$ and $D(6, y_0)$,

   find y_0 if   *(a) $l_1 \parallel l_2$.   (b) $l_1 \perp l_2$.

7. Given the lines

   l_1 through $A(2, 3)$ and $B(7, 6)$, l_2 through $C(1, 3)$ and $D(x_0, 1)$

   find x_0 if   (a) $l_1 \parallel l_2$.   *(b) $l_1 \perp l_2$.

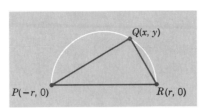

Figure 3.3.5

8. Show that every triangle inscribed in a semicircle is a right triangle. HINT: Take
    P, Q, R, as in Figure 3.3.5 and show that

$$\text{(the slope of } \overline{PQ})\text{(the slope of } \overline{RQ}) = -1.$$

Optional | **Supplement to Section 3.3**

PROOF OF (3.3.2). In Figure 3.3.6 we have drawn two nonvertical lines. By
elementary geometry,

$$l_1 \parallel l_2 \quad \text{iff} \quad A = C \text{ and } B = D,$$

$$\text{iff} \quad \text{the two corresponding triangles are similar,}$$

$$\text{iff} \quad m_1 = \frac{b}{a} = \frac{d}{c} = m_2. \quad \square$$

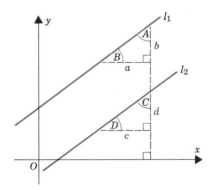

Figure 3.3.6

PROOF OF (3.3.3). In Figure 3.3.7 you see two nonvertical lines that
intersect at a point $P_0(x_0, y_0)$.

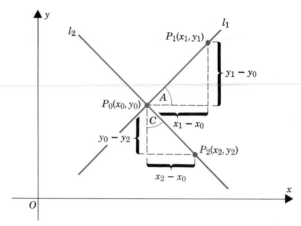

Figure 3.3.7

If $l_1 \perp l_2$, then $A = C$† and the right triangles in the figure are similar. It then follows that the comparable sides are proportional:

$$\frac{y_1 - y_0}{x_2 - x_0} = \frac{x_1 - x_0}{y_0 - y_2}.$$

This gives

$$\frac{y_1 - y_0}{x_1 - x_0} \frac{y_0 - y_2}{x_2 - x_0} = 1,$$

and therefore

$$\frac{y_1 - y_0}{x_1 - x_0} \frac{y_2 - y_0}{x_2 - x_0} = -1. \qquad \text{(we replaced } y_0 - y_2 \text{ by } y_2 - y_0\text{)}$$

The first fraction on the left is m_1; the second fraction is m_2. As you can see,

$$m_1 m_2 = -1.$$

We have shown that

$$\text{if} \quad l_1 \perp l_2, \qquad \text{then} \quad m_1 m_2 = -1.$$

---

† If $l_1 \perp l_2$, then

$$A + B = 90°, \qquad C + B = 90°$$

and therefore

$$A = C.$$

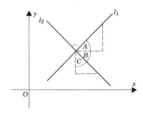

Optional | By retracing the steps of the proof, you can show that
$$\text{if} \quad m_1 m_2 = -1, \qquad \text{then} \quad l_1 \perp l_2. \quad \square$$

3.4 Equations for Lines

Vertical Lines

If l is vertical, then l intersects the x-axis at some point $Q(a, 0)$. We call a the *x-intercept*. A point $P(x, y)$ lies on l iff

(3.4.1)
$$\boxed{x = a.}$$

Nonvertical Lines

If l is nonvertical (Figure 3.4.1), then l has some slope m and intersects the y-axis at some point $Q(0, b)$. We call b the *y-intercept*.

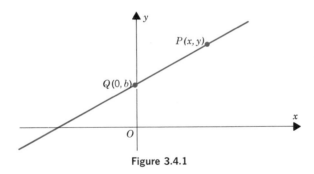

Figure 3.4.1

A point $P(x, y)$ different from $Q(0, b)$ lies on l iff

$$\frac{y - b}{x - 0} = m.$$

[The left-hand side is the slope of the line through $Q(0, b)$ and $P(x, y)$. For $P(x, y)$ to lie on l, this slope must be m, the slope of l.]

We now multiply by x and write the equation as

(3.4.2)
$$\boxed{y = mx + b.}$$

This equation is satisfied by the coordinates of all the points on l, including Q.
The equation

$$y = mx + b$$

is said to be written in *slope-intercept* form: it represents the line with slope m and y-intercept b. If the line is horizontal, then $m = 0$, and the equation reduces to

$$y = b.$$

If the line passes through the origin, then $b = 0$, and the equation takes the form

$$y = mx.$$

We illustrate the various possibilities in Figure 3.4.2.

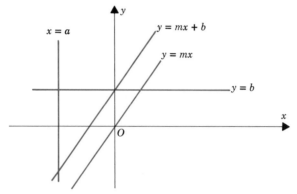

Figure 3.4.2

Examples

1. Equations

$$x = -1, \qquad x = 0, \qquad x = 1 \qquad \text{(Figure 3.4.3)}$$

all represent vertical lines. The first equation represents the line one unit to the left of the y-axis; the second, the y-axis; the third, the line one unit to the right of the y-axis.

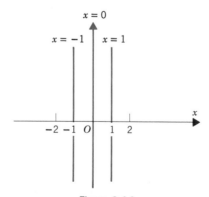

Figure 3.4.3

2. Equations

$$y = -1, \qquad y = 0, \qquad y = 1 \qquad \text{(Figure 3.4.4)}$$

all represent horizontal lines. The first equation represents the line one unit below the x-axis; the second, the x-axis; the third, the line one unit above the x-axis.

3. Equations

$$y = -2x, \qquad y = \tfrac{1}{2}x, \qquad y = 2x \qquad \text{(Figure 3.4.5)}$$

all represent lines that pass through the origin. The first line has slope -2; the second has slope $\tfrac{1}{2}$; the third one has slope 2.

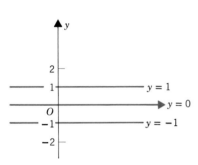

Figure 3.4.4

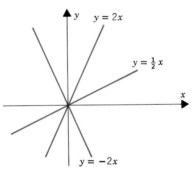

Figure 3.4.5

4. Equations

$$y = \tfrac{2}{5}x - 1, \qquad y = \tfrac{2}{5}x, \qquad y = \tfrac{2}{5}x + 1 \qquad \text{(Figure 3.4.6)}$$

all represent parallel lines, each with slope $\tfrac{2}{5}$. The first line has y-intercept -1; the second one passes through the origin; the third one has y-intercept 1.   □

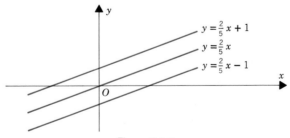

Figure 3.4.6

An equation of the form

$$Ax + By + C = 0 \qquad \text{with } A \text{ and } B \text{ not both } 0$$

is called a *linear equation in x and y.* Here is the reason:

Theorem   3.4.3

Every line has an equation of the form

$$Ax + By + C = 0 \qquad \text{with } A \text{ and } B \text{ not both } 0,$$

and every such equation represents a line.

PROOF. We begin with a line l. If l is vertical, then l has an equation of the form

$$x = a.$$

We can rewrite this as

$$1 \cdot x + 0 \cdot y - a = 0. \qquad (A = 1, B = 0, C = -a)$$

If l is not vertical, then l has an equation of the form

$$y = mx + b.$$

We can rewrite this as

$$mx - y + b = 0. \qquad (A = m, B = -1, C = b)$$

So far we have shown that every line has an equation of the right form. Now we go the other way. We take an equation

$$Ax + By + C = 0 \qquad \text{with } A \text{ and } B \text{ not both } 0$$

and show that it represents a line.

If $B = 0$, then we have

$$Ax + C = 0$$

and, since $A \neq 0$,

$$x = -\frac{C}{A}.$$

This gives a vertical line.

If $B \neq 0$, then we can divide the initial equation by B and obtain

$$\frac{A}{B}x + y + \frac{C}{B} = 0,$$

which we can rewrite as

$$y = -\frac{A}{B}x - \frac{C}{B}.$$

This equation is of the form $y = mx + b$. It represents a line with slope $-A/B$ and y-intercept $-C/B$. ☐

Problem.   Find the slope and the y-intercept of

$$l_1: 2x + 5y - 3 = 0 \quad \text{and} \quad l_2: 2x - 3 = 0.$$

SOLUTION.   The equation of l_1 can be rewritten as

$$5y = -2x + 3$$

and therefore as

$$y = -\tfrac{2}{5}x + \tfrac{3}{5}.$$

This is in the form $y = mx + b$. The slope m is $-\tfrac{2}{5}$, and the y-intercept is $\tfrac{3}{5}$.
The equation of l_2 can be written

$$x = \tfrac{3}{2}.$$

The line is vertical and has no slope; and since it does not cross the y-axis, it has no y-intercept. See Figure 3.4.7. □

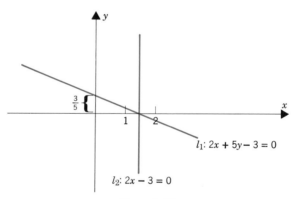

Figure 3.4.7

Exercises

1.   Find the slope and the y-intercept of each of the following lines:
  *(a) $y = x + 4$.   (b) $y = 3x - 1$.
  *(c) $4x = 1$.   (d) $3y = x + 2$.
  *(e) $x + y + 1 = 0$.   (f) $4x - 2y + 6 = 0$.
  *(g) $7x + 4y + 4 = 0$.   (h) $7x - 4y + 4 = 0$.
  *(i) $\dfrac{x}{a} + \dfrac{y}{b} = 1$.   (j) $\dfrac{x}{a} - \dfrac{y}{b} = 1$.
  *(k) $2y + 5 = 0$.   (l) $\tfrac{1}{2}x + \tfrac{1}{3}y = \tfrac{1}{6}$.

2.   Write an equation $y = mx + b$ for the line
   *(a) with slope 5 and y-intercept 2.
   (b) with slope -5 and y-intercept 2.
   *(c) with slope 5 and y-intercept -2.
   (d) with slope -5 and y-intercept -2.

3.   Write an equation for the horizontal line
   *(a) 3 units above the x-axis.
   (b) 3 units below the x-axis.

4.   Write an equation for the vertical line
   (a) 3 units to the right of the y-axis.
   *(b) 3 units to the left of the y-axis.

5.   Given
$$l: 2x + y + 4 = 0,$$
   find an equation $Ax + By + C = 0$ for the line that is
   *(a) parallel to l and has y-intercept 1.
   (b) parallel to l and has y-intercept -3.
   *(c) perpendicular to l and has y-intercept 1.
   (d) perpendicular to l and has y-intercept -3.

3.5   More on Lines

When we write the equation of a nonvertical line in its slope-intercept form

$$y = mx + b,$$

we are featuring the slope m and the y-intercept b. There are other ways of writing the equation of such a line. *The point-slope form* features the slope and a point on the line; *the two-point form* features two points; *the two-intercept form* features the two intercepts.

Theorem   3.5.1   The Point-Slope Form

If a line passes through a point $P_0(x_0, y_0)$ with slope m, then we can write its equation as

$$y - y_0 = m(x - x_0).$$

PROOF.   Since the slope is m, we can write the equation of the line in the form

(1) $y = mx + b.$

Since the point $P_0(x_0, y_0)$ lies on the line, its coordinates must satisfy the equation. Thus we know that

$$y_0 = mx_0 + b.$$

This gives

$$b = y_0 - mx_0.$$

Substituting this value of b into Equation (1), we have

$$y = mx + y_0 - mx_0$$

and thus

$$y - y_0 = m(x - x_0). \quad \square$$

Examples

1. The equation

$$y - 1 = 2(x - 3)$$

represents the line that passes through the point $P(3, 1)$ with slope 2.

2. The equation

$$y - 6 = \tfrac{1}{3}(x + 2)$$

represents the line that passes through the point $Q(-2, 6)$ with slope $\tfrac{1}{3}$. $\quad \square$

Theorem   3.5.2   The Two-Point Form

If a line passes through two points $P_0(x_0, y_0)$ and $P_1(x_1, y_1)$ with $x_0 \neq x_1$, then its equation can be written

$$y - y_0 = \frac{y_1 - y_0}{x_1 - x_0}(x - x_0).$$

PROOF. Simply note that

$$m = \frac{y_1 - y_0}{x_1 - x_0}$$

and use the point-slope form. $\quad \square$

Problem. Find an equation for the line that passes through the points $P(2, 5)$ and $Q(6, -3)$.

SOLUTION. The slope of the line is

$$m = \frac{-3 - 5}{6 - 2} = -2.$$

Taking P as P_0, we have

$$y - 5 = -2(x - 2).$$

Taking Q as P_0, we have

$$y + 3 = -2(x - 6).$$

The equations look different, but they are really both the same. Each can be simplified to read

$$y = -2x + 9. \quad \square$$

Theorem   3.5.3   The Two-Intercept Form

If a line has x-intercept a and y-intercept b with a and b both different from 0, then we can write its equation as

$$\frac{x}{a} + \frac{y}{b} = 1.$$

PROOF. Such a line passes through $P(a, 0)$ and $Q(0, b)$. The slope is

$$m = \frac{b - 0}{0 - a} = -\frac{b}{a},$$

and the equation is therefore

$$y = -\frac{b}{a}x + b. \qquad \text{(slope-intercept form)}$$

Dividing by b, we have

$$\frac{y}{b} = -\frac{x}{a} + 1,$$

and therefore

$$\frac{x}{a} + \frac{y}{b} = 1. \quad \square$$

Examples

1. The equation

$$\frac{x}{2} + \frac{y}{3} = 1$$

represents the line with x-intercept 2 and y-intercept 3.

2. The equation

$$\frac{x}{2} - \frac{y}{3} = 1$$

can be written

$$\frac{x}{2} + \frac{y}{-3} = 1.$$

The x-intercept is 2, and the y-intercept is -3. $\square$

Problem. Find the intercepts of the line

$$3x + 4y + 7 = 0.$$

SOLUTION. To find the x-intercept, we set $y = 0$ and solve for x:

$$3x + 7 = 0$$
$$x = -\tfrac{7}{3}.$$

The x-intercept is $-\tfrac{7}{3}$.
    To find the y-intercept, we set $x = 0$ and solve for y:

$$4y + 7 = 0$$
$$y = -\tfrac{7}{4}.$$

The y-intercept is $-\tfrac{7}{4}$. $\square$

Problem. Find an equation for the line l_2 that is perpendicular to

$$l_1: x - 5y + 5 = 0$$

and passes through the point $P(2, -4)$.

SOLUTION. We can rewrite the equation of l_1 as

$$5y = x + 5$$

and thus as

$$y = \tfrac{1}{5}x + 1.$$

The slope of l_1 is $\tfrac{1}{5}$. The slope of l_2 is therefore -5. [*Remember:* For perpendicular lines, $m_1 m_2 = -1$.]
    Since l_2 passes through $P(2, -4)$ with slope -5, we can write its equation as

$$y + 4 = -5(x - 2). \quad \square$$

Summary on Lines   3.5.4

General form: $Ax + By + C = 0$ with A and B not both 0.

Vertical line: $x = a$.

Slope-intercept form: $y = mx + b$.

Point-slope form: $y - y_0 = m(x - x_0)$.

Two-point form: $y - y_0 = \dfrac{y_1 - y_0}{x_1 - x_0}(x - x_0)$.

Two-intercept form: $\dfrac{x}{a} + \dfrac{y}{b} = 1$.

Parallel nonvertical lines: $m_1 = m_2$.

Perpendicular nonvertical lines: $m_1 m_2 = -1$.

Exercises

1. Write an equation in point-slope form for the line
    *(a) through $P(1, 3)$ with slope 5.
    (b) through $P(3, 1)$ with slope 5.
    *(c) through $P(1, 3)$ with slope -5.
    (d) through $P(3, 1)$ with slope -5.
    *(e) through $P(-1, 3)$ with slope 5.
    (f) through $P(1, -3)$ with slope 5.
    *(g) through $P(-1, -3)$ with slope 5.
    (h) through $P(-1, -3)$ with slope -5.

2. Find an equation $Ax + By + C = 0$ for the line through
    *(a) $P(1, 2)$, $Q(2, 1)$.   (b) $P(1, -2)$, $Q(3, 2)$.
    *(c) $P(-1, 2)$, $Q(-2, 5)$.   (d) $P(-1, -1)$, $Q(2, -4)$.

3. Write an equation in two-intercept form for the line
    *(a) with x-intercept 3, y-intercept 5.
    (b) with x-intercept 3, y-intercept -5.
    *(c) with x-intercept -3, y-intercept 5.
    (d) with x-intercept -3, y-intercept -5.

4. Find an equation $Ax + By + C = 0$ for the line through $P(2, 7)$ that is
    *(a) parallel to the x-axis.
    (b) parallel to the y-axis.
    *(c) parallel to the line $x + y + 1 = 0$.

(d) perpendicular to the line $x + y + 1 = 0$.
(e) parallel to the line $3x - 2y + 6 = 0$.
*(f) perpendicular to the line $3x - 2y + 6 = 0$.

5. Take the equation
$$3y + 6x - 12 = 0$$

and put it in its
*(a) slope-intercept form.
(b) point-slope form using the point $P(1, 2)$.
*(c) two-intercept form.

*6. Write an equation in point-slope form for the perpendicular bisector of the line segment that joins $P(2, -1)$ and $Q(4, 7)$.

3.6 Intersections of Lines and Circles

Two distinct lines either intersect at a single point, or they are parallel and don't intersect at all.
     If two lines
$$l_1: A_1 x + B_1 y + C_1 = 0 \quad \text{and} \quad l_2: A_2 x + B_2 y + C_2 = 0$$

intersect at the point Q, then the coordinates of Q must satisfy both equations. We can find these coordinates by solving the two equations simultaneously.

Problem. Find the point Q at which the lines
$$2x - 3y + 4 = 0, \qquad 5x + 2y - 1 = 0$$
intersect.

SOLUTION. We can solve the two equations simultaneously by solving one of the equations for one variable in terms of the other and then substituting the result into the other equation. In this case let's solve the first equation for y in terms of x:
$$2x - 3y + 4 = 0$$
$$3y = 2x + 4$$
(∗)
$$y = \tfrac{1}{3}(2x + 4).$$
Substituting $\tfrac{1}{3}(2x + 4)$ for y into the second equation, we have
$$5x + \tfrac{2}{3}(2x + 4) - 1 = 0$$
$$15x + 4x + 8 - 3 = 0$$
$$19x = -5$$
$$x = -\tfrac{5}{19}.$$

Substituting this value of x into equation (∗), we have

$$y = \tfrac{1}{3}(-\tfrac{10}{19} + 4) = \tfrac{1}{3}(\tfrac{66}{19}) = \tfrac{22}{19}.$$

According to our calculations, the two lines intersect at the point $Q(-\tfrac{5}{19}, \tfrac{22}{19})$.
We can check the result by substituting $x = -\tfrac{5}{19}$ and $y = \tfrac{22}{19}$ in the initial equations:

$$2(-\tfrac{5}{19}) - 3(\tfrac{22}{19}) + 4 = -\tfrac{10}{19} - \tfrac{66}{19} + 4 = -\tfrac{76}{19} + 4 \overset{\checkmark}{=} 0,$$

$$5(-\tfrac{5}{19}) + 2(\tfrac{22}{19}) - 1 = -\tfrac{25}{19} + \tfrac{44}{19} - 1 = \tfrac{19}{19} - 1 \overset{\checkmark}{=} 0. \quad \square$$

REMARK. Substitution is not the only way to solve linear equations simultaneously. Let's return to the equations

$$2x - 3y + 4 = 0 \qquad \text{and} \qquad 5x + 2y - 1 = 0.$$

We can eliminate y from both equations by multiplying the first equation by 2, the second by 3, and then adding the resulting equations:

$$2(2x - 3y + 4) = 0,$$
$$3(5x + 2y - 1) = 0;$$

$$4x - 6y + 8 = 0,$$
$$15x + 6y - 3 = 0;$$

$$19x + 5 = 0,$$
$$x = -\tfrac{5}{19}.$$

We can eliminate x from both equations by multiplying the first equation by 5, the second by 2, and then subtracting:

$$5(2x - 3y + 4) = 0,$$
$$2(5x + 2y - 1) = 0;$$

$$10x - 15y + 20 = 0,$$
$$10x + 4y - 2 = 0;$$

$$-19y + 22 = 0,$$
$$y = \tfrac{22}{19}.$$

These are the same solutions we obtained before. $\quad \square$

If two lines

$$l_1: A_1 x + B_1 y + C_1 = 0 \qquad \text{and} \qquad l_2: A_2 x + B_2 y + C_2 = 0$$

are parallel, there is no point of intersection and no simultaneous solution to their equations. This happens when $A_1 B_2 = A_2 B_1$: if B_1 and B_2 are both zero, then the two lines are vertical, and, if not, then both lines have the same slope.†

The next problems involve lines and circles.

Problem.   Find the points, if any, at which the lines

$$l_1: 4x - 3y + 2 = 0, \qquad l_2: 3x - 4y - 20 = 0, \qquad l_3: x - 7 = 0$$

intersect the circle

$$(x - 1)^2 + (y - 2)^2 = 25.$$

SOLUTION.   We can find the points where a line intersects a circle by solving their equations simultaneously. The equation for l_1 can be written

$$(*) \qquad\qquad y = \tfrac{1}{3}(4x + 2).$$

Substituting $\tfrac{1}{3}(4x + 2)$ for y into the equation for the circle we have

$$(x - 1)^2 + [\tfrac{1}{3}(4x + 2) - 2]^2 = 25,$$

which simplifies to

$$(x - 1)^2 + \tfrac{16}{9}(x - 1)^2 = 25.$$

This last equation gives

$$9(x - 1)^2 + 16(x - 1)^2 = 225$$
$$25(x - 1)^2 = 225$$
$$(x - 1)^2 = 9$$
$$x - 1 = \pm 3$$
$$x = -2, 4.$$

Substitution of $x = -2$ into $(*)$ gives $y = -2$; substitution of $x = 4$ gives $y = 6$. The line l_1 intersects the circle at the points $P(-2, -2)$ and $Q(4, 6)$.

We check these calculations by verifying that the pairs $x = -2, y = -2$ and $x = 4, y = 6$ satisfy both the original equation for l_1 and the equation for the circle:

$$4(-2) - 3(-2) + 2 = -8 + 6 + 2 \overset{\checkmark}{=} 0,$$
$$(-2 - 1)^2 + (-2 - 2)^2 = 9 + 16 \overset{\checkmark}{=} 25;$$

$$4(4) - 3(6) + 2 = 16 - 18 + 2 \overset{\checkmark}{=} 0,$$
$$(4 - 1)^2 + (6 - 2)^2 = 9 + 16 \overset{\checkmark}{=} 25.$$

---

† You know that $m_1 = -A_1/B_1$ and $m_2 = -A_2/B_2$. If $A_1 B_2 = A_2 B_1$, then $A_1/B_1 = A_2/B_2$ and therefore $m_1 = m_2$.

To find where l_2 intersects the circle we rewrite its equation in the form

(**) $$y = \tfrac{1}{4}(3x - 20).$$

We now substitute $\tfrac{1}{4}(3x - 20)$ for y into the equation for the circle. This gives

$$(x - 1)^2 + [\tfrac{1}{4}(3x - 20) - 2]^2 = 25.$$

We leave it to you to simplify this equation and verify that $x = 4$ is the only solution. Substitution of $x = 4$ into (**) gives $y = -2$. The line l_2 intersects the circle at the point $R(4, -2)$. You can check this by verifying that $x = 4$ and $y = -2$ satisfy both the original equation for l_2 and the equation for the circle.

To find where l_3 intersects the circle we write its equation as $x = 7$ and substitute 7 for x into the equation for the circle. This gives

$$36 + (y - 2)^2 = 25$$

$$(y - 2)^2 = -11.$$

Since this equation has no solution (explain), we conclude that the line l_3 doesn't intersect the circle at all.

The circle and the lines l_1, l_2, l_3 have been sketched in Figure 3.6.1. ☐

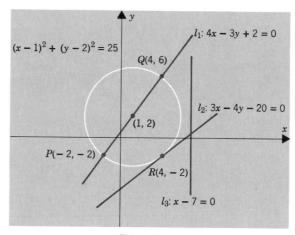

Figure 3.6.1

A line is said to be *tangent* to a circle iff it intersects the circle at one and only one point. Thus, for example, the line l_2 in Figure 3.6.1 is tangent to the circle at $(4, -2)$.

Figure 3.6.2 shows a circle centered at a point Q and a line tangent to the circle. Observe that the tangent line is perpendicular to the line segment $\overline{QP}$ where P is the point of tangency.

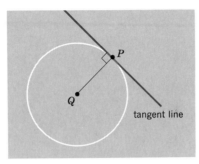

Figure 3.6.2

Problem. Find an equation for the line tangent to the circle

$$(x - 2)^2 + (y + 8)^2 = 169$$

at the point $P(7, 4)$.

SOLUTION. Call the tangent line l. The center of the circle is the point $Q(2, -8)$. We know that l passes through the point $P(7, 4)$ and is perpendicular to the line segment $\overline{QP}$. Since

$$\text{the slope of } \overline{QP} = \frac{-8 - 4}{2 - 7} = \frac{-12}{-5} = \frac{12}{5},$$

the slope of l is $-\frac{5}{12}$. $(m_1 m_2 = -1)$ As an equation for l we can write

$$y - 4 = -\tfrac{5}{12}(x - 7)$$

or, more simply,

$$5x + 12y - 83 = 0. \quad \square$$

Exercises

Determine whether the equations represent the same line, parallel lines, or intersecting lines. If they represent intersecting lines, find the point of intersection.

*1. $2x + y - 2 = 0,$ $x + 2y + 2 = 0.$

2. $2x + y - 2 = 0,$ $6x + 3y - 1 = 0.$

*3. $3x - y + 4 = 0,$ $x + 2y - 3 = 0.$

4. $7x - 2y + 1 = 0,$ $2x + y - 1 = 0.$

*5. $3x - 5y + 1 = 0,$ $12x - 20y + 4 = 0.$

6. $4x - y = 0$, $x + 3y + 3 = 0$.

*7. $y = -\frac{2}{3}(x + 5)$, $2x + 3y + 2 = 0$.

8. $x - 4y - 6 = 0$, $2x + 5y + 1 = 0$.

*9. $\frac{1}{2}x - \frac{1}{3}y - \frac{1}{6} = 0$, $\frac{3}{2}x + \frac{1}{3}y + \frac{1}{6} = 0$.

10. $16x - 5y + 2 = 0$, $16x - 8y + 14 = 0$.

Find the points, if any, where the line intersects the circle $x^2 + y^2 = 8$.

*11. $x - y = 0$.   12. $x + y = 0$.

*13. $2x + y = 0$.   14. $2x - y = 0$.

*15. $x - 2\sqrt{2} = 0$.   16. $y + 2\sqrt{2} = 0$.

*17. $x + 3 = 0$.   18. $y - 1 = 0$.

*19. $x + y - 2 = 0$.   20. $x + y - 4 = 0$.

21. Find an equation $Ax + By + C = 0$ for the line tangent to the circle

$$(x - 4)^2 + (y - 5)^2 = 25$$

at the point   *(a) $P(0, 2)$.   (b) $P(4, 10)$.   *(c) $P(9, 5)$.   (d) $P(7, 9)$.

3.7 Additional Exercises

*1. For each of the points given, find the coordinates of the point that is symmetric
to it about (i) the x-axis,   (ii) the y-axis,   (iii) the origin.
(a) $P(7, 7)$.   (b) $P(\sqrt{2}, -\sqrt{2})$.   (c) $P(-1, 0)$.   (d) $P(-\pi, \pi)$.
(e) $P(-3, -4)$.   (f) $P(0, -d)$.   (g) $P(\frac{1}{3}, 0.33)$.   (h) $P(a - b, a + b)$.

*2. Sketch the line segment with the indicated endpoints and find the coordinates
of the midpoint.
(a) $P(0, 1)$, $Q(1, 0)$.   (b) $P(-\pi, 1)$, $Q(\pi, -1)$.
(c) $P(\sqrt{2}, -\sqrt{2})$, $Q(3\sqrt{2}, 3\sqrt{2})$.   (d) $P(\frac{1}{2}, 2)$, $Q(-2, \frac{1}{2})$.

*3. Given that the point $Q(1, 2)$ bisects the line segment $\overline{P_1 P_2}$, find the coordinates
of P_2 given that P_1 has coordinates (a, b).

*4. Find the distance between P and Q:
(a) $P(a, a)$, $Q(a + \sqrt{2}, a + \sqrt{2})$.   (b) $P(1, 2)$, $Q(2, 2 + \sqrt{3})$.
(c) $P(-1, -1)$, $Q(1, 1)$.   (d) $P(-1, 3)$, $Q(4, 15)$.

*5. Give all possible coordinates for the third vertex of an equilateral triangle
given that $P(-1, 0)$ and $Q(1, 0)$ are two of the vertices.

*6. Find the slope of the line through the given points:
     (a) $O, Q(\frac{1}{2}, \frac{1}{2}\sqrt{3})$.   (b) $P(a, a), Q(b, b)$.   (c) $P(-\pi, 10), Q(\frac{2}{3}, 10)$.

*7. Write an equation $Ax + By + C = 0$ for the line with y-intercept 5 and slope
     (a) 1.   (b) -2.   (c) $\frac{1}{3}$.   (d) -1.   (e) $\sqrt{3}$.

*8. Write an equation $Ax + By + C = 0$ for the line
     (a) through $P(a, a)$ with slope -1.
     (b) through $P(\frac{1}{2}, \frac{1}{2}\sqrt{3})$ and $Q(1, \sqrt{3})$.
     (c) with x-intercept 2 and y-intercept 6.
     (d) with x-intercept $\frac{3}{5}$ and y-intercept $-\frac{2}{3}$.

*9. Write an equation $Ax + By + C = 0$ for the line through $P(1, \sqrt{3})$ that is
     (a) parallel to $y = -x$.
     (b) parallel to $\sqrt{3}y + x + 2\sqrt{3} = 0$.
     (c) perpendicular to $y = \sqrt{3}x + 1$.

*10. Determine all possible second coordinates for a point on the unit circle
     $x^2 + y^2 = 1$ given that the first coordinate is
     (a) 0.   (b) $\frac{1}{2}\sqrt{2}$.   (c) $-\frac{1}{2}$.   (d) 1.

*11. Write an equation for the circle given that the diameter has endpoints
     (a) $(-3, 1), (-1, 3)$.   (b) $(a, b), (-a, -b)$.

*12. Find the radius and center of the following circles:
     (a) $x^2 + y^2 - 6x - 8y = 0$.   (b) $x^2 + y^2 + 4x + 3 = 0$.

*13. An isosceles triangle has a base with endpoints $(\frac{5}{2}, 2)$ and $(\frac{11}{2}, 6)$. The third
     vertex has coordinates $(0, b)$. Find the area of the triangle.

*14. The base of an isosceles triangle has endpoints $P(3, -5)$ and $Q(6, -1)$. Find
     all possible positions for the third vertex if the area of the triangle is 25.

*15. Write an equation in two-intercept form for the line through $P(2, 1)$ that forms
     an isosceles right triangle with the coordinate axes in the first quadrant.

*16. Write an equation in two-intercept form for each line through $P(2, 1)$ that
     forms a triangle of area 6 with the coordinate axes in the first quadrant.

*17. Show that the perpendicular bisectors of the sides of a triangle meet at a point.
     [HINT: Place the vertices at $P(a, 0)$, $Q(-a, 0)$, $R(b, c)$ with a, b, c positive.]

*18. Sketch the graph of $4x^2 - 9y^2 = 0$. [HINT: Factor.]

*19. Determine whether the lines are parallel. If they are not parallel, give the point
     of intersection.
     (a) $3x + 4y - 5 = 0$, $2x + y = 0$.
     (b) $3x + 4y - 5 = 0$, $\frac{1}{4}x + \frac{1}{3}y + 1 = 0$.
     (c) $2x - 13 = 0$, $6y - 1 = 0$.
     (d) $\sqrt{2}x - \sqrt{3}y + 1 = 0$, $\sqrt{3}x + \sqrt{2}y - 2 = 0$.

*20. Find the points, if any, where the line intersects the circle

$$(x - 4)^2 + (y - 1)^2 = 25.$$

(a) $x - 7 = 0$.   (b) $y - 6 = 0$.   (c) $x - 10 = 0$.

(d) $x - y + 1 = 0$.   (e) $3x + 4y - 41 = 0$.   (f) $x - 7y + 28 = 0$.

*21. The lines l_1 and l_2 are tangent to the unit circle at the points $(\frac{1}{2}\sqrt{2}, \frac{1}{2}\sqrt{2})$ and $(\frac{1}{2}, -\frac{1}{2}\sqrt{3})$, respectively. Where do these lines intersect?

FUNCTIONS

4

The fundamental processes of calculus (called *differentiation* and *integration*) are processes applied to functions. To understand these processes and to be able to carry them out, you have to be thoroughly familiar with functions. You have to understand what is meant by *domain, range, graph, composition, one-to-one, inverse, polynomial,* etc. Here we discuss such notions.

4.1 Domain and Range

We begin with a set D of real numbers.

(4.1.1)
> By a *function f on D* we mean a rule that assigns exactly one real number $f(x)$ to each number x in D.†

In this context, the set D is called the *domain* of the function, and the set of assignments that the function makes (the set of values that the function takes on) is called the *range*.

Example 1. We begin with the squaring function

$$f(x) = x^2, \qquad x \text{ real.}$$

---

† $f(x)$ is read "f of x." Some of you may have been taught the more formal ordered-pair definition of function; accordingly, a function f with domain D would be the set of all ordered pairs $(x, f(x))$ with $x \in D$. From our point of view f is the rule that assigns $f(x)$ to each x in D. The two approaches are not so different.

Its value at 0 is 0:

$$f(0) = 0^2 = 0.$$

Its value at -1 and 1 is 1:

$$f(-1) = (-1)^2 = 1 = 1^2 = f(1).$$

In general, its value at $-x$ is the same as its value at x:

$$f(-x) = (-x)^2 = x^2 = f(x).$$

The domain of f is explicitly given as the set of real numbers. As x runs through all the real numbers, x^2 runs through all the nonnegative numbers. The range is therefore $[0, \infty)$. For short, we write

$$\operatorname{dom}(f) = (-\infty, \infty), \qquad \operatorname{ran}(f) = [0, \infty). \quad \square$$

Example 2.   Now consider the function

$$g(x) = \sqrt{x + 4}, \qquad x \in [0, 5].$$

The domain of g is given as the closed interval $[0, 5]$. At 0, g takes on the value 2:

$$g(0) = \sqrt{0 + 4} = \sqrt{4} = 2.$$

At 5, g takes on the value 3:

$$g(5) = \sqrt{5 + 4} = \sqrt{9} = 3.$$

As x runs through all the numbers from 0 to 5, $g(x)$ runs through all the numbers from 2 to 3. The range of g is therefore the closed interval $[2, 3]$. In abbreviated form,

$$\operatorname{dom}(g) = [0, 5], \qquad \operatorname{ran}(g) = [2, 3]. \quad \square$$

Example 3.   The absolute value function,

$$f(x) = |x|,$$

assigns to each real number its absolute value. The domain of f is the set of real numbers, but the range is only $[0, \infty)$. $\quad \square$

Example 4.   The Dirichlet function

$$g(x) = \begin{cases} 1, & x \text{ rational} \\ 0, & x \text{ irrational} \end{cases}$$

assigns the number 1 to each rational number and the number 0 to each irrational number. While the domain is the set of all real numbers, the range consists only of 0 and 1:

$$\operatorname{dom}(g) = (-\infty, \infty), \qquad \operatorname{ran}(g) = \{0, 1\}. \quad \square$$

Sometimes the domain of a function is not explicitly given. Thus, for example, we might write

$$f(x) = x^3 \quad \text{or} \quad f(x) = \sqrt{x}$$

without further explanation. In such cases take as the domain the maximal set of real numbers x for which $f(x)$ is itself a real number. For the cubing function take dom $(f) = (-\infty, \infty)$ and for the square root function take dom $(f) = [0, \infty)$.

Problem. Find the domain and range of

$$f(x) = x^2 + 6x + 4.$$

SOLUTION.   The domain is clearly $(-\infty, \infty)$: we can form the expression $x^2 + 6x + 4$ for all real x.

To find the range, we complete the square:

$$f(x) = (x^2 + 6x +) + 4$$
$$= (x^2 + 6x + 9) + 4 - 9$$
$$= (x + 3)^2 - 5.$$

Since $(x + 3)^2$ takes on all nonnegative values, $(x + 3)^2 - 5$ takes on all nonnegative values minus 5. The range is $[-5, \infty)$.   □

Problem. Find the domain and range of

$$f(x) = \frac{1}{\sqrt{2 - x}} + 5.$$

SOLUTION.   First we look for the domain. To be able to form $\sqrt{2 - x}$ we need $2 - x \geq 0$ and therefore $x \leq 2$. But at $x = 2$, $\sqrt{2 - x}$ is 0, and its reciprocal $1/\sqrt{2 - x}$ is not defined. We must therefore restrict x to $x < 2$. The domain is thus $(-\infty, 2)$.

Now we look for the range. As x runs through $(-\infty, 2)$, $\sqrt{2 - x}$ runs through all the positive numbers and so does its reciprocal. Therefore the range of f consists of all positive numbers plus 5. In short, ran $(f) = (5, \infty)$.   □

Exercises

Find $f(0)$, $f(\tfrac{1}{2})$, and $f(1)$ if defined:

*1. $f(x) = 1 - x$.

2. $f(x) = \sqrt{1 - x^2}$.

*3. $f(x) = \dfrac{1}{x}$.

4. $f(x) = \dfrac{2}{1 - x}$.

Find the number(s), if any, where f takes on the value 1:

*5. $f(x) = \sqrt{1 + x}$.   6. $f(x) = |2 - x|$.

*7. $f(x) = x^2 + 4x + 5$.   8. $f(x) = 4 + 10x - x^2$.

Express $f(x - 1)$ and $f(x + 1)$:

*9. $f(x) = x^2$.   10. $f(x) = 2x^2 + 3x + 1$.

*11. $f(x) = (x + 1)^2$.   12. $f(x) = x(x - 1)^2 + (x + 2)(x - 1)$.

Find the domain and the range:

*13. $f(x) = 1 + |x|$.   14. $g(x) = x^2 - 1$.

*15. $F(t) = 2t - 1$.   16. $G(z) = \sqrt{z} - 1$.

*17. $f(x) = \dfrac{1}{x^2}$.   18. $g(x) = \dfrac{1}{x}$.

*19. $f(t) = \sqrt{1 - t}$.   20. $g(t) = \sqrt{t - 1}$.

*21. $f(x) = \sqrt{1 - x} - 1$.   22. $g(x) = \sqrt{x - 1} - 1$.

*23. $h(x) = \dfrac{1}{\sqrt{1 - x}}$.   24. $F(x) = \dfrac{1}{\sqrt{1 + x^2}}$.

*25. $g(t) = \sqrt{1 - t^2}$.   26. $\phi(x) = 2 - \sqrt{1 + x^2}$.

*27. $f(x) = 2 - |1 - x^2|$.   28. $g(x) = x^2 + 8x - 5$.

*29. $f(x) = 5 - 3x + x^2$.   30. $P(x) = 2x^2 - 4x + 9$.

*31. $f(x) = -\dfrac{1}{x^2 + 1}$.   32. $g(x) = \dfrac{1}{x^2 + 6x + 24}$.

*33. $f(x) = \sqrt{x^2 - 5x + 6}$.   34. $g(x) = \sqrt{x^2 + 5x + 4}$.

4.2 The Graph of a Function

In the usual pictorial representation of a function, the domain is pictured on the x-axis and the range on the y-axis. The *graph* of a function f is the set of all points $P(x, y)$ such that $y = f(x)$.

In Figure 4.2.1 we have drawn a graph of a function f, and on that graph we have marked some points.

1. The domain of f is the closed interval $[a, e]$, pictured on the x-axis; the range of f is the closed interval $[A, D]$, pictured on the y-axis.

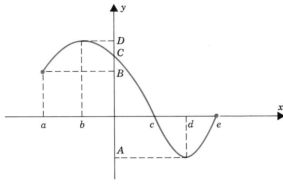

Figure 4.2.1

2. From the figure,

$$f(a) = B, \quad f(b) = D, \quad f(0) = C, \quad f(c) = 0, \quad f(d) = A, \quad f(e) = 0.$$

3. The maximum value of the function is D, and the minimum value is A. These values are taken on by f at b and d, respectively.

4. The function increases on $[a, b]$, decreases on $[b, d]$, then increases again on $[d, e]$.†

5. The function is positive on $[a, c)$, zero at c, negative on (c, e), and zero at e.   □

Next we turn to some specific examples. All of them are important and worth remembering.

Example 1. The graph of a *constant* function

$$f(x) = c$$

is the horizontal line

$$y = c. \quad □ \qquad\qquad \text{(Figure 4.2.2)}$$

---

†It's probably obvious what we mean here. Nevertheless here is a formal definition.

Definition 4.2.1

A function f *increases* on an interval I iff for every two numbers x_1, x_2 in I

$$x_1 < x_2 \quad \text{implies} \quad f(x_1) < f(x_2);$$

it *decreases* on an interval I iff for every two numbers x_1, x_2 in I

$$x_1 < x_2 \quad \text{implies} \quad f(x_1) > f(x_2).$$

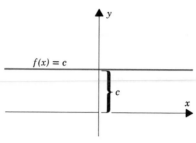

Figure 4.2.2

Example 2. The graph of the *identity* function

$$f(x) = x$$

is the line

$$y = x. \hspace{4cm} \text{(Figure 4.2.3)}$$

This line bisects the first and third quadrants. More generally, the graph of a *linear* function

$$f(x) = mx + b$$

is the line

$$y = mx + b. \hspace{3cm} \text{(Figure 4.2.3)}$$

It has slope m and y-intercept b. ☐

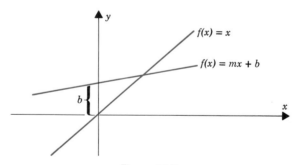

Figure 4.2.3

Example 3. The graph of the *squaring* function

$$f(x) = x^2$$

is the curve $y = x^2$ sketched in Figure 4.2.4.† You can arrive at such a curve by making a table of values and plotting the corresponding points.   (Figure 4.2.5)

---

†This curve is a *parabola*. Parabolas are discussed in detail in Section 6.2.

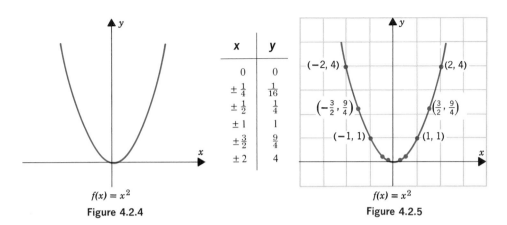

$f(x) = x^2$

Figure 4.2.4

$f(x) = x^2$

Figure 4.2.5

The squaring function decreases on $(-\infty, 0]$ and increases on $[0, \infty)$. It has a minimum value 0 at $x = 0$ but no maximum value. $\quad\square$

Example 4. The graph of the cubing function

$$f(x) = x^3$$

is the curve $y = x^3$ sketched in Figure 4.2.6. You can obtain such a curve by making a table of values and plotting the corresponding points. (Figure 4.2.7)

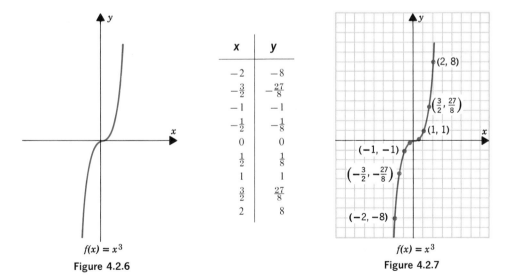

$f(x) = x^3$

Figure 4.2.6

$f(x) = x^3$

Figure 4.2.7

In contrast to the squaring function, the cubing function increases throughout. It has no minimum value and no maximum value. $\quad\square$

Example 5. The graph of the absolute value function

$$f(x) = |x|$$

is easy to visualize. Since

$$|x| = \begin{cases} x, & x \ge 0 \\ -x, & x \le 0 \end{cases},$$

we need no table of values. For $x \ge 0$, the graph follows the line $y = x$; for $x \le 0$, the graph follows the line $y = -x$. The graph is thus the V pictured in Figure 4.2.8. □

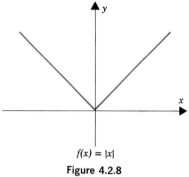

$f(x) = |x|$

Figure 4.2.8

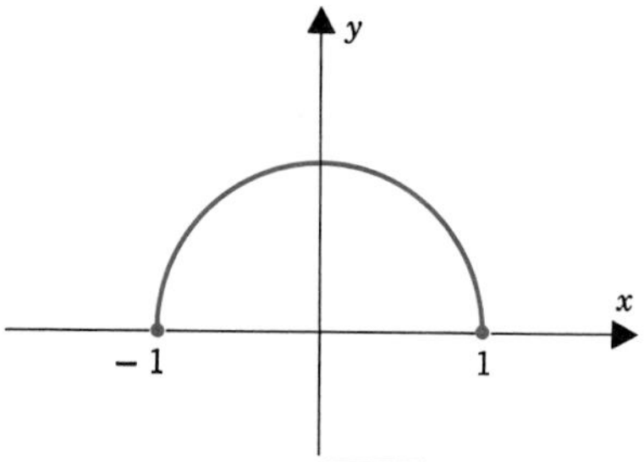

$f(x) = \sqrt{1 - x^2}$

Figure 4.2.9

Example 6. The graph of the function

$$f(x) = \sqrt{1 - x^2}$$

is the curve

$$y = \sqrt{1 - x^2}. \qquad\qquad \text{(Figure 4.2.9)}$$

This is the upper half of the unit circle

$$x^2 + y^2 = 1.$$

The domain is the closed interval $[-1, 1]$ and the range is the closed interval $[0, 1]$. The minimum value 0 is taken on by f at $x = -1$ and $x = 1$. The maximum value 1 is taken on at $x = 0$. □

Example 7. The function

$$f(x) = \begin{cases} x, & x \le -1 \\ x^2, & -1 < x < 1 \\ -x, & x > 1 \end{cases}$$

is defined everywhere except at $x = 1$. The graph consists of three pieces:

$$y = x, \qquad\qquad x \le -1. \qquad\qquad \text{(a half-line)}$$

$$y = x^2, \qquad -1 < x < 1. \qquad \text{(a piece of the curve } y = x^2\text{)}$$

$$y = -x, \qquad\qquad x > 1. \qquad \text{(a half-line without its endpoint)}$$

You can see a sketch of the graph in Figure 4.2.10. □

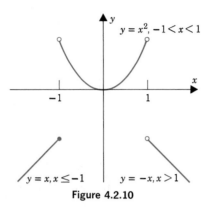

Figure 4.2.10

Exercises

For the first eight exercises take f as graphed in Figure 4.2.11.

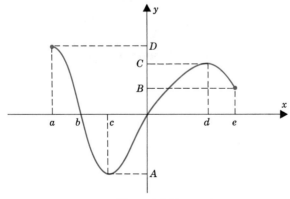

Figure 4.2.11

*1.   Find $f(a), f(b), f(c), f(0), f(d), f(e)$.

*2.   What is the domain of f?

*3. What is the range of f?

*4. Where does f take on the value 0?

*5. What is the maximum value of f? Where does f take on this value?

*6. What is the minimum value of f? Where does f take on this value?

*7. On what intervals does f increase? decrease?

*8. On what intervals is f positive? negative?

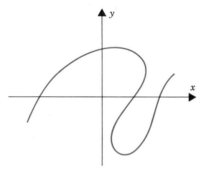

Figure 4.2.12

9. Explain why the curve displayed in Figure 4.2.12 is not the graph of a function.

Sketch the graph:

*10. $f(x) = 1$. \qquad\qquad 11. $f(x) = -1$.

*12. $f(x) = 2x$. \qquad\qquad 13. $f(x) = 2x + 1$.

*14. $f(x) = \frac{1}{2}x$. \qquad\qquad 15. $f(x) = -\frac{1}{2}x$.

*16. $f(x) = \frac{1}{2}x + 2$. \qquad\qquad 17. $f(x) = -\frac{1}{2}x - 3$.

*18. $f(x) = \sqrt{4 - x^2}$. \qquad\qquad 19. $f(x) = \sqrt{9 - x^2}$.

*20. Given that

$$f(x) = \sqrt{x}$$

construct a table of values based on $x = 0, \frac{1}{9}, \frac{1}{4}, 1, \frac{16}{9}, \frac{9}{4}, 4, 9$. Plot the corresponding points and then sketch the graph.

21.   Given that
$$f(x) = 1/\sqrt{x}$$
construct a table of values based on $x = \frac{1}{16}, \frac{1}{9}, \frac{1}{4}, 1, \frac{16}{9}, \frac{9}{4}, 4, 9$. Plot the corresponding points and then sketch the graph.

Sketch the graph and specify the domain and range:

*22. $f(x) = \begin{cases} -1, & x < 0 \\ 1, & x > 0 \end{cases}$.

23. $f(x) = \begin{cases} x^2, & x \le 0 \\ 1 - x, & x > 0 \end{cases}$.

*24. $f(x) = \begin{cases} 1 + x, & 0 \le x \le 1 \\ x, & 1 < x < 2 \\ \frac{1}{2}x + 1, & x \ge 2 \end{cases}$.

25. $f(x) = \begin{cases} x^2, & x < 0 \\ -1, & 0 < x < 2 \\ x, & x > 2 \end{cases}$.

4.3   Displacements of a Graph; Reflections in the x-Axis

Vertical Displacements

In Figure 4.3.1, you see the graphical results of adding a constant to a function. The graph of $g(x) = x^2 + 1$ is the graph of $f(x) = x^2$ raised one unit; the graph of $h(x) = x^2 - 1$ is the graph of $f(x) = x^2$ lowered one unit.

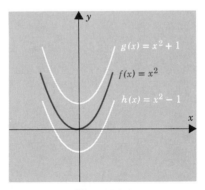

Figure 4.3.1

In Figure 4.3.2, we started with the function $f(x) = \sqrt{1 - x^2}$. Its graph is the upper half of the unit circle. The graph of $g(x) = \sqrt{1 - x^2} + 2$ is that same curve raised two units; the graph of $h(x) = \sqrt{1 - x^2} - 2$ is that same curve lowered two units.

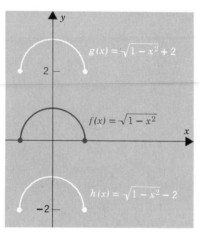

Figure 4.3.2

(4.3.1)

In general, if $C > 0$, then the graph of
$$g(x) = f(x) + C$$
is the graph of f raised C units, and the graph of
$$h(x) = f(x) - C$$
is the graph of f lowered C units.

Lateral Displacements

In Figure 4.3.3, you can see the graph of $f(x) = x^2$ displaced one unit to the right and also one unit to the left. The value that f takes on at x is taken on by g at $x + 1$ and by h at $x - 1$.

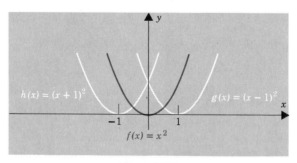

Figure 4.3.3

In Figure 4.3.4, you can see the graph of $f(x) = \sqrt{1 - x^2}$ displaced one-half unit to the right and also one-half unit to the left. The value that f takes on at x is taken on by g at $x + \frac{1}{2}$ and by h at $x - \frac{1}{2}$.

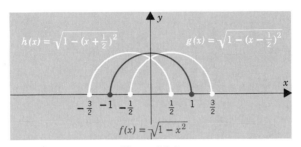

Figure 4.3.4

(4.3.2)

> In general, if $c > 0$, then the graph of
> $$g(x) = f(x - c)$$
> is the graph of f displaced c units to the right, and the graph of
> $$h(x) = f(x + c)$$
> is the graph of f displaced c units to the left.

Reflections in the x-Axis

Figures 4.3.5 and 4.3.6 show the effect of changing the sign of a function. In general, if $g(x) = -f(x)$, then for each point $P(x, y)$ on the graph of f, there is a corresponding point $Q(x, -y)$ on the graph of g. The graph of g is thus a mirror image of the graph of f. We speak of the graph of g as the graph of f *reflected in the x-axis.*

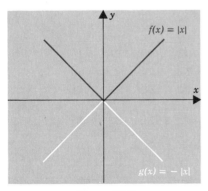

Figure 4.3.5

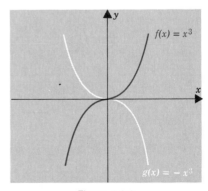

Figure 4.3.6

It's time to use the ideas that we've been discussing.

Problem.   Sketch the graph of

$$F(x) = (x - 1)^2 + 2.$$

SOLUTION.   The graph of F is the graph of

$$f(x) = x^2$$

displaced one unit to the right and raised two units. (Figure 4.3.7)   □

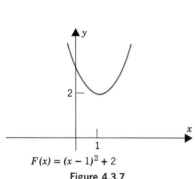

$F(x) = (x - 1)^2 + 2$

Figure 4.3.7

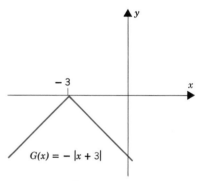

$G(x) = -|x + 3|$

Figure 4.3.8

Problem.   Sketch the graph of

$$G(x) = -|x + 3|.$$

SOLUTION.   Here

$$G(x) = -f(x + 3)$$

with

$$f(x) = |x|.$$

The graph of G is the graph V of the absolute value function displaced three units to the left and then reflected in the x-axis. (Figure 4.3.8)   □

Problem.   Sketch the graph of

$$H(x) = |x^2 - 1|.$$

SOLUTION.   Figure 4.3.9 shows the graph of

$$g(x) = x^2 - 1.$$

It is the graph of the squaring function lowered one unit.

Since $g(x)$ is nonnegative for $x \leq -1$, negative for $-1 < x < 1$, and nonnegative again for $x \geq 1$, you can see that

$$H(x) = |g(x)| = \begin{cases} g(x), & x \leq -1 \\ -g(x), & -1 < x < 1 \\ g(x), & 1 \leq x \end{cases}.$$

The graph of H agrees with that of g except between -1 and 1. There the graph of H is the graph of g reflected in the x-axis. (Figure 4.3.10)   □

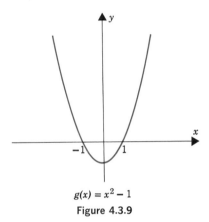

$g(x) = x^2 - 1$

Figure 4.3.9

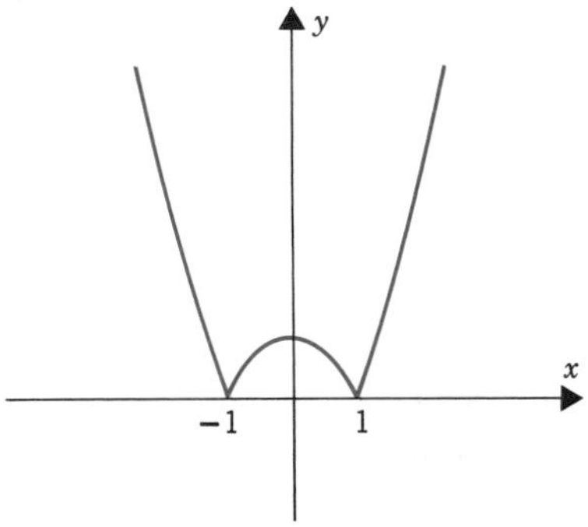

$H(x) = |x^2 - 1|$

Figure 4.3.10

Exercises

*1. Draw a figure displaying the graphs of
$$f(x) = -2x, \qquad g(x) = -2x + 1, \qquad h(x) = -2x - 1.$$

2. Draw a figure displaying the graphs of
$$f(x) = x^2, \qquad g(x) = x^2 + 3, \qquad h(x) = x^2 - 3.$$

*3. Draw a figure displaying the graphs of
$$f(x) = x^3, \qquad g(x) = (x - 2)^3, \qquad h(x) = (x + 2)^3.$$

Sketch the graph:

*4. $F(x) = (x - 2)^2.$ 5. $G(x) = (x + 2)^2.$

*6. $F(x) = (x - 2)^2 + 3.$ 7. $G(x) = (x - 2)^2 - 3.$

*8. $F(x) = -(x - 2)^2.$ 9. $G(x) = 3 - (x - 2)^2.$

*10. $F(x) = (x + 1)^3$.

11. $G(x) = (x - 1)^3$.

*12. $F(x) = 1 - (x - 1)^3$.

13. $G(x) = 1 + (x + 1)^3$.

*14. $F(x) = 1 - |x|$.

15. $G(x) = 2 + |x|$.

*16. $F(x) = |x + 4|$.

17. $G(x) = 2 - |x + 4|$.

*18. $F(x) = 2 - \sqrt{1 - x^2}$.

19. $G(x) = 1 - \sqrt{1 - (x - 2)^2}$.

*20. $F(x) = |x^3 - 1|$.

21. $G(x) = 2 - |x^3 - 1|$.

4.4 Reciprocals; Boundedness; Odd and Even Functions

Reciprocals

We begin with the *reciprocal* function

$$f(x) = \frac{1}{x}.$$

For $x > 0$, $f(x) > 0$; at $x = 0$, f is not defined; for $x < 0$, $f(x) < 0$. This tells us that the graph consists of two branches: one branch in the first quadrant (both coordinates positive) and one branch in the third quadrant (both coordinates negative).

To graph the reciprocal function we make a table of values and plot the corresponding points (Figure 4.4.1).

x	y		x	y
$\frac{1}{4}$	4		$-\frac{1}{4}$	-4
$\frac{1}{3}$	3		$-\frac{1}{3}$	-3
$\frac{1}{2}$	2		$-\frac{1}{2}$	-2
1	1		-1	-1
2	$\frac{1}{2}$		-2	$-\frac{1}{2}$
3	$\frac{1}{3}$		-3	$-\frac{1}{3}$
4	$\frac{1}{4}$		-4	$-\frac{1}{4}$

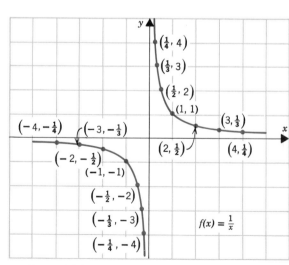

Figure 4.4.1

As x approaches 0, the curve becomes steeper and steeper and approaches the y-axis as a limiting position. We call the y-axis a *vertical asymptote.*†

As x moves away from 0, the curve becomes flatter and flatter and approaches the x-axis as a limiting position. We call the x-axis a *horizontal asymptote.*

Figure 4.4.2 shows the graphs of

$$f(x) = \frac{1}{x} \quad \text{and} \quad g(x) = \frac{1}{|x|}$$

side by side.

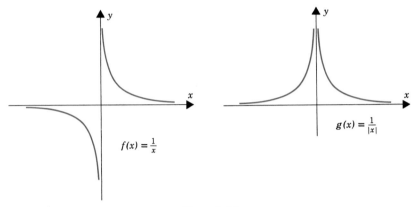

Figure 4.4.2

To the right of the origin, the two graphs agree:

$$\text{for } x > 0, \quad g(x) = f(x).$$

To the left of the origin, the graph of g is the graph of f reflected in the x-axis:

$$\text{for } x < 0, \quad g(x) = -f(x). \quad \square$$

Problem. Sketch the graphs of

$$F(x) = \frac{1}{x-1} + 2 \quad \text{and} \quad G(x) = 2 - \frac{1}{|x|}.$$

SOLUTION. Applying the techniques of the last section, you can see that the graph of F must be the graph of the reciprocal function

$$f(x) = \frac{1}{x}$$

displaced one unit to the right and then raised two units. (Figure 4.4.3)

---

†From a Greek word meaning "to fall together."

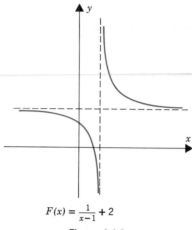

$$F(x) = \frac{1}{x-1} + 2$$

Figure 4.4.3

We can obtain the graph of G by starting with the graph of

$$g(x) = \frac{1}{|x|},$$

reflecting it in the x-axis (this gives $-(1/|x|)$) and then raising the result two units. (Figure 4.4.4) ☐

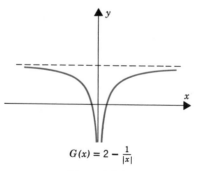

$$G(x) = 2 - \frac{1}{|x|}$$

Figure 4.4.4

Boundedness

We have already discussed boundedness and unboundedness for sets of real numbers. Here we apply these notions to functions.

Definition   4.4.1   Boundedness for Functions

A function is said to be

bounded above   iff its range is   bounded above,
bounded below   iff its range is   bounded below,
bounded   iff its range is   bounded above and below.

Thus, for a function f to be bounded above, there must exist a real number M such that

$$f(x) \leq M \qquad \text{for all} \quad x \in \text{dom}(f).$$

For f to be bounded below, there must exist a real number m such that

$$m \leq f(x) \qquad \text{for all} \quad x \in \text{dom}(f).$$

For f to be bounded, there must exist real numbers m and M such that

$$m \leq f(x) \leq M \qquad \text{for all} \quad x \in \text{dom}(f).$$

Examples

1. The squaring function

$$f(x) = x^2$$

   is bounded below by 0:

$$0 \leq x^2 \qquad \text{for all real } x.$$

   It is not bounded above: there is no real number M such that

$$x^2 \leq M \qquad \text{for all real } x.\dagger$$

   The graph (Figure 4.2.4) does not reach below the x-axis, but it does reach above any horizontal line that we could draw.   ☐

2. The cubing function

$$f(x) = x^3$$

$\dagger$ The inequality

$$x^2 \leq M$$

cannot possibly hold for all real numbers x since, no matter how M is chosen,

$$(|M| + 1)^2 = M^2 + 2|M| + 1 \nleq M.$$

is unbounded above: there is no number M such that

$$x^3 \le M \qquad \text{for all real } x.$$

It is also unbounded below: there is no number m such that

$$m \le x^3 \qquad \text{for all real } x.$$

The graph of the cubing function (Figure 4.2.6) reaches above and below any fixed horizontal line. ☐

3. The function

$$f(x) = \sqrt{1 - x^2}$$

is bounded (both above and below):

$$0 \le \sqrt{1 - x^2} \le 1 \qquad \text{for all } x \text{ in the domain.}$$

The graph (Figure 4.2.9) lies entirely between the horizontal lines $y = 0$ and $y = 1$. ☐

4. The function

$$f(x) = 1 - \frac{1}{x^2}, \qquad x \ne 0$$

is bounded above by 1:

$$1 - \frac{1}{x^2} < 1 \qquad \text{for all } \quad x \ne 0.$$

It is not bounded below: there is no number m such that

$$m \le 1 - \frac{1}{x^2} \qquad \text{for all } \quad x \ne 0.\dagger \quad ☐$$

Odd and Even Functions

If n is an even integer, then

$$(-x)^n = x^n, \qquad [\text{for example, } (-x)^2 = x^2]$$

but, if n is an odd integer, then

$$(-x)^n = -x^n. \qquad [\text{for example, } (-x)^3 = -x^3]$$

These ideas can be carried over to functions in general.

---

† You are asked to show this in Exercise 26.

Definition 4.4.2   Odd and Even Functions

A function f is said to be *even* iff

$$f(-x) = f(x) \qquad \text{for all} \quad x \in \text{dom}(f).$$

It is said to be *odd* iff

$$f(-x) = -f(x) \qquad \text{for all} \quad x \in \text{dom}(f).$$

As already suggested, the even powers of x are even functions and the odd powers are odd functions. There are, of course, other examples.

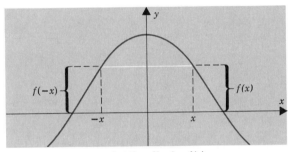

even function: $f(-x) = f(x)$
graph is symmetric about the y-axis
Figure 4.4.5

First some pictures. The function graphed in Figure 4.4.5 is even: if it takes on one value to one side of the origin, it takes on that same value the same number of units to the other side of the origin. The graph is therefore *symmetric about the y-axis*.
    The function displayed in Figure 4.4.6 is *odd*: if it takes on one value to one side

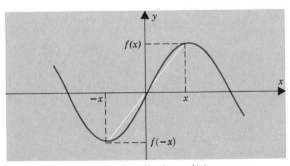

odd function: $f(-x) = -f(x)$
graph is symmetric about the origin
Figure 4.4.6

of the origin, it takes on the same value with an opposite sign the same number of units to the other side of the origin. The graph is *symmetric about the origin.*
  Now let's look at some examples defined analytically.

Examples

1. The absolute value function

$$f(x) = |x|$$

is an even function:

$$f(-x) = |-x| = |x| = f(x).$$

The graph (Figure 4.2.8) is symmetric about the y-axis.   ☐

2. The reciprocal function

$$f(x) = \frac{1}{x}$$

is an odd function:

$$f(-x) = \frac{1}{-x} = -\frac{1}{x} = -f(x).$$

The graph (Figure 4.4.1) is symmetric about the origin.   ☐

3. The function

$$f(x) = \frac{1}{x-1} + 2$$

is neither even nor odd. Its graph (Figure 4.4.3) is not symmetric about the y-axis nor about the origin, but about the point $P(1, 2)$.   ☐

4. The Dirichlet function

$$f(x) = \begin{cases} 1, & x \text{ rational} \\ 0, & x \text{ irrational} \end{cases}$$

is an even function:
  (a) If x is rational, so is $-x$, and therefore

$$f(-x) = 1 = f(x).$$

  (b) If x is irrational, so is $-x$, and therefore again

$$f(-x) = 0 = f(x).$$

The graph (though impossible to draw) is symmetric about the y-axis.   ☐

5. As final examples, take the functions

$$f(x) = x|x| \quad \text{and} \quad g(x) = x^2(4x^2 + 1).$$

The first function is odd:

$$f(-x) = -x|-x| = -x|x| = -f(x).$$

The second is even:

$$g(-x) = (-x)^2[4(-x)^2 + 1] = x^2(4x^2 + 1) = g(x). \quad \square$$

Exercises

Which of the following functions are even? odd?

*1. $f(x) = 1$.

2. $g(x) = x^5 + 1$.

*3. $f(x) = x(x - 1)$.

4. $g(x) = x(x^2 - 1)$.

*5. $f(x) = \sqrt{25 - x^2}$.

6. $g(x) = x^3|x|$.

*7. $f(x) = x + \dfrac{1}{x}$.

8. $g(x) = \dfrac{2}{x^2} - 5x^2 + 1$.

Which of the following functions are bounded above? bounded below? bounded?

*9. $f(x) = 2 + |x|$.

10. $g(x) = \sqrt{1 - x}$.

*11. $f(x) = x + \dfrac{1}{x}$.

12. $g(x) = x^2 - 8x + 1$.

*13. $f(x) = 2 - x^2$.

14. $g(x) = (x - 2)^3$.

*15. $f(x) = \dfrac{x^2}{x^2 + 1}$.

16. $g(x) = -\dfrac{1}{|x - 2|}$.

Sketch the graph of each of the following functions:

17. $f(x) = \dfrac{1}{x} + 1$.

*18. $g(x) = \dfrac{1}{|x|} + 1$.

19. $f(x) = \dfrac{1}{x - 2}$.

20. $g(x) = \dfrac{1}{|x - 2|}$.

21. $f(x) = \dfrac{1}{x - 2} - 1$.

22. $g(x) = \dfrac{1}{x - 3} + 2$.

*23. $f(x) = -\dfrac{1}{|x - 1|}$.

24. $g(x) = \dfrac{1}{|x - 5|} - 1$.

25.   Take f as a function defined for all real numbers and set

$$g(x) = \tfrac{1}{2}[f(x) + f(-x)], \qquad h(x) = \tfrac{1}{2}[f(x) - f(-x)].$$

Show that g is even and h is odd.

26.   Show that the function

$$f(x) = 1 - \frac{1}{x^2}$$

is not bounded below.

4.5   Algebraic Combinations of Functions

If we have two functions f and g *with a common domain*, then we can form their *sum $f + g$*:

(4.5.1)

$$(f + g)(x) = f(x) + g(x);$$

their *difference $f - g$*:

(4.5.2)

$$(f - g)(x) = f(x) - g(x);$$

and their *product fg*:

(4.5.3)

$$(fg)(x) = f(x)g(x).$$

Taking

$$f(x) = x^2 \quad \text{and} \quad g(x) = 3x + 1,$$

we have

$$(f + g)(x) = f(x) + g(x) = x^2 + 3x + 1,$$
$$(f - g)(x) = f(x) - g(x) = x^2 - 3x - 1,$$
$$(fg)(x) = f(x)g(x) = x^2(3x + 1) = 3x^3 + x^2. \quad \square$$

Taking

$$f(x) = x + \frac{1}{x} \quad \text{and} \quad g(x) = x - \frac{1}{x},$$

we have

$$(f + g)(x) = f(x) + g(x) = \left(x + \frac{1}{x}\right) + \left(x - \frac{1}{x}\right) = 2x,$$

$$(f - g)(x) = f(x) - g(x) = \left(x + \frac{1}{x}\right) - \left(x - \frac{1}{x}\right) = \frac{2}{x},$$

$$(fg)(x) = f(x)g(x) = \left(x + \frac{1}{x}\right)\left(x - \frac{1}{x}\right) = x^2 - \frac{1}{x^2}. \quad \square$$

If $g(x) \neq 0$, then we can form the *reciprocal* $1/g$:

(4.5.4)
$$\boxed{\frac{1}{g}(x) = \frac{1}{g(x)}}$$

and also the *quotient* f/g:

(4.5.5)
$$\boxed{\frac{f}{g}(x) = \frac{f(x)}{g(x)}.}$$

With

$$f(x) = x^2 \quad \text{and} \quad g(x) = x^4 + 1,$$

we have

$$\frac{1}{g}(x) = \frac{1}{g(x)} = \frac{1}{x^4 + 1}$$

and

$$\frac{f}{g}(x) = \frac{f(x)}{g(x)} = \frac{x^2}{x^4 + 1}. \quad \square$$

We can also multiply a function f by a real number α and thereby form the *scalar multiple* αf:

(4.5.6)
$$\boxed{(\alpha f)(x) = \alpha f(x).}$$

With

$$f(x) = 2x^2 - 1,$$

we have

$$(4f)(x) = 4f(x) = 4(2x^2 - 1) = 8x^2 - 4$$

and

$$(-3f)(x) = -3f(x) = -3(2x^2 - 1) = -6x^2 + 3. \quad \square$$

By combining scalar multiples we can form *linear combinations* $\alpha f + \beta g$:

(4.5.7)
$$\boxed{(\alpha f + \beta g)(x) = \alpha f(x) + \beta g(x).}$$

With

$$f(x) = \frac{2}{x} - x^2 \quad \text{and} \quad g(x) = \frac{5}{x^2} - x^3,$$

we have

$$(3f + 2g)(x) = 3f(x) + 2g(x)$$

$$= 3\left(\frac{2}{x} - x^2\right) + 2\left(\frac{5}{x^2} - x^3\right)$$

$$= \frac{6}{x} - 3x^2 + \frac{10}{x^2} - 2x^3$$

$$= -2x^3 - 3x^2 + \frac{6}{x} + \frac{10}{x^2},$$

and, as you can verify,

$$(3f - 2g)(x) = 2x^3 - 3x^2 + \frac{6}{x} - \frac{10}{x^2}. \quad \square$$

Functions that are defined piecewise require special care:

Problem.   Given that

$$f(x) = \begin{cases} 1 - x^2, & x \le 0 \\ x, & x > 0 \end{cases} \quad \text{and} \quad g(x) = \begin{cases} -2x, & x < 1 \\ 1 - x, & x \ge 1 \end{cases},$$

find $f + g$, $f - g$, and fg.

SOLUTION.   We begin by breaking up the domain of both functions into pieces over which neither f nor g changes in defining formula. The formula for f changes at $x = 0$; the formula for g changes at $x = 1$. On the pieces

$$x \le 0, \qquad 0 < x < 1, \qquad x \ge 1$$

there is no formula change. Writing

$$f(x) = \begin{cases} 1 - x^2, & x \le 0 \\ x, & 0 < x < 1 \\ x, & x \ge 1 \end{cases}, \qquad g(x) = \begin{cases} -2x, & x \le 0 \\ -2x, & 0 < x < 1 \\ 1 - x, & x \ge 1 \end{cases},$$

we can combine the functions piece by piece:

$$(f + g)(x) = f(x) + g(x) = \begin{cases} 1 - 2x - x^2, & x \le 0 \\ -x, & 0 < x < 1 \\ 1, & x \ge 1 \end{cases},$$

$$(f - g)(x) = f(x) - g(x) = \begin{cases} 1 + 2x - x^2, & x \le 0 \\ 3x, & 0 < x < 1 \\ -1 + 2x, & x \ge 1 \end{cases},$$

$$(fg)(x) = f(x)g(x) = \begin{cases} -2x + 2x^3, & x \le 0 \\ -2x^2, & 0 < x < 1 \\ x - x^2, & x \ge 1 \end{cases}. \quad \square$$

Exercises

*1. Given that

$$f(x) = x^2 - a^2, \qquad g(x) = x^2 + a^2,$$

find (a) $f + g$, (b) $f - g$, (c) fg, (d) f/g.

2. Given that

$$f(x) = 2x^2 + 5x - 1, \qquad g(x) = 3x^2 - 4x + 4,$$

find (a) $3f - 2g$, (b) $4f + 5g$, (c) $4f + g$.

*3. Given that

$$f(x) = \sqrt{x} - \frac{1}{\sqrt{x}}, \qquad g(x) = \sqrt{x} + \frac{2}{\sqrt{x}},$$

find (a) $6f + 3g$, (b) fg, (c) f/g.

4. Given that

$$f(x) = \begin{cases} x, & x \le 0 \\ -1, & x > 0 \end{cases}, \qquad g(x) = \begin{cases} -x, & x < 1 \\ x^2, & x \ge 1 \end{cases},$$

find (a) $f + g$, (b) $f - g$, (c) fg.

*5. Given that

$$f(x) = \begin{cases} 1 - x, & x \le 1 \\ 2x - 1, & x > 1 \end{cases}, \qquad g(x) = \begin{cases} 0, & x < 2 \\ -1, & x \ge 2 \end{cases},$$

find (a) $f + g$, (b) $f - g$, (c) fg.

Sketch the graphs of the following functions with f and g as in Figure 4.5.1.

6. $2f$.	*7. $\frac{1}{2}f$.	8. $-f$.
*9. $2g$.	10. $\frac{1}{2}g$.	*11. $-2g$.

12. $f + g.$ *13. $f - g.$ 14. $f + 2g.$

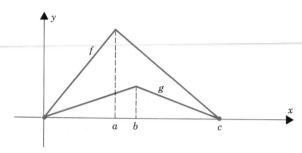

Figure 4.5.1

*15.   Is the product of two odd functions odd, even, or neither?

16.   Is the product of two even functions odd, even, or neither?

*17.   Is the product of an odd function and an even function odd, even, or neither?

18.   Show that every function defined for all real numbers can be written as the sum of an even function and an odd function.   HINT: Recall Exercise 25, Section 4.4.

4.6   The Composition of Functions

In the last section, you saw how to combine functions algebraically. There is another way to combine functions, called *composition*. To describe it, we begin with two functions f and g and a number x in the domain of g. By applying g to x, we get the number $g(x)$. If $g(x)$ is in the domain of f, then we can apply f to $g(x)$ and thereby obtain the number $f(g(x))$.

What is $f(g(x))$? It is the result of first applying g to x and then applying f to $g(x)$. The idea is portrayed schematically in Figure 4.6.1.

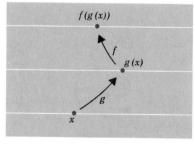

Figure 4.6.1

If the range of g is completely contained in the domain of f (namely, if each value that g takes on is in the domain of f), then we can form $f(g(x))$ *for each x in the domain of g* and in this manner create a new function. This new function—it takes each x in the domain of g and assigns to it the value $f(g(x))$—is called the *composition* of f and g and is denoted by the symbol $f \circ g$. What is this function $f \circ g$? It is the function that results from first applying g and then f. (Figure 4.6.2)

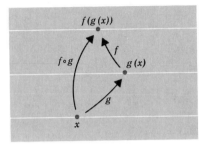

Figure 4.6.2

Definition   4.6.1   Composition

If the range of the function g is contained in the domain of the function f, then the composition $f \circ g$ (read "f circle g") is the function defined on the domain of g by setting

$$(f \circ g)(x) = f(g(x)).$$

Examples

1. Suppose that

$$g(x) = x^2 \qquad \text{(the squaring function)}$$

and

$$f(x) = x + 1. \qquad \text{(the function that adds 1)}$$

Then

$$(f \circ g)(x) = f(g(x)) = g(x) + 1 = x^2 + 1.$$

In other words, $f \circ g$ is the function that *first* squares and *then* adds 1. □

2. If, on the other hand,

$$g(x) = x + 1 \qquad \text{(g adds 1)}$$

and

$$f(x) = x^2, \qquad \text{(f squares)}$$

then
$$(f \circ g)(x) = f(g(x)) = [g(x)]^2 = (x + 1)^2.$$

Here $f \circ g$ is the function that *first* adds 1 and *then* squares. □

3. If
$$g(x) = \frac{1}{x} + 1 \qquad (g \text{ adds 1 to the reciprocal})$$

and
$$f(x) = x^2 - 5, \quad (f \text{ subtracts 5 from the square})$$
then
$$(f \circ g)(x) = f(g(x)) = [g(x)]^2 - 5 = \left(\frac{1}{x} + 1\right)^2 - 5.$$

Here $f \circ g$ *first* adds 1 to the reciprocal and *then* subtracts 5 from the square. □

4. If
$$f(x) = \frac{1}{x} + x^2 \quad \text{and} \quad g(x) = \frac{x^2 + 1}{x^4 + 1},$$
then
$$(f \circ g)(x) = f(g(x)) = \frac{1}{g(x)} + [g(x)]^2 = \frac{x^4 + 1}{x^2 + 1} + \left(\frac{x^2 + 1}{x^4 + 1}\right)^2. \quad \square$$

It is possible to form the composition of more than two functions. For example, the triple composition $f \circ g \circ h$ consists of first h, then g, and then f:
$$(f \circ g \circ h)(x) = f(g(h(x))).$$

Obviously we can go on in this manner with as many functions as we like.

Example. If
$$f(x) = \frac{1}{x}, \qquad g(x) = x + 2, \qquad h(x) = (x^2 + 1)^2,$$
then
$$(f \circ g \circ h)(x) = f(g(h(x)))$$
$$= \frac{1}{g(h(x))}$$
$$= \frac{1}{h(x) + 2}$$
$$= \frac{1}{(x^2 + 1)^2 + 2}. \quad \square$$

To apply some of the techniques of calculus you will need to be able to recognize composites. You will need to be able to start with a function, say $F(x) = (x + 1)^5$, and recognize how it is a composition.

Problem.   Find functions f and g such that $f \circ g = F$ if

$$F(x) = (x + 1)^5.$$

A SOLUTION. The function consists of first adding 1 and then taking the fifth power. We can therefore set

$$g(x) = x + 1 \qquad\qquad \text{(adding 1)}$$

and

$$f(x) = x^5. \qquad\qquad \text{(taking the fifth power)}$$

As you can see,

$$(f \circ g)(x) = f(g(x)) = [g(x)]^5 = (x + 1)^5. \quad \square$$

Problem.   Find functions f and g such that $f \circ g = F$ if

$$F(x) = \frac{1}{x} - 6.$$

A SOLUTION. F takes the reciprocal and then subtracts 6. We can therefore set

$$g(x) = \frac{1}{x} \qquad\qquad \text{(taking the reciprocal)}$$

and

$$f(x) = x - 6. \qquad\qquad \text{(subtracting 6)}$$

As you can check,

$$(f \circ g)(x) = f(g(x)) = g(x) - 6 = \frac{1}{x} - 6. \quad \square$$

Problem.   Find three functions f, g, h such that $f \circ g \circ h = F$ if

$$F(x) = \frac{1}{|x| + 3}.$$

A SOLUTION. F takes the absolute value, adds 3, and then inverts. Let h take the absolute value:

$$\text{set} \quad h(x) = |x|.$$

Let g add 3:

$$\text{set} \quad g(x) = x + 3.$$

Let f do the inverting:

$$\text{set} \quad f(x) = \frac{1}{x}.$$

With this choice of f, g, h we have

$$(f \circ g \circ h)(x) = f(g(h(x)))$$

$$= \frac{1}{g(h(x))}$$

$$= \frac{1}{h(x) + 3}$$

$$= \frac{1}{|x| + 3}. \quad \square$$

In the next problem we start with the same function

$$F(x) = \frac{1}{|x| + 3},$$

but this time we are asked to break it up as the composition of only two functions.

Problem. Find two functions f and g such that $f \circ g = F$ if

$$F(x) = \frac{1}{|x| + 3}.$$

SOME SOLUTIONS. F takes the absolute value, adds 3, and then inverts. We can let g do the first two things by setting

$$g(x) = |x| + 3$$

and then let f do the inverting by setting

$$f(x) = \frac{1}{x}.$$

Or, we could let g just take the absolute value,

$$g(x) = |x|,$$

and then have f add 3 and invert:

$$f(x) = \frac{1}{x + 3}.$$

There are still other possible choices for f and g. For example, we could set

$$g(x) = x \quad \text{and} \quad f(x) = \frac{1}{|x| + 3}$$

or

$$g(x) = \frac{1}{|x| + 3} \quad \text{and} \quad f(x) = x,$$

but these last choices of f and g don't advance us at all. They are much like factoring 12 into $12 \cdot 1$ or $1 \cdot 12$. $\square$

Exercises

Form the composition $f \circ g$:

*1. $f(x) = 2x + 5$, $g(x) = x^2$.

2. $f(x) = x^2$, $g(x) = 2x + 5$.

*3. $f(x) = \sqrt{x}$, $g(x) = x^2 + 5$.

4. $f(x) = x^2 + x$, $g(x) = \sqrt{x}$.

*5. $f(x) = \dfrac{1}{x}$, $g(x) = \dfrac{1}{x}$.

6. $f(x) = x^2$, $g(x) = \dfrac{1}{x - 1}$.

*7. $f(x) = x - 1$, $g(x) = \dfrac{1}{x}$.

8. $f(x) = x^2 - 1$, $g(x) = x(x - 1)$.

*9. $f(x) = \dfrac{1}{\sqrt{x - 1}}$, $g(x) = (x^2 + 2)^2$.

10. $f(x) = \dfrac{1}{x} - \dfrac{1}{x + 1}$, $g(x) = \dfrac{1}{x^2}$.

Form the composition $f \circ g \circ h$:

*11. $f(x) = 4x$, $g(x) = x - 1$, $h(x) = x^2$.

12. $f(x) = x - 1$, $g(x) = 4x$, $h(x) = x^2$.

*13. $f(x) = x^2$, $g(x) = x - 1$, $h(x) = x^4$.

14. $f(x) = x - 1$, $g(x) = x^2$, $h(x) = 4x$.

*15. $f(x) = \dfrac{1}{x}$, $g(x) = \dfrac{1}{2x + 1}$, $h(x) = x^2$.

16. $f(x) = \dfrac{x + 1}{x}$, $g(x) = \dfrac{1}{2x + 1}$, $h(x) = x^2$.

17. Find f such that $f \circ g = F$ given that

*(a) $g(x) = x^2$ and $F(x) = ax^2 + b$.

(b) $g(x) = x^2$ and $F(x) = \sqrt{x^4 + 1}$.

*(c) $g(x) = x^2 + 5$ and $F(x) = \dfrac{1}{(x^2 + 5)^3}$.

18. Find g such that $f \circ g = F$ given that

(a) $f(x) = x^3$ and $F(x) = \left(1 - \dfrac{1}{x^4}\right)^3$.

*(b) $f(x) = x + \dfrac{1}{x}$ and $F(x) = a^2x^2 + \dfrac{1}{a^2x^2}$.

(c) $f(x) = x^5 + 1$ and $F(x) = (2x^3 - 1)^5 + 1$.

4.7   Fractional Exponents

This is a bit of an aside but useful for the next section.

You have seen the square root of a nonnegative number x written as $\sqrt{x}$. It is sometimes also written as $\sqrt[2]{x}$. This *radical* notation extends to other roots as well. We can write the cube root of x as $\sqrt[3]{x}$, the fourth root of x as $\sqrt[4]{x}$, etc. In general, for $x \geq 0$, we write $\sqrt[n]{x}$ for the unique positive number which has the property that

$$(\sqrt[n]{x})^n = x.$$

Thus

$$(\sqrt[2]{x})^2 = x, \qquad (\sqrt[3]{x})^3 = x, \qquad (\sqrt[4]{x})^4 = x, \qquad \text{etc.}$$

There is another notation for nth roots that lends itself better to the formulas of calculus. Instead of writing $\sqrt[n]{x}$, we write $x^{1/n}$. The square root of x becomes $x^{1/2}$, the cube root $x^{1/3}$, and so on.

Definition   4.7.1

If $x \geq 0$ and n is a positive integer, then $x^{1/n}$, *the nth root of x,* is the unique nonnegative number with the property that

$$(x^{1/n})^n = \underbrace{x^{1/n} \cdot x^{1/n} \cdot \ \cdots \ \cdot x^{1/n}}_{n} = x.$$

Our definition raises two questions:

1. How do we know that every nonnegative number has an nth root?
2. Granted that it has an nth root, how can we be sure that it does not have more than one?

The first question is easy to answer with calculus,† difficult to answer without it. To answer the second question, it is enough to show that

$$\text{if} \quad 0 \leq a < b, \qquad \text{then} \quad a^n < b^n.$$

We do this in Section 7.5, where we have induction at our disposal.

---

† See, for example, S. L. Salas and E. Hille, *Calculus: One and Several Variables,* 3rd ed. (New York: Wiley, 1978), page 83, Exercise 36.

Examples of nth roots

1. Since

$$5^2 = 25, \qquad 5^3 = 125, \qquad 5^4 = 625,$$

you can see that

$$25^{1/2} = 5, \qquad 125^{1/3} = 5, \qquad 625^{1/4} = 5. \quad \square$$

2. Since

$$2^2 = 4, \qquad 2^3 = 8, \qquad 2^4 = 16, \qquad 2^5 = 32,$$

we have

$$4^{1/2} = 2, \qquad 8^{1/3} = 2, \qquad 16^{1/4} = 2, \qquad 32^{1/5} = 2. \quad \square$$

3. As you can see by raising 0.1 to the appropriate power,

$$(0.01)^{1/2} = 0.1, \qquad (0.001)^{1/3} = 0.1, \qquad (0.0001)^{1/4} = 0.1. \quad \square$$

So far we have considered $x^{1/n}$ only for $x \geq 0$.

Definition 4.7.2

If $x < 0$ and n is an odd integer, then $x^{1/n}$ is the unique negative number with the property that

$$(x^{1/n})^n = x.$$

The questions about existence and uniqueness that came up before come up again here. They can be answered in a similar manner.

Here are some examples of odd roots of negative numbers†:

$$(-8)^{1/3} = -2 \quad \text{since} \quad (-2)^3 = (-1)^3 2^3 = -8.$$

$$(-0.001)^{1/3} = -0.1 \quad \text{since} \quad (-0.1)^3 = (-1)^3 (0.1)^3 = -0.001.$$

$$(-100{,}000)^{1/5} = -10 \quad \text{since} \quad (-10)^5 = (-1)^5 10^5 = -100{,}000. \quad \square$$

We come now to positive fractional exponents. If r is a positive rational number, then r can be written p/q with p and q positive integers, p/q in lowest terms.

---

† We have avoided discussion of the even roots of negative numbers because these are not real numbers. They are *complex numbers,* and these lie outside the scope of our present discussion. Complex numbers are discussed in Chapter 7.

Definition   4.7.3

Let p and q be positive integers, p/q in lowest terms. If $x^{1/q}$ is defined, then we set

$$x^{p/q} = (x^{1/q})^p.$$

From this definition you can see that

$$8^{2/3} = (8^{1/3})^2 = 2^2 = 4,$$

$$4^{5/2} = (4^{1/2})^5 = 2^5 = 32,$$

$$(-8)^{4/3} = [(-8)^{1/3}]^4 = (-2)^4 = 16,$$

$$(10,000)^{3/4} = (10,000^{1/4})^3 = 10^3 = 1000. \quad \square$$

Finally we come to negative fractional exponents.

Definition   4.7.4

Let p and q be positive integers, p/q in lowest terms. If $x^{p/q}$ is defined and nonzero, then we set

$$x^{-p/q} = \frac{1}{x^{p/q}}.$$

Here are some examples:

$$10^{-5} = \frac{1}{10^5} = \frac{1}{100,000} = 0.00001.$$

$$8^{-2/3} = \frac{1}{8^{2/3}} = \frac{1}{(8^{1/3})^2} = \frac{1}{2^2} = \frac{1}{4}.$$

$$4^{-5/2} = \frac{1}{4^{5/2}} = \frac{1}{(4^{1/2})^5} = \frac{1}{2^5} = \frac{1}{32}.$$

$$(-8)^{-4/3} = \frac{1}{(-8)^{4/3}} = \frac{1}{[(-8)^{1/3}]^4} = \frac{1}{(-2)^4} = \frac{1}{16}. \quad \square$$

As a special convention we set

(4.7.5)
$$\boxed{x^0 = 1 \quad \text{for all } x \neq 0.}$$

Thus $(-5)^0 = 1$ and $10^0 = 1$ but 0^0 is left undefined. Note also that

(4.7.6)

$$1^r = 1 \quad \text{for all rational exponents } r.$$

The laws of exponents that we gave before for positive integer exponents (page 10) apply now to all rational exponents (positive, negative, or zero) provided that the *base* (the x in $x^{p/q}$) is positive.

Theorem 4.7.7   Laws of Exponents

Let x and y be positive numbers. If r and s are rational numbers, then

$$\text{(i) } x^r \cdot x^s = x^{r+s}, \qquad \text{(ii) } \frac{x^r}{x^s} = x^{r-s},$$

$$\text{(iii) } (x^r)^s = x^{rs}, \qquad \text{(iv) } (xy)^r = x^r y^r.$$

You can prove this theorem by expressing r and s in fractional form:

$$r = \frac{a}{b}, \qquad s = \frac{c}{d},$$

but the details are tedious and not very rewarding. $\square$

We go on to computations.

Problem. Simplify

$$8^{5/3} + 2(\tfrac{4}{9})^{3/2}.$$

SOLUTION

$$8^{5/3} + 2(\tfrac{4}{9})^{3/2} = (8^{1/3})^5 + 2[(\tfrac{4}{9})^{1/2}]^3$$
$$= 2^5 + 2(\tfrac{2}{3})^3$$
$$= 32 + 2(\tfrac{8}{27})$$
$$= 32\tfrac{16}{27}. \quad \square$$

Problem. Simplify

$$(\tfrac{1}{9})^{-1/2} + (\tfrac{8}{27})^{-4/3}.$$

SOLUTION

$$(\tfrac{1}{9})^{-1/2} + (\tfrac{8}{27})^{-4/3} = 9^{1/2} + (\tfrac{27}{8})^{4/3}$$

$$= 3 + (\tfrac{27}{8})^{1/3}(\tfrac{27}{8}) \qquad \text{(explain)}$$

$$= 3 + (\tfrac{3}{2})(\tfrac{27}{8})$$

$$= 3 + \tfrac{81}{16}$$

$$= \tfrac{129}{16}. \quad \square$$

Problem.   Simplify

$$\frac{a^{1/2}}{a^{-1/2} - 5a^{-3/2}}. \qquad (a > 0)$$

SOLUTION.   We begin by removing the fractional exponents from the denominator. We can do this by multiplying through by $a^{3/2}$.

$$\frac{a^{1/2}}{a^{-1/2} - 5a^{-3/2}} = \left(\frac{a^{3/2}}{a^{3/2}}\right)\left(\frac{a^{1/2}}{a^{-1/2} - 5a^{-3/2}}\right)$$

$$= \frac{a^2}{a^1 - 5a^0}$$

$$= \frac{a^2}{a - 5}. \quad \square$$

Problem.   Simplify

$$(27a^{-3}b^{3/5})^{-2/3}. \qquad (a > 0, \, b > 0)$$

SOLUTION

$$(27a^{-3}b^{3/5})^{-2/3} = (27)^{-2/3}(a^{-3})^{-2/3}(b^{3/5})^{-2/3}$$

$$= \left(\frac{1}{27^{2/3}}\right)(a^2)\left(\frac{1}{b^{2/5}}\right)$$

$$= \frac{1}{9}a^2 \left(\frac{1}{b^{2/5}}\right)$$

$$= \frac{a^2b^{3/5}}{9b}.$$

In the last step we removed the fractional exponent from the denominator by multiplying numerator and denominator by $b^{3/5}$. $\square$

Problem. Simplify

$$\frac{2 + x^{-1}}{x^{-1} - 2x^{-2}}. \qquad\qquad (x \neq 0)$$

SOLUTION. We remove the negative exponents by multiplying through by x^2:

$$\frac{2 + x^{-1}}{x^{-1} - 2x^{-2}} = \left(\frac{x^2}{x^2}\right)\left(\frac{2 + x^{-1}}{x^{-1} - 2x^{-2}}\right)$$

$$= \frac{2x^2 + x}{x - 2}. \quad \square$$

Problem. Simplify

$$\sqrt[3]{1^{8/5} + \sqrt{49}}.$$

SOLUTION. Remember that

$$\sqrt[n]{x} = x^{1/n} \quad \text{and} \quad 1^r = 1 \quad \text{for all rational } r.$$

Here

$$\sqrt[3]{1^{8/5} + \sqrt{49}} = (1 + \sqrt{49})^{1/3} = (1 + 7)^{1/3} = 2. \quad \square$$

Problem. Taking $a \geq 0$ and $b \geq 0$ simplify

$$(a^{3/2} + 2a^{3/4}b^{3/4} + b^{3/2})^{1/2} - (a^{3/2} - 2a^{3/4}b^{3/4} + b^{3/2})^{1/2}.$$

SOLUTION. Since

$$a^{3/2} + 2a^{3/4}b^{3/4} + b^{3/2} = (a^{3/4} + b^{3/4})^2$$

and

$$a^{3/2} - 2a^{3/4}b^{3/4} + b^{3/2} = (a^{3/4} - b^{3/4})^2,$$

the expression reduces to

$$(a^{3/4} + b^{3/4}) - (a^{3/4} - b^{3/4}) = 2b^{3/4}. \quad \square$$

Exercises

Write each expression as an integer or the quotient of two integers.

*1. $125^{2/3}$.

2. $125^{-2/3}$.

*3. $8^{-5/3}$.

4. $3^2 \cdot 8^{-2/3} \cdot 49^{-1/2}$.

*5. $(-2)^4(-27)^{2/3}$.

6. $(-27)^{-5/3}$.

*7. $9^{3/2} - 27^{2/3}$.

8. $2(8^{2/3} + 9^{-1/2})$.

*9. $[2(4^{3/2})]^{3/4}$.

10. $\sqrt[3]{54}\,\sqrt[3]{4}$.

*11. $\dfrac{2^{2n+1} + 4^n}{4^{n+1}}$.

12. $\dfrac{121^{-1/2} - 144^{-1/2}}{9^{-1/2}}$.

*13. $\dfrac{(4^3)^{5/6} - (3^4)^{1/2}}{16^{-1/4}}$.

14. $3^{5/6} \cdot 6^{1/3} \cdot 18^{2/3} \cdot 27^{1/2}$.

Simplify to an expression without negative exponents, fractional exponents, or radicals. All letters represent positive numbers.

*15. $(64a^6 b^9)^{1/3}$.

16. $\sqrt{2}(8y^{2/3})^{3/2}$.

*17. $\left(\dfrac{a^4 b^{4/3}}{a^{8/3}}\right)^{3/4}$.

18. $\dfrac{(8y^7)^{1/3} + y^{1/3}}{(27y)^{1/3}}$.

*19. $\dfrac{x^{-1} - y^{-1}}{y - x}$.

20. $\dfrac{x^{-4} - y^{-4}}{x^{-2} - y^{-2}}$.

*21. $\dfrac{a^{4/3} - a^{-2/3}}{a^{1/3} + a^{-2/3}}$.

22. $\dfrac{\sqrt{x^2(2 + x^2 + x^{-2})y^4}}{y^2(x^2 + 1)}$.

*23. $\dfrac{a^{1/3}}{5a^{-2/3} - 3a^{-5/3}}$.

24. $\left(\dfrac{a^{2/3}}{a^{5/3} - a^{8/3}}\right)^{-1}$.

*25. $\left[\dfrac{(u^{-2} - v^{-2})(u^{-1} - v^{-1})}{u^{-1} + v^{-1}}\right]^{-1/2}$.

26. $\left[4\left(\dfrac{ab}{\sqrt{a} + \sqrt{b}}\right)^{-1} - 3(ab)^{-1/2}\left(\dfrac{1}{\sqrt{b}} + \dfrac{1}{\sqrt{a}}\right)\right](\sqrt{a} - \sqrt{b})$.

4.8 One-to-One Functions; Inverses

One-to-One Functions

A function can take on the same value at different points of its domain. Thus, for example, the squaring function takes on the same value at $-c$ as it does at c. In the case of

$$f(x) = (x - 3)(x - 5),$$

we have not only

$$f(3) = 0 \quad \text{but also} \quad f(5) = 0.$$

Functions for which this kind of repetition does *not* occur are called *one-to-one*.

Definition   4.8.1   One-to-One Function

A function is said to be *one-to-one* iff there are no two points at which it takes on the same value.

As examples take

$$f(x) = x^3 \quad \text{and} \quad g(x) = \sqrt{x}.$$

The cubing function is one-to-one because no two numbers have the same cube. The square root function is one-to-one because no two numbers have the same square root.

In the case of a one-to-one function, no horizontal line intersects the graph more than once. (Figure 4.8.1) If a horizontal line intersects the graph more than once, then the function is not one-to-one. (Figure 4.8.2)

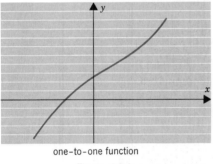

one-to-one function

Figure 4.8.1

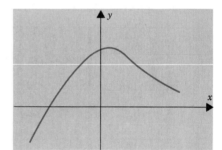

function that is not one-to-one

Figure 4.8.2

Inverses

I am thinking of a real number and its cube is 8. Can you figure out the number? Obviously yes. It must be 2.

Now I am thinking of another number and its square is 9. Can you figure out the number? No, not unless you are clairvoyant. It could be 3 but it could also be -3.

The difference between the two situations can be phrased this way: In the first case we are dealing with a one-to-one function—no two numbers have the same cube. So, knowing the cube, we can work back to the original number. In the second case we are dealing with a function that is not one-to-one—different numbers can have the same square. Knowing the square we still cannot figure out what the original number was.

Throughout the rest of this section we shall be looking at functional values and tracing them back to where they came from. So as to be able to trace these values back, we shall restrict ourselves entirely to functions that are one-to-one.

We begin with a theorem about one-to-one functions.

Theorem 4.8.2

If f is a one-to-one function, then there is one and only one function g that is defined on the range of f and satisfies the equation

$$f(g(x)) = x \quad \text{for all } x \text{ in the range of } f.$$

PROOF.   The proof is easy. If x is in the range of f, then f must take on the value x at some number. Since f is one-to-one, there can be only one such number. We have called it $g(x)$. $\square$

The function that we have named g in the theorem is called the *inverse* of f and is usually denoted by f^{-1}.

Definition 4.8.3   Inverse Function

Let f be a one-to-one function. The *inverse* of f, denoted by f^{-1}, is the unique function that is defined on the range of f and satisfies the equation

$$f(f^{-1}(x)) = x \quad \text{for all } x \text{ in the range of } f.$$

Problem.   Find the inverse of f for

$$f(x) = x^3.$$

SOLUTION.   We begin by setting

$$f(f^{-1}(x)) = x.$$

Since f is the cubing function, we must have

$$f(f^{-1}(x)) = [f^{-1}(x)]^3$$

so that

$$[f^{-1}(x)]^3 = x.$$

Taking cube roots, we have

$$f^{-1}(x) = x^{1/3}.$$

What we've found is that the cube root function is the inverse of the cubing function.

As you can check,

$$f(f^{-1}(x)) = [f^{-1}(x)]^3 = [x^{1/3}]^3 = x. \quad \square$$

Problem. Find the inverse of f for

$$f(x) = 3x - 5.$$

SOLUTION. Set

$$f(f^{-1}(x)) = x.$$

With our choice of f this gives

$$3f^{-1}(x) - 5 = x,$$

which we can now solve for $f^{-1}(x)$:

$$3f^{-1}(x) = x + 5,$$
$$f^{-1}(x) = \tfrac{1}{3}x + \tfrac{5}{3}.$$

As you can check,

$$f(f^{-1}(x)) = 3f^{-1}(x) - 5 = 3(\tfrac{1}{3}x + \tfrac{5}{3}) - 5 = (x + 5) - 5 = x. \quad \square$$

Problem. Find the inverse of f for

$$f(x) = (1 - x^3)^{1/5} + 2.$$

SOLUTION. Once again we begin by setting

$$f(f^{-1}(x)) = x.$$

Since

$$f(f^{-1}(x)) = \{1 - [f^{-1}(x)]^3\}^{1/5} + 2,$$

we must have

$$\{1 - [f^{-1}(x)]^3\}^{1/5} + 2 = x.$$

We now solve for $f^{-1}(x)$:

$$\{1 - [f^{-1}(x)]^3\}^{1/5} = x - 2$$
$$1 - [f^{-1}(x)]^3 = (x - 2)^5$$
$$[f^{-1}(x)]^3 = 1 - (x - 2)^5$$
$$f^{-1}(x) = [1 - (x - 2)^5]^{1/3}. \quad \square$$

By definition f^{-1} satisfies the equation

(4.8.4) $$f(f^{-1}(x)) = x \quad \text{for all } x \text{ in the range of } f.$$

It is also true that

(4.8.5) $$f^{-1}(f(x)) = x \quad \text{for all } x \text{ in the domain of } f.$$

PROOF.   We rewrite (4.8.4) replacing the letter x by the letter t:

(∗) $$f(f^{-1}(t)) = t \quad \text{for all } t \text{ in the range of } f.$$

If x is in the domain of f, then $f(x)$ is in the range of f, and therefore by (∗)

$$f(f^{-1}(f(x))) = f(x). \qquad\qquad [\text{let } t = f(x)]$$

This tells us that f takes on the same value at $f^{-1}(f(x))$ as it does at x. With f one-to-one, this can only happen if

$$f^{-1}(f(x)) = x. \quad \square$$

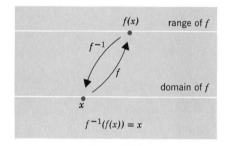

Figure 4.8.3

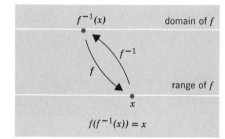

Figure 4.8.4

Equation 4.8.5 tells us that f^{-1} "undoes" the work of f:

if f takes x to $f(x)$,   then f^{-1} takes $f(x)$ back to x.   (Figure 4.8.3)

Equation 4.8.4 tells us that f "undoes" the work of f^{-1}:

if f^{-1} takes x to $f^{-1}(x)$,   then f takes $f^{-1}(x)$ back to x.   (Figure 4.8.4)

The Graphs of f and f⁻¹

There is an important relation between the graph of a one-to-one function f and the graph of f^{-1}. (See Figure 4.8.5) The graph of f consists of points of the form $P(x, f(x))$. Since f^{-1} takes on the value x at $f(x)$, the graph of f^{-1} consists of points of the form $Q(f(x), x)$. Since, for each x, P and Q are symmetric with respect to the line $y = x$, the graph of f^{-1} is the graph f reflected in the line $y = x$.

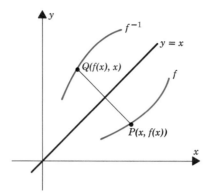

Figure 4.8.5

Problem. Given the graph of f in Figure 4.8.6, sketch the graph of f^{-1}.

SOLUTION. First draw in the line $y = x$. Then reflect the graph of f in that line. See Figure 4.8.7. □

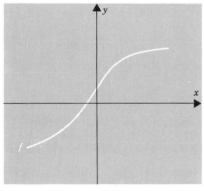

Figure 4.8.6

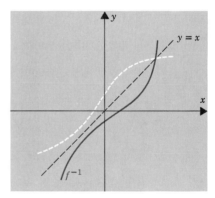

Figure 4.8.7

Exercises

Determine whether or not the given function is one-to-one and, if so, find its inverse.

*1. $f(x) = 5x + 3$.

2. $f(x) = 3x + 5$.

*3. $f(x) = 4x - 7$.

4. $f(x) = 7x - 4$.

*5. $f(x) = 1 - x^2$.

6. $f(x) = x^5$.

*7. $f(x) = x^5 + 1$.

8. $f(x) = x^2 - 3x + 2$.

*9. $f(x) = 1 + 3x^3$.

10. $f(x) = x^3 - 1$.

*11. $f(x) = (1 - x)^3$.

12. $f(x) = (1 - x)^4$.

*13. $f(x) = (x + 1)^3 + 2$.

14. $f(x) = (4x - 1)^3$.

*15. $f(x) = x^{3/5}$.

16. $f(x) = 1 - (x - 2)^{1/3}$.

*17. $f(x) = (2 - 3x)^3$.

18. $f(x) = (2 - 3x^2)^3$.

*19. $f(x) = \dfrac{1}{x}$.

20. $f(x) = \dfrac{1}{1 - x}$.

*21. $f(x) = x + \dfrac{1}{x}$.

22. $f(x) = \dfrac{x}{|x|}$.

*23. $f(x) = \dfrac{1}{x^3 + 1}$.

24. $f(x) = \dfrac{1}{1 - x} - 1$.

*25. What is the relation between f and $(f^{-1})^{-1}$?

26. If $f(x) = x^2$ for all $x \geq 0$, then f is one-to-one and its inverse is the square root function:

$$f^{-1}(x) = x^{1/2}, \qquad x \geq 0.$$

Draw a figure displaying both graphs.

*27. If $f(x) = x^3$ for all real x, then f is one-to-one and its inverse is the cube root function:

$$f^{-1}(x) = x^{1/3}, \qquad \text{all real } x.$$

Draw a figure displaying both graphs.

28. Sketch the graph of f^{-1} given the graph of f in (a) Figure 4.8.8. (b) Figure 4.8.9. (c) Figure 4.8.10.

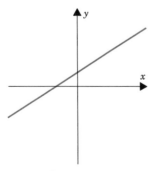

Figure 4.8.8

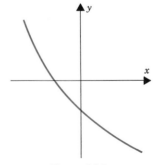

Figure 4.8.9

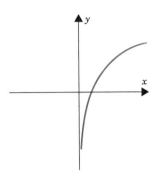

Figure 4.8.10

Optional   29. (a) Show that the composition $f \circ g$ of one-to-one functions is one-to-one.

           (b) Express $(f \circ g)^{-1}$ in terms of f^{-1} and g^{-1}.

     30. Show that the inverse of an odd one-to-one function is odd.

     31. (a) Show that the sum $f + g$ of one-to-one functions need not be one-to-one.

           (b) Show that the product fg of one-to-one functions need not be one-to-one.

4.9   Polynomials

You are familiar with the integral powers:

$$x^0 = 1, \quad x^1 = x, \quad x^2, x^3, \cdots, x^n, \cdots.$$

Any function that can be expressed as a finite linear combination of such powers is called a *polynomial*. For example,

$$P(x) = 3x^2 - x + 7, \qquad Q(x) = -\tfrac{1}{2}x^7 + 4x^5 + \sqrt{2}, \qquad R(x) = 5$$

are polynomials. So are

$$f(x) = x(x - 1)(x - 2) \quad \text{and} \quad g(x) = (2x^2 + 5)(x^3 - 1),$$

as you can see by carrying out the indicated multiplications:

$$f(x) = x(x - 1)(x - 2) = x(x^2 - 3x + 2) = x^3 - 3x^2 + 2x,$$

$$g(x) = (2x^2 + 5)(x^3 - 1) = 2x^5 + 5x^3 - 2x^2 - 5.\dagger \quad \square$$

---

†The polynomials discussed in this book are all *real polynomials*; they are *real* in the sense that all the coefficients are *real* numbers. It is also possible to form polynomials with *complex* coefficients, but these lie outside the scope of our present discussion.

The functions

$$r(x) = \frac{x}{x^2 + 1}, \qquad A(x) = |x|, \qquad t(x) = \sqrt{x + 1}$$

are not polynomials. There is no way that we can write these functions as finite linear combinations of the integral powers. □

Linear Combinations and Products

The set of polynomials is *closed* under linear combinations:

if P and Q are polynomials, so is $\alpha P + \beta Q$;†

and it is *closed* under multiplication:

if P and Q are polynomials, so is PQ.

Thus, for example, if

$$P(x) = x^3 - 2x + 1 \quad \text{and} \quad Q(x) = 2x^4 - x^3 + 7,$$

then

$$
\begin{aligned}
(3P - 2Q)(x) &= 3P(x) - 2Q(x) \\
&= 3(x^3 - 2x + 1) - 2(2x^4 - x^3 + 7) \\
&= 3x^3 - 6x + 3 - 4x^4 + 2x^3 - 14 \\
&= -4x^4 + 5x^3 - 6x - 11
\end{aligned}
$$

and

$$
\begin{aligned}
(PQ)(x) &= P(x)Q(x) \\
&= (x^3 - 2x + 1)(2x^4 - x^3 + 7) \\
&= 2x^7 - x^6 + 7x^3 - 4x^5 + 2x^4 - 14x + 2x^4 - x^3 + 7 \\
&= 2x^7 - x^6 - 4x^5 + 4x^4 + 6x^3 - 14x + 7. \quad \square
\end{aligned}
$$

The quotient of two polynomials is in general not a polynomial: with P and Q as above,

$$\frac{P}{Q}(x) = \frac{P(x)}{Q(x)} = \frac{x^3 - 2x + 1}{2x^4 - x^3 + 7},$$

and this cannot be written as a polynomial.‡ □

---

† Here α and β are arbitrary real numbers. Linear combinations were defined in (4.5.7).
‡ Just as the quotient of two integers is called a *rational number,* the quotient of two polynomials is called a *rational function.*

Degree

Definition 4.9.1   The Degree of a Polynomial

A polynomial

$$P(x) = a_0 x^n + a_1 x^{n-1} + \cdots + a_{n-1} x + a_n,$$

with the leading coefficient $a_0 \neq 0$, is said to have *degree n*.
    A constant polynomial

$$P(x) = c x^0 = c \qquad (c \neq 0)$$

is said to have *degree* 0.
    The zero polynomial

$$P(x) = 0$$

is *not assigned a degree.*

For example,

$$P(x) = 3x^7 - x^2 + 1 \qquad \text{has degree 7,}$$

$$Q(x) = -2x + 1 \qquad \text{has degree 1,}$$

$$R(x) = 2 \qquad \text{has degree 0,}$$

$$S(x) = 0 \qquad \text{has no degree.}$$

When multiplying powers of x, we add the exponents:

$$(x^p)(x^q) = x^{p+q}.$$

As a result, when a polynomial of degree m is multiplied by a polynomial of degree n, the result is a polynomial of degree $m + n$:

$$P(x) = 3x^7 - x^2 + 1 \qquad \text{has degree 7,}$$

$$Q(x) = -2x + 1 \qquad \text{has degree 1,}$$

and therefore the product has degree $7 + 1 = 8$:

$$(PQ)(x) = P(x)Q(x) = (3x^7 - x^2 + 1)(-2x + 1)$$

$$= -6x^8 + 3x^7 + 2x^3 - x^2 - 2x + 1. \quad \square$$

Factors

You've seen that some polynomials can be factored:

$$P(x) = x^2 - 5x + 6 = (x - 2)(x - 3),$$

$$Q(x) = 3x^3 + 9x^2 + 27x = 3x(x^2 + 3x + 9),$$

$$R(x) = x^5 + 2x^4 - x - 2 = (x - 1)(x + 1)(x + 2)(x^2 + 1).$$

In these examples, and (more to the point) in general,

(4.9.2) | *the degree of a polynomial is the sum of the degrees of its factors.* |

Problem.   Verify that $2x^4 + 1$ is a factor of

$$P(x) = 2x^6 + 2x^5 + 6x^4 + x^2 + x + 3.$$

SOLUTION.   We can do this by long division:

$$
\begin{array}{r}
x^2 + x + 3 \\
2x^4 + 1 \overline{\smash{\big)}\ 2x^6 + 2x^5 + 6x^4 + x^2 + x + 3} \\
\underline{2x^6 + x^2} \\
2x^5 + 6x^4 + x + 3 \\
\underline{2x^5 + x} \\
6x^4 + 3 \\
\underline{6x^4 + 3} \\
0.
\end{array}
$$

Since the remainder is 0,

$$P(x) = (2x^4 + 1)(x^2 + x + 3). \quad \square$$

We come now to a key result about polynomials.

Theorem   4.9.3   The Factor Theorem

If P is a polynomial and c is a real number, then

$$P(c) = 0 \quad \text{iff} \quad x - c \text{ is a factor of } P(x).$$

Optional | PROOF.   Observe, first of all, that for each positive integer k,

$$x^k - c^k = (x - c)(x^{k-1} + cx^{k-2} + c^2x^{k-3} + \cdots + c^{k-2}x + c^{k-1}).$$

You can check this by carrying out the multiplication on the right. The product is a sum that collapses to $x^k - c^k$. All the other terms cancel out.
    Getting back to the theorem, take

$$P(x) = a_0x^n + a_1x^{n-1} + \cdots + a_{n-1}x + a_n.$$

Optional | Substituting c for x, we have

$$P(c) = a_0c^n + a_1c^{n-1} + \cdots + a_{n-1}c + a_n.$$

If $P(c) = 0$, then

$$P(x) = P(x) - P(c)$$
$$= a_0(x^n - c^n) + a_1(x^{n-1} - c^{n-1}) + \cdots + a_{n-1}(x - c).$$

The term $x - c$ is a factor of $P(x)$ because it is a factor of each of the terms enclosed by the parentheses.

Conversely, suppose now that $x - c$ is a factor of $P(x)$. Then,

$$P(x) = (x - c)Q(x),$$

where Q is a polynomial. Thus,

$$P(c) = (c - c)Q(c) = 0Q(c) = 0$$

no matter what $Q(c)$ is.   □

Take, for example, the polynomial

$$P(x) = 2x^3 - 7x^2 + 7x - 2.$$

If somehow you know that $x - 2$ is a factor of $P(x)$, then you can be sure that $P(2) = 0$:

$$P(2) = 2(2)^3 - 7(2)^2 + 7(2) - 2 = 16 - 28 + 14 - 2 = 0.$$

Conversely, if somehow you know that $P(1) = 0$, then you can be sure that $x - 1$ is a factor:

$$
\begin{array}{r}
2x^2 - 5x \;\; + 2 \\
x - 1\overline{\smash{\big)}\,2x^3 - 7x^2 + 7x - 2} \\
\underline{2x^3 - 2x^2} \\
-5x^2 + 7x \\
\underline{-5x^2 + 5x} \\
2x - 2 \\
\underline{2x - 2} \\
0
\end{array}
$$

so that

$$P(x) = (x - 1)(2x^2 - 5x + 2).   □$$

The following is a direct consequence of the factor theorem. The proof is left to you as an optional exercise.

Theorem 4.9.4   The Remainder Theorem

If P is a polynomial of positive degree n and c is a real number, then
$$P(x) = (x - c)Q(x) + P(c)$$
with Q a polynomial of degree $n - 1$.

As you can check, for
$$P(x) = x^3 + 4x - 1,$$
we have

$$P(x) = (x - 3)(x^2 + 3x + 13) + 38, \qquad P(3) = 27 + 12 - 1 = 38$$

$$P(x) = (x - 2)(x^2 + 2x + 8) + 15, \qquad P(2) = 8 + 8 - 1 = 15$$

$$P(x) = (x - 1)(x^2 + x + 5) + 4, \qquad P(1) = 1 + 4 - 1 = 4$$

$$P(x) = (x - 0)(x^2 + 4) - 1$$
$$= x(x^2 + 4) - 1, \qquad\qquad P(0) = 0 + 0 - 1 = -1$$

$$P(x) = (x + 1)(x^2 - x + 5) - 6$$
$$= [x - (-1)](x^2 - x + 5) - 6, \qquad P(-1) = -1 - 4 - 1 = -6. \quad \square$$

Exercises

*1.   Which of these functions are polynomials?

(a) $f(x) = (x^4 + 1)(2x - 3)$.　　　(b) $f(x) = \dfrac{1}{x}$.

(c) $f(x) = x + \sqrt{x}$.　　　　　　(d) $f(x) = \dfrac{x^3 + x}{x^2 + 1}$.

2.   Determine the polynomials $3P - 2Q$ and PQ given that
　*(a) $P(x) = x^5 - x + 4, \quad Q(x) = 2x^2 - 1$.
　(b) $P(x) = 2x^2 - 3x + 5, \quad Q(x) = 3x^3 + 2x^2 + x - 1$.

3.   Determine the degree of each polynomial:
　*(a) $P(x) = 4x - \frac{3}{2}$.　　　　　(b) $Q(x) = 4$.

　*(c) $P(x) = \frac{7}{3}x^8 - x^5 + 2x^2 + 1$.　(d) $Q(x) = (x^2 - 1)(x^4 + \sqrt{5})$.

　*(e) $P(x) = x(x^3 - 5x + 1) + x^5$.　(f) $Q(x) = x^2(x - 1)^3(x + 2)^4$.

4.   Determine whether or not $x^3 - x - 2$ is a factor:
　*(a) $P(x) = 2x^4 + x^3 - 2x^2 - 5x - 2$.
　(b) $Q(x) = x^6 - 2x^4 - 4x^3 + x^2 + 4x + 4$.

    (c) $R(x) = x^6 - x^2 + 4x - 4$.

*(d) $S(x) = 2x^6 + x^5 - 2x^4 - 4x^3 - 2x^2 - x - 2$.

5. Determine the polynomial P with leading coefficient 1 given that
   *(a) the degree of P is 3, $P(1) = 0$, and $x^2 + 5x + 4$ is a factor of $P(x)$.
   (b) the degree of P is 3, $P(2) = 0$, $P(3) = 0$, and $x + 2$ is a factor of $P(x)$.
   *(c) the degree of P is 4, $P(-2) = 0$, $P(-1) = 0$, and x^2 is factor of $P(x)$.
   (d) the degree of P is 4, $P(-2) = 0$, $P(-1) = 0$, $P(1) = 0$, $P(2) = 0$.
   *(e) the degree of P is 4, $P(0) = 0$, $P(-1) = 1$, and $x^2 - 3$ is a factor of $P(x)$.

6. Use the remainder theorem to evaluate
   *(a) $P(x) = 3x^5 - 4x^4 + x - 6$ at $x = 2$.
   (b) $P(x) = 3x^5 - 4x^4 + x - 6$ at $x = -2$.
   *(c) $P(x) = 2x^4 - 17x^3 + 2x^2 + 200$ at $x = 3$.
   (d) $P(x) = x^5 - 2x^4 + 3x^2 - 50$ at $x = 5$.

Optional | 7. Prove the remainder theorem.   HINT: Consider the polynomial

$$R(x) = P(x) - P(c).$$

4.10 The Zeros of a Polynomial

The *real zeros* of a polynomial are the real numbers at which the polynomial takes on the value 0:

(4.10.1) | a real number c is called a *zero* of P iff $P(c) = 0$. |

    The real zeros of

$$P(x) = (x + 3)^2(x - 1)^3(x - 2)$$

are -3, 1, and 2. The polynomial

$$P(x) = x^2 + 1$$

has no real zeros.
    If c is a real zero of a polynomial P, then by the factor theorem, $x - c$ divides [is a factor of] $P(x)$. It may happen that $(x - c)^2$ and $(x - c)^3$ also divide $P(x)$. The highest power of $x - c$ that divides $P(x)$ is called the *multiplicity* of c:

(4.10.2) | a real number c is a zero of multiplicity k

                             iff

    $(x - c)^k$ divides $P(x)$ but $(x - c)^{k+1}$ does not divide $P(x)$. |

In the case of

$$P(x) = (x - \tfrac{1}{2})^3(x - 4)(x^2 + 3),$$

$\tfrac{1}{2}$ is a zero of multiplicity three and 4 is a zero of multiplicity one—called a *simple* zero. There are no other real zeros.

Problem.   Find the real zeros of

$$Q(x) = 2x^4 - x^3 - 3x^2$$

and determine their multiplicity.

SOLUTION.   We begin by factoring:

$$Q(x) = 2x^4 - x^3 - 3x^2$$
$$= x^2(2x^2 - x - 3)$$
$$= x^2(2x - 3)(x + 1).$$

Since we can write

$$2x - 3 = 2(x - \tfrac{3}{2}),$$

we have

$$Q(x) = 2x^2(x - \tfrac{3}{2})(x + 1).$$

The number 0 is a zero of multiplicity 2; $\tfrac{3}{2}$ and -1 are simple zeros (multiplicity one). $\square$

Problem.   Find the real zeros of

$$P(x) = x^5 + 4x^4 + 4x^3 - x^2 - 4x - 4$$

and determine their multiplicity given that $x + 2$ is a factor.

SOLUTION.   We know that $x + 2$ is a factor. Let's check whether

$$(x + 2)^2 = x^2 + 4x + 4$$

is also a factor. We can do this by long division:

$$
\begin{array}{r}
x^3 - 1 \\
x^2 + 4x + 4 \overline{\smash{\big)}\, x^5 + 4x^4 + 4x^3 - x^2 - 4x - 4} \\
\underline{x^5 + 4x^4 + 4x^3} \\
-x^2 - 4x - 4 \\
\underline{-x^2 - 4x - 4} \\
0.
\end{array}
$$

We have just shown that

$$P(x) = (x^2 + 4x + 4)(x^3 - 1) = (x + 2)^2(x^3 - 1).$$

Since

$$x^3 - 1 = (x - 1)(x^2 + x + 1),$$

we have

$$P(x) = (x + 2)^2(x - 1)(x^2 + x + 1).$$

With x real, the quadratic at the end is never 0:

$$x^2 + x + 1 = (x^2 + x + \tfrac{1}{4}) + \tfrac{3}{4} = (x + \tfrac{1}{2})^2 + \tfrac{3}{4} > 0.$$

The polynomial P takes on the value 0 only at -2 and at 1; -2 is a zero of multiplicity 2, and 1 is a simple zero.   □

Exercises

1. Find the real zeros of each polynomial and determine the multiplicity of each:
    *(a) $P(x) = (x - 1)(x - 2)(x - 3)$.
    (b) $Q(x) = 2(x - 1)^2(x - 2)(x - 3)^3$.
    *(c) $P(x) = (2x^2 + 1)(2x + 5)^4$.
    (d) $Q(x) = (x^4 - 1)(3x - 2)$.
    *(e) $P(x) = (x^4 - 1)^2(x^4 + 1)$.
    (f) $Q(x) = (x^2 + 3x + 2)(x^2 + 4x + 3)$.
    *(g) $P(x) = (x + 2)(x^6 - x^2)^2$.
    (h) $Q(x) = 2 + (x^2 + 1)(x^4 + 1)$.

2. Find the real zeros of each polynomial and determine the multiplicity of each given that $x - 1$ is a factor:
    *(a) $P(x) = x^3 - 3x^2 + 3x - 1$.
    (b) $P(x) = x^4 - 3x^3 - 17x^2 + 39x - 20$.
    *(c) $P(x) = x^6 - x^5 + x^4 - x^3$.
    (d) $P(x) = x^5 - 11x^4 + 35x^3 - 25x^2$.

Optional | 3. Prove that a polynomial of degree n cannot have more than n real zeros.

4.11   Exponentials and Logarithms

Exponentials

We begin with a positive number b different from 1. In Section 4.7 we defined b^r for r rational, but so far we've attached no meaning to irrational powers of b:

$$b^{\sqrt{2}}, \quad b^{\sqrt{3}}, \quad b^{\pi}, \quad \text{etc.}$$

We can define b^z for z irrational by imposing the following betweenness condition:

(4.11.1)   for any pair of rational numbers r_1 and r_2
if z lies between r_1 and r_2, then b^z lies between b^{r_1} and b^{r_2}.

This betweenness condition determines b^z completely. Thus, for example,

$1.7 < \sqrt{3} < 1.8,$ $2^{1.7} < 2^{1.8},$ and therefore $2^{1.7} < 2^{\sqrt{3}} < 2^{1.8}$

$1.73 < \sqrt{3} < 1.74,$ $2^{1.73} < 2^{1.74},$ and therefore $2^{1.73} < 2^{\sqrt{3}} < 2^{1.74}$

$1.731 < \sqrt{3} < 1.732,$ $2^{1.731} < 2^{1.732},$ and therefore $2^{1.731} < 2^{\sqrt{3}} < 2^{1.732}$, etc.

The laws of exponents,

(4.11.2)

(i) $b^x \cdot b^y = b^{x+y},$ (ii) $\dfrac{b^x}{b^y} = b^{x-y},$

(iii) $(b^x)^y = b^{xy},$ (iv) $(bc)^x = b^x c^x,$

now hold for all real exponents x and y, irrational as well as rational. You will see a proof of this when you study calculus. □

Thus, for example,

$$10^{\sqrt{2}} \cdot 10^{\sqrt{3}} = 10^{\sqrt{2}+\sqrt{3}}, \qquad \frac{10^{\sqrt{2}}}{10^{\sqrt{3}}} = 10^{\sqrt{2}-\sqrt{3}},$$

$$(10^{\sqrt{2}})^{\sqrt{3}} = 10^{\sqrt{6}}, \qquad (10\pi)^{\sqrt{2}} = 10^{\sqrt{2}}\pi^{\sqrt{2}}. \quad □$$

The function defined by setting

(4.11.3) $f(x) = b^x$ for all real x

is called the *exponential function with base b*.

In elementary mathematics we usually take $b = 2$ or $b = 10$. Figure 4.11.1 shows the graph of

$$f(x) = 2^x,$$

the exponential function with base 2. As you can read from the figure, the function is positive and increasing. The graph passes through the point $(0, 1)$:

$$f(0) = 2^0 = 1.$$

If the x-coordinate is increased by 1, the y-coordinate is doubled:

$$f(x + 1) = 2^{x+1} = 2x \cdot 2^1 = 2 \cdot 2^x = 2f(x).$$

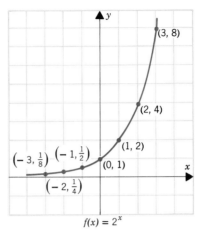

Figure 4.11.1

Logarithms

We continue with a positive number b different from 1. If x is also a positive number, then there is a unique number c such that

$$b^c = x.$$

This number is called the *logarithm of x to the base b*. In symbols,

(4.11.4) $c = \log_b x$ iff $b^c = x.$

As an immediate consequence of this definition, you can see that

(4.11.5) $\log_b b^t = t$ for all real numbers $t.$

Examples

1. $\log_{10} 1000 = \log_{10} 10^3 = 3.$

2. $\log_{10} \frac{1}{100} = \log_{10} 10^{-2} = -2.$

3. $\log_2 64 = \log_2 2^6 = 6.$

4. $\log_2 \frac{1}{32} = \log_2 2^{-5} = -5.$ ☐

The characteristic property of logarithms is that they transform multiplication into addition:

the log of a product = the sum of the logs.

Theorem 4.11.6

If x_1 and x_2 are positive,

$$\log_b (x_1 x_2) = \log_b x_1 + \log_b x_2.$$

PROOF. Set

$$c_1 = \log_b x_1 \quad \text{and} \quad c_2 = \log_b x_2.$$

Then

$$x_1 = b^{c_1} \quad \text{and} \quad x_2 = b^{c_2}.$$

Consequently,

$$\log_b (x_1 x_2) = \log_b (b^{c_1} b^{c_2})$$
$$= \log_b (b^{c_1 + c_2})$$
$$= c_1 + c_2$$
$$= \log_b x_1 + \log_b x_2. \quad \square$$

Here are some other properties worth noting:

(4.11.7)

(i) $\log_b 1 = 0.$

(ii) $\log_b (x_1/x_2) = \log_b x_1 - \log_b x_2.$ $\quad (x_1 > 0, x_2 > 0)$

(iii) $\log_b x^t = t \log_b x.$ $\quad (x > 0, \quad \text{all real } t)$

PROOFS

(i) $\log_b 1 = \log_b b^0 = 0.$ $\quad \square$

(ii) $\log_b (x_1/x_2) = \log_b (x_1 \cdot x_2^{-1})$
$$= \log_b x_1 + \log_b x_2^{-1} = \log_b x_1 - \log_b x_2. \quad \square$$

(iii) Set

$$x = b^u \quad \text{so that} \quad \log_b x = u.$$

Then

$$\log_b x^t = \log_b (b^u)^t = \log_b b^{tu} = tu = t \log_b x. \quad \square$$

Table 4.11.1 gives the logarithms to the base 10 of the integers 2 through 9 rounded off to four decimal places.

Problem. Use Table 4.11.1 to estimate the following logarithms:
(a) $\log_{10} 0.2.$ (b) $\log_{10} 2.4.$ (c) $\log_{10} \sqrt{24}.$ (d) $\log_{10} 810.$

Table 4.11.1

n	$\log_{10} n$
2	0.3010
3	0.4771
4	0.6021
5	0.6990
6	0.7782
7	0.8451
8	0.9031
9	0.9542

SOLUTION

(a) $\log_{10} 0.2 = \log_{10} \left(\frac{2}{10}\right)$

$\qquad = \log_{10} 2 - \log_{10} 10$

$\qquad \cong 0.3010 - 1 = -0.6990.$ ☐

(b) $\log_{10} 2.4 = \log_{10} \left[\dfrac{(3)(8)}{10}\right]$

$\qquad = \log_{10} 3 + \log_{10} 8 - \log_{10} 10$

$\qquad \cong 0.4771 + 0.9031 - 1 = 0.3802.$ ☐

(c) $\log_{10} \sqrt{24} = \log_{10} (24)^{1/2}$

$\qquad = \frac{1}{2}\log_{10} 24$

$\qquad = \frac{1}{2}\log_{10} [(3)(8)]$

$\qquad = \frac{1}{2}(\log_{10} 3 + \log_{10} 8)$

$\qquad \cong \frac{1}{2}(0.4771 + 0.9031) = 0.6901.$ ☐

(d) $\log_{10} 810 = \log_{10} [(10)(9)^2]$

$\qquad = \log_{10} 10 + \log_{10} 9^2$

$\qquad = 1 + 2\log_{10} 9$

$\qquad \cong 1 + 2(0.9542) = 2.9084.$ ☐

Although largely displaced by the calculator and the computer, logarithms to the base 10 are still used to carry out numerical computations, particularly those which involve complicated products, quotients, powers, or roots. In what follows we give some sample computations. We base these on the table of logarithms that appears at the end of the book. The table gives the logarithm of numbers between 1 and 10 to four decimal places. To find the logarithm of a number less than 1 or greater than 10, we use the relation

(4.11.8) $$\boxed{\log_{10} 10^n \, x = n + \log_{10} x.}$$ (check this out)

Thus, for example,

$$\log_{10} 0.375 = -1 + \log_{10} 3.75 = -1 + 0.5740 = -0.4260,$$
$$\log_{10} 37.50 = 1 + \log_{10} 3.75 = 1 + 0.5740 = 1.5740,$$
$$\log_{10} 3750 = 3 + \log_{10} 3.75 = 3 + 0.5740 = 3.5740.$$

Problem.   Use logarithms to the base 10 to estimate $\sqrt{237}$.

SOLUTION

$$\log_{10} \sqrt{237} = \tfrac{1}{2} \log_{10} 237$$
$$= \tfrac{1}{2}(2 + \log 2.37)$$
$$\cong \tfrac{1}{2}(2 + 0.3747)$$
$$\cong 1.1874.$$

Reading backward from the table, you can see that
$$\sqrt{237} \cong 10(1.54) = 15.4. \quad \square$$

Problem.   Use logarithms to the base 10 to estimate
$$t = \frac{(478)(6.19)^2}{75.2}.$$

SOLUTION

$$\log t = \log 478 + \log (6.19)^2 - \log 75.2$$
$$= (2 + \log 4.78) + 2 \log 6.19 - (1 + \log 7.52)$$
$$\cong 2 + 0.6794 + 2(0.7917) - 1 - 0.8762$$
$$\cong 2.3866$$
$$t \cong 100(2.44) = 244. \quad \square$$

Problem.   Use logarithms to the base 10 to estimate
$$u = \frac{\sqrt{87}\sqrt[3]{70}}{\sqrt[5]{600}}.$$

SOLUTION

$$\log u = \tfrac{1}{2}\log 87 + \tfrac{1}{3}\log 70 - \tfrac{1}{5}\log 600$$

$$= \tfrac{1}{2}(1 + \log 8.7) + \tfrac{1}{3}(1 + \log 7) - \tfrac{1}{5}(2 + \log 6)$$

$$= \tfrac{1}{2}(1 + 0.9395) + \tfrac{1}{3}(1 + 0.8451) - \tfrac{1}{5}(2 + 0.7782)$$

$$\cong 1.0291$$

$$u \cong 10(1.07) = 10.7. \quad \square$$

Any positive number other than 1 can be used as a base for a system of logarithms. Sometimes it is useful to change from one base to another. Here is the formula for changing from base b to base B:

(4.11.9)
$$\log_B x = \frac{\log_b x}{\log_b B}.$$

The verification of this formula is left to you as an exercise. $\quad \square$

Problem. Estimate the following logarithms:
(a) $\log_2 10$.   (b) $\log_2 3$.   (c) $\log_2 0.79$.

SOLUTION. We first change these logarithms to base 10.

$$\text{(a) } \log_2 10 = \frac{\log_{10} 10}{\log_{10} 2} = \frac{1}{\log_{10} 2} \cong \frac{1}{0.3010} \cong 3.3223.$$

$$\text{(b) } \log_2 3 = \frac{\log_{10} 3}{\log_{10} 2} \cong \frac{0.4771}{0.3010} \cong 1.5850.$$

$$\text{(c) } \log_2 0.79 = \frac{\log_{10} 0.79}{\log_{10} 2} \cong \frac{-1 + 0.8976}{0.3010} \cong -0.3402. \quad \square$$

The function

(4.11.10)
$$g(x) = \log_b x, \qquad x > 0$$

is called the *logarithm function with base b*.
     Now set

$$f(x) = b^x.$$

Since

$$f(g(x)) = b^{g(x)} = b^{\log_b x} = x \qquad \text{for all } x > 0,$$
$$\underset{(4.11.4)}{\nearrow}$$

(4.11.11)

the logarithm function with base b is the inverse of the exponential function with base b and therefore its graph is the graph of the exponential function reflected in the line $y = x$.

(Section 4.8)

Figure 4.11.2 displays both graphs taking $b = 2$.

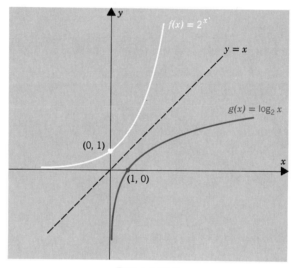

Figure 4.11.2

In our brief explanation of logarithms we have concentrated on base 10 and base 2. These are the most widely used in elementary mathematics. There is, however, another base that is more "natural" and much more important in advanced work. This other base is called e after the great Swiss mathematician Leonard Euler (1707–1783). The number e can be characterized as the unique number that satisfies the condition

$$\left(1 + \frac{1}{n}\right)^n < e < \left(1 + \frac{1}{n}\right)^{n+1} \qquad \text{for all positive integers } n.$$

Rounded off to eight decimal places

$$e \cong 2.71828182.$$

The advantages of basing the system of logarithms on the number e will be clear to you after you have studied some calculus.

Exercises

Simplify using the laws of exponents.

*1. $3^{\sqrt{2}} \cdot 3^{2\sqrt{2}}$.

2. $3^{1-\sqrt{2}} \cdot 3^{2\sqrt{2}-1}$.

*3. $3^{1-\sqrt{2}}/3^{2-\sqrt{2}}$.

*4. $(3^{1-\sqrt{2}})^{1+\sqrt{2}}$.

5. $(3^{\sqrt{3}}/3^{\sqrt{2}})^{\sqrt{3}+\sqrt{2}}$.

*6. $(3^{2+\sqrt{2}}/9)^{\sqrt{2}/2}$.

Find the following logarithms:

*7. $\log_{10} \frac{1}{10}$.

8. $\log_3 27$.

*9. $\log_5 \frac{1}{125}$.

*10. $\log_2 4^3$.

11. $\log_9 \sqrt{3}$.

*12. $\log_{100} 10^{-4/5}$.

*13. $\log_{10} 100^{-4/5}$.

14. $\log_8 32$.

*15. $\log_{16} \frac{1}{8}$.

Use Table 4.11.1 to estimate the following logarithms:

*16. $\log_{10} 15$.

17. $\log_{10} 2.7$.

*18. $\log_{10} 0.016$.

*19. $\log_{10} \sqrt[3]{25}$.

20. $\log_{10} \sqrt{140}$.

*21. $\log_{10} 14.4$.

*22. $\log_{10} 0.9$.

23. $\log_{10} \frac{108}{625}$.

*24. $\log_{10} \sqrt[4]{36}$.

Use the table of logarithms at the end of the book to estimate the following:

*25. $\sqrt{110}$.

26. $\sqrt[3]{350}$.

*27. $\sqrt[5]{1620}$.

28. $\dfrac{(0.622)(0.738)}{1.40}$.

*29. $\dfrac{(63.9)^2(7.2)^3}{(11.5)^4}$.

30. $\dfrac{\sqrt{40}\sqrt[3]{75}}{\sqrt[4]{69}}$.

*31. Show the validity of the formula for changing bases:

$$\log_B x = \frac{\log_b x}{\log_b B}.$$

Estimate the following logarithms. Use the formula for changing bases and the base 10 table at the end of the book.

*32. $\log_5 10$.

33. $\log_5 3$.

*34. $\log_5 100$.

35. $\log_5 \frac{1}{2}$.

Solve the following equations for x:

*36. $(2 - \log_{10} x) \log_{10} x = 0$.

37. $\frac{1}{2} \log_{10} x = \log_{10} (2x - 1)$.

*38. $\log_{2x} 640 = 5$.

39. $\log_5 b \log_b x = 4$.

*40. Sketch the graph of
$$f(x) = (\tfrac{1}{2})^x = 2^{-x}$$
plotting the points with $x = -3, -2, -1, 0, 1, 2, 3$.

41. Sketch the graphs of
$$f(x) = 3^x \quad \text{and} \quad g(x) = (\tfrac{1}{3})^x = 3^{-x}$$
plotting the points with $x = -2, -1, 0, 1, 2$.

*42. Explain how the graph of
$$g(x) = \left(\frac{1}{b}\right)^x = b^{-x}$$
can be obtained from the graph of
$$f(x) = b^x.$$

43. Draw a figure displaying the graphs of
*(a) $f(x) = 3^x$ and $g(x) = \log_3 x$. (b) $f(x) = (\tfrac{1}{2})^x$ and $g(x) = \log_{\frac{1}{2}} x$.

4.12   Additional Exercises

*1. Which of the following functions are one-to-one?
(a) $f(x) = x^{3/8}$.   (b) $f(x) = x^{8/3}$.   (c) $f(x) = x^2 + x$.

*2. Which of the following functions are even?
(a) $f(x) = 4$.   (b) $f(x) = (x^2 + 1)^{1/3}$.   (c) $f(x) = (x^{1/3} + 1)^2$.

*3. Which of the following functions are odd?
(a) $f(x) = 3x^2 - 5x + 1$.

(b) $f(x) = \left\{ \begin{array}{ll} 1, & x > 0 \\ -1, & x < 0 \end{array} \right]$.

(c) $f(x) = (1 + x^{1/3})^2 + (1 - x^{1/3})^2$.
(d) $f(x) = (1 + x^{1/3})^2 - (1 - x^{1/3})^2$.

Find the real zeros of the polynomial and determine their multiplicity:

*4. $P(x) = x^6 - 1$.   *5. $P(x) = x^2(x^3 - 1)$.

*6. $P(x) = 2x^5 - x^4 - 10x^3$.   *7. $P(x) = x^2(x^2 - 1)(x^2 - 2)^2$.

*8. Given that
$$f(x) = (x - 2)^2 \quad \text{and} \quad g(x) = x + \frac{1}{x},$$

form the following:

(a) $f(x - 2)$.   (b) $f(x + 2)$.   (c) $g(x^2)$.
(d) $g(1/x)$.   (e) $f(g(x))$.   (f) $f(f(x))$.

Find the domain and the range:

*9. $f(x) = (x^{1/3} - 1)^{1/2}$.   *10. $f(x) = \dfrac{x}{|x|}$.

*11. $f(x) = 1 - \dfrac{1}{x^2}$.   *12. $f(x) = \dfrac{1}{5 + \sqrt{x}}$.

Find the inverse:

*13. $f(x) = 1 - x^3$.   *14. $f(x) = 3x^{1/5}$.   *15. $f(x) = \dfrac{2}{x} - 1$.

*16. Show that if g is even and $f \circ g$ is defined, then $f \circ g$ is even.

*17. Show that if f and g are bounded on a common domain, then $f + g$ and fg are bounded.

*18. Find the slope of the line that cuts the graph of the function f at the points $(x, f(x))$ and $(x + h, f(x + h))$.

Sketch the graph:

*19. $f(x) = 3|x|$.   *20. $f(x) = -3|x|$.

*21. $f(x) = 3 + |x|$.   *22. $f(x) = 3 - |x|$.

*23. $f(x) = |x - 3|$.   *24. $f(x) = |x + 3|$.

*25. $f(x) = x|x|$.   *26. $f(x) = x + |x|$.

*27. $f(x) = \begin{cases} x^2, & x \le 1 \\ 2 - x, & x > 1 \end{cases}$.   *28. $f(x) = \begin{cases} 2 - x, & x \le 1 \\ x^2, & x > 1 \end{cases}$.

Sketch the graph of each of the following functions, taking the graph of f as given in Figure 4.12.1, page 174:

*29. $g(x) = f(x) + 2$.   *30. $g(x) = f(x) - 2$.

*31. $g(x) = -f(x)$.   *32. $g(x) = f(x + 2)$.

*33. $g(x) = f(x - 2)$.   *34. $g(x) = 2 - f(x)$.

*35. $g(x) = |f(x)|$.   *36. $g(x) = f(|x|)$.

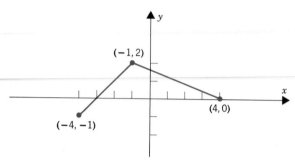

Figure 4.12.1

Draw a figure displaying the two graphs.

*37. $f(x) = 5^x$, $g(x) = 5^{-x}$.

*38. $f(x) = 5^x$, $g(x) = \log_5 x$.

5

TRIGONOMETRY

The word *trigonometry* comes from the Greek: *trigōnon* means triangle and *metron* means measure. Long the tool of the surveyor and the navigator, elementary trigonometry is concerned with the relations that hold between the sides of a triangle and its angles. The importance of trigonometric ideas to calculus and to science in general is much more far reaching. The trigonometric functions play a prominent role in the study of all periodic phenomena. Wherever there are waves to analyze (sound waves, radio waves, light waves, etc.), the trigonometric functions hold the center stage.

In this chapter we define the trigonometric functions and examine their properties. We begin with a discussion of degrees and radians, taking as our setting the unit circle.

5.1 Degrees and Radians

Degrees

In high school geometry you probably measured angles in degrees. A complete revolution was 360 degrees, a straight angle 180 degrees, a right angle 90 degrees.

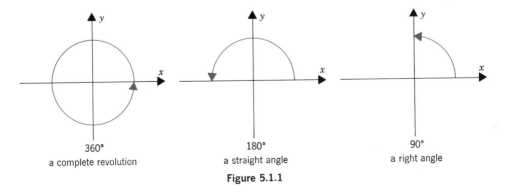

360°
a complete revolution

180°
a straight angle

90°
a right angle

Figure 5.1.1

The trouble with degree measurement is that it's artificial. Why 360 degrees for a complete revolution? Why not 400? or 100?

There is another way to measure angles that is more natural and lends itself much better to the methods of calculus, and that is to measure angles in *radians*.

Radians

The circumference of a circle of radius r (the distance around it) is given by the formula

$$C = 2\pi r.$$

In the case of a unit circle, a circle of radius 1, the circumference is simply 2π.

Since the total distance around a unit circle is 2π, the distance around half of it is π, the distance around a quarter of it is $\frac{1}{2}\pi$, and so on proportionally.

To measure an angle in radians, we draw a unit circle with the center at the vertex and measure the length of the subtended arc. (Figure 5.1.2) The *radian measure* of the angle is the length of this subtended arc. (Figure 5.1.3)

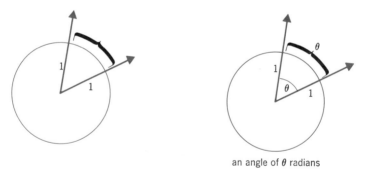

an angle of θ radians

Figure 5.1.2   **Figure 5.1.3**

Figure 5.1.4 illustrates the angles shown in Figure 5.1.1, this time measured in radians.

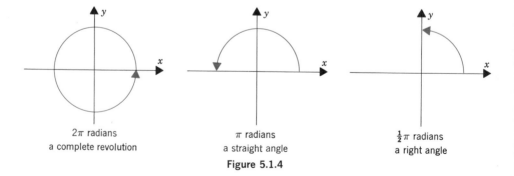

2π radians
a complete revolution

π radians
a straight angle

$\frac{1}{2}\pi$ radians
a right angle

Figure 5.1.4

The Number π

The number π is irrational and therefore cannot be represented as a finite decimal or as a repeating decimal. It is possible, however, to approximate π by finite decimals as closely as you may wish. Here, for example, is π to twenty decimal places:

$$\pi \cong 3.14159\ 26535\ 89793\ 23846.$$

For our purpose the estimate

$$\pi \cong 3.14$$

is close enough.

Conversion Formulas

Since

$$2\pi \text{ radians} = \text{one complete revolution} = 360 \text{ degrees,}$$

you can see that

$$1 \text{ radian} = \frac{180}{\pi} \text{ degrees} \cong 57.30 \text{ degrees}$$

and

$$1 \text{ degree} = \tfrac{1}{180}\pi \text{ radians} \cong 0.0175 \text{ radians.}$$

Problem. How many degrees are there in an angle of $\frac{1}{8}\pi$ radians?

SOLUTION. Since

$$1 \text{ radian} = \frac{180}{\pi} \text{ degrees,}$$

you can see that

$$\tfrac{1}{8}\pi \text{ radians} = \tfrac{1}{8}\pi \left(\frac{180}{\pi}\right) \text{ degrees} = 22.5 \text{ degrees.} \quad \square$$

Problem. How many radians are there in an angle of 60 degrees?

SOLUTION

$$1 \text{ degree} = \tfrac{1}{180}\pi \text{ radians.}$$

Consequently,

$$60 \text{ degrees} = 60(\tfrac{1}{180}\pi) \text{ radians} = \tfrac{1}{3}\pi \text{ radians.} \quad \square$$

Table 5.1.1 gives some important angles measured in degrees and in radians.

Table 5.1.1

degrees	0	30	45	60	90	120	135	150	180
radians	0	$\frac{1}{6}\pi$	$\frac{1}{4}\pi$	$\frac{1}{3}\pi$	$\frac{1}{2}\pi$	$\frac{2}{3}\pi$	$\frac{3}{4}\pi$	$\frac{5}{6}\pi$	π

Sectors

In Figure 5.1.5 you can see a circle of radius r with a sector shaded in. Given that the central angle has θ radians,
(i) What is the length of the subtended arc?
(ii) What is the area of the sector?

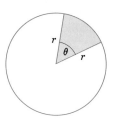

Figure 5.1.5

Both of these questions are easy to answer by proportionality. The length of the subtended arc is to the total circumference $2\pi r$ what θ is to a complete revolution 2π:

$$\frac{\text{length of the subtended arc}}{2\pi r} = \frac{\theta}{2\pi}.$$

Clearing fractions, we have

(5.1.1) $\boxed{\text{length of the subtended arc} = r\theta.}$

This answers (i). To answer (ii), observe that the area of the sector is to the total area πr^2 what θ is to a complete revolution 2π:

$$\frac{\text{area of the sector}}{\pi r^2} = \frac{\theta}{2\pi}.$$

Clearing fractions again, we have

(5.1.2) $\boxed{\text{area of the sector} = \tfrac{1}{2}r^2\theta.}$

Exercises

1. Convert to radians:
   *(a) 270°.   (b) 240°.   *(c) 300°.   (d) 25°.

2. Convert to degrees:
   *(a) 3π radians.   (b) $\frac{3}{2}\pi$ radians.

   (c) $\frac{3}{8}\pi$ radians.   *(d) $\frac{13}{8}\pi$ radians.

3. What is the length of the arc subtended by a central angle of 135° in a circle of radius 4? What is the area of the sector?

*4. What is the length of the arc subtended by a central angle of $\frac{1}{6}\pi$ radians in a circle of area 4π? What is the area of the sector?

5. What central angle subtends an arc of length 3π in a circle of area 25π? What is the area of the corresponding sector?

*6. Compare the radii of two circles given that a sector of angle θ in the first circle has the same area as a sector of angle 3θ in the second circle?

5.2 Sines and Cosines

In Figure 5.2.1 you can see a *unit radius vector* (a pointing needle of length 1 anchored at the origin) resting along the positive *x*-axis.

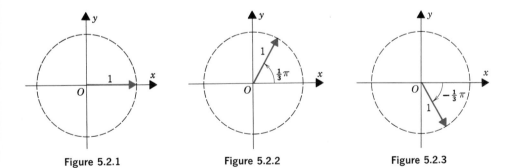

Figure 5.2.1   Figure 5.2.2   Figure 5.2.3

If we rotate this vector about the origin, its tip traces out the unit circle

$$x^2 + y^2 = 1.$$

We can describe these rotations by real numbers t.

Figure 5.2.2 shows the position of the radius vector after the rotation $\frac{1}{3}\pi$; Figure 5.2.3 shows its position after the rotation $-\frac{1}{3}\pi$. There are of course other rotations that take the radius vector to these same positions.

Problem. Determine all the rotations that take the radius vector to the position shown in Figure 5.2.2.

SOLUTION. To the rotation $\frac{1}{3}\pi$ we can add any number of complete revolutions: all rotations of the form

$$(*) \qquad\qquad \tfrac{1}{3}\pi \pm 2n\pi \qquad\qquad (n \text{ a positive integer})$$

take the radius vector to the same position.

We can also reach this position by rotating $\frac{5}{3}\pi$ radians in the clockwise direction and then adding any number of complete revolutions; namely, by rotations of the form

$$(**) \qquad\qquad -\tfrac{5}{3}\pi \pm 2n\pi.$$

But (**) offers nothing new. Every rotation of type (**) can be written as a rotation of type (*): for example,

$$-\tfrac{5}{3}\pi + 8\pi = (-\tfrac{5}{3} + 2)\pi + 6\pi = \tfrac{1}{3}\pi + 6\pi,$$
$$-\tfrac{5}{3}\pi - 8\pi = (-\tfrac{5}{3} + 2)\pi - 10\pi = \tfrac{1}{3}\pi - 10\pi. \quad \square$$

The Cosine Function and The Sine Function

The arrow $\overrightarrow{OP}$ of Figure 5.2.4 shows the unit radius vector after the rotation t:

(5.2.2) $\qquad \cos t = x\text{-coordinate of } P, \qquad \sin t = y\text{-coordinate of } P.$

The function

$$f(t) = \cos t, \qquad \text{all real } t$$

is called the *cosine function,* and the function

$$g(t) = \sin t, \qquad \text{all real } t$$

is called the *sine function.*

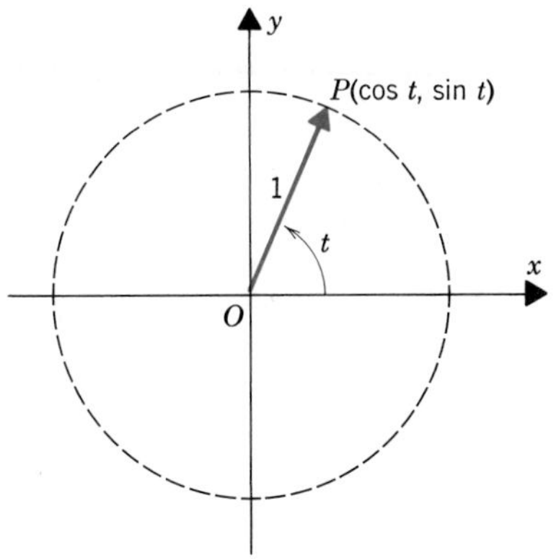

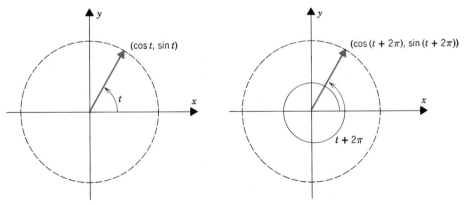

Figure 5.2.4

Properties of cos t and sin t

1. Both the cosine function and the sine function have *period* 2π: for all real t,

(5.2.3) $\boxed{\cos (t + 2\pi) = \cos t, \qquad \sin (t + 2\pi) = \sin t.}$

This property is illustrated in Figure 5.2.5. The rotation $t + 2\pi$ takes the radius vector to the same position as the rotation t.

Figure 5.2.5

2. While the addition of 2π to t has no effect on the values of the cosine and the sine, the addition of π to t reverses the sign of each function: for all real t,

(5.2.4) $\boxed{\cos (t + \pi) = -\cos t, \qquad \sin (t + \pi) = -\sin t.}$

We illustrate this in Figure 5.2.6. The rotation t takes the tip of the radius vector to the point $(\cos t,\ \sin t)$; π units more and the tip of the vector reaches the point diametrically opposite.

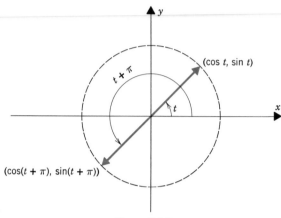

Figure 5.2.6

3. The cosine function is even and the sine function is odd: for all real t,

(5.2.5) $$\cos\left(-t\right) = \cos t, \qquad \sin\left(-t\right) = -\sin t.$$

This is evident in Figure 5.2.7.

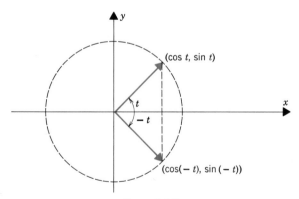

Figure 5.2.7

4. For all real t, we have

$$(\cos t)^2 + (\sin t)^2 = 1,$$

usually written

(5.2.6) $\boxed{\cos^2 t + \sin^2 t = 1.}$

This identity comes from the fact that each point $(\cos t, \sin t)$ lies on the unit circle and each point (x, y) on the unit circle satisfies the equation

$$x^2 + y^2 = 1. \quad \square$$

In what follows we'll be making frequent use of 30, 60, 90 degree triangles and 45, 45, 90 degree triangles. Figure 5.2.8 shows the relative dimensions of the three sides. In each triangle the length of the shortest side is marked a.

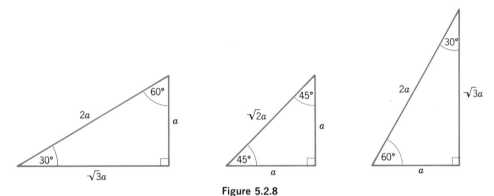

Figure 5.2.8

Some Special Values

The rotation 0 (0 radians) leaves the tip of the unit radius vector at its initial position $(1, 0)$. This gives

$$\cos 0 = 1, \qquad \sin 0 = 0. \qquad \text{(Figure 5.2.9)}$$

The rotation $\frac{1}{6}\pi$, being a counterclockwise rotation of 30 degrees, takes the tip of the unit radius vector to the point $(\frac{1}{2}\sqrt{3}, \frac{1}{2})$. (See Figure 5.2.10. Here $a = \frac{1}{2}$.) It follows that

$$\cos \tfrac{1}{6}\pi = \tfrac{1}{2}\sqrt{3}, \qquad \sin \tfrac{1}{6}\pi = \tfrac{1}{2}.$$

Figure 5.2.11 shows that

$$\cos \tfrac{1}{4}\pi = \tfrac{1}{2}\sqrt{2}, \qquad \sin \tfrac{1}{4}\pi = \tfrac{1}{2}\sqrt{2}. \qquad (a = \tfrac{1}{2}\sqrt{2})$$

Figure 5.2.12 shows that

$$\cos \tfrac{1}{3}\pi = \tfrac{1}{2}, \qquad \sin \tfrac{1}{3}\pi = \tfrac{1}{2}\sqrt{3}. \qquad (a = \tfrac{1}{2})$$

Figure 5.2.13 shows that

$$\cos \tfrac{1}{2}\pi = 0, \qquad \sin \tfrac{1}{2}\pi = 1. \quad \square$$

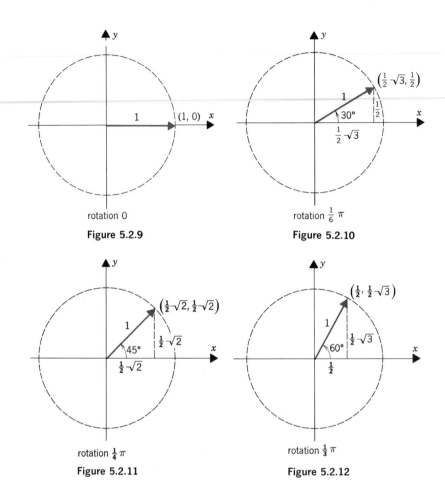

rotation 0

Figure 5.2.9

rotation $\frac{1}{6}\pi$

Figure 5.2.10

rotation $\frac{1}{4}\pi$

Figure 5.2.11

rotation $\frac{1}{3}\pi$

Figure 5.2.12

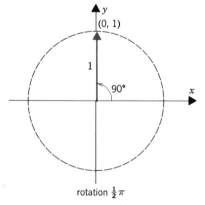

rotation $\frac{1}{2}\pi$

Figure 5.2.13

Problem. Calculate $\cos \frac{9}{4}\pi$ and $\sin \frac{11}{6}\pi$.

SOLUTION

$$\cos \tfrac{9}{4}\pi = \cos\left(\tfrac{1}{4}\pi + 2\pi\right) = \cos \tfrac{1}{4}\pi = \tfrac{1}{2}\sqrt{2}$$
$$\underset{\cos(t+2\pi)=\cos t}{}$$

and

$$\sin \tfrac{11}{6}\pi = \sin\left(-\tfrac{1}{6}\pi + 2\pi\right) = \sin\left(-\tfrac{1}{6}\pi\right) = -\sin \tfrac{1}{6}\pi = -\tfrac{1}{2}. \quad \square$$
$$\underset{\sin(t+2\pi)=\sin t}{} \qquad\qquad \underset{\sin(-t)=-\sin t}{}$$

Problem. Calculate $\cos\left(-\tfrac{7}{6}\pi\right)$ and $\sin\left(-\tfrac{5}{4}\pi\right)$.

SOLUTION

$$\cos\left(-\tfrac{7}{6}\pi\right) = \cos \tfrac{7}{6}\pi = \cos\left(\tfrac{1}{6}\pi + \pi\right) = -\cos \tfrac{1}{6}\pi = -\tfrac{1}{2}\sqrt{3}$$
$$\underset{\cos(-t)=\cos t}{} \qquad\qquad \underset{\cos(t+\pi)=-\cos t}{}$$

and

$$\sin\left(-\tfrac{5}{4}\pi\right) = -\sin \tfrac{5}{4}\pi = -\sin\left(\tfrac{1}{4}\pi + \pi\right) = \sin \tfrac{1}{4}\pi = \tfrac{1}{2}\sqrt{2}. \quad \square$$
$$\underset{\sin(-t)=-\sin t}{} \qquad\qquad \underset{\sin(t+\pi)=-\sin t}{}$$

Problem. What are the possible values of $\sin t$ given that $\cos t = \tfrac{1}{3}$?

SOLUTION. In general

$$\cos^2 t + \sin^2 t = 1.$$

With $\cos t = \tfrac{1}{3}$, we have

$$\tfrac{1}{9} + \sin^2 t = 1,$$
$$\sin^2 t = \tfrac{8}{9},$$
$$\sin t = \pm\tfrac{2}{3}\sqrt{2}. \quad \square$$

Problem. Find all possible values for $\cos t$ given that $2 \sin t - 3 \cos t = 1$.

SOLUTION. Again we use the identity

(∗) $$\cos^2 t + \sin^2 t = 1.$$

The condition $2 \sin t - 3 \cos t = 1$ gives

$$\sin t = \tfrac{1}{2}(1 + 3 \cos t).$$

Substituting $\tfrac{1}{2}(1 + 3 \cos t)$ for $\sin t$ in (∗) we have

$$\cos^2 t + \tfrac{1}{4}(1 + 3 \cos t)^2 = 1$$
$$\cos^2 t + \tfrac{1}{4}(1 + 6 \cos t + 9 \cos^2 t) = 1$$

$$4 \cos^2 t + 1 + 6 \cos t + 9 \cos^2 t = 4$$

$$13 \cos^2 t + 6 \cos t - 3 = 0.$$

This is a quadratic equation in $\cos t$. By the general quadratic formula (2.3.1)

$$\cos t = \frac{-6 \pm \sqrt{36 + 156}}{26} = \frac{-6 \pm 8\sqrt{3}}{26} = \frac{-3 \pm 4\sqrt{3}}{13}. \quad \square$$

Exercises

Determine all positive rotations t that take the unit radius vector to the same position as the given rotation:

*1. $\frac{1}{4}\pi$.　　　　　　2. $-\frac{1}{6}\pi$.　　　　　*3. π.　　　　　　4. $-\frac{5}{12}\pi$.

Determine all negative rotations t that take the unit radius vector to the same position as the given rotation:

*5. $\frac{1}{4}\pi$.　　　　　　6. $-\frac{1}{6}\pi$.　　　　　*7. π.　　　　　　8. $-\frac{5}{12}\pi$.

Calculate the following values:

*9. $\cos \pi$.　　　　　10. $\sin \pi$.　　　　　*11. $\sin \frac{3}{2}\pi$.　　　　12. $\cos \frac{3}{2}\pi$.

*13. $\cos(-\pi)$.　　　　14. $\sin(-\pi)$.　　　　*15. $\sin 2\pi$.　　　　16. $\cos 2\pi$.

*17. $\cos(-\frac{1}{3}\pi)$.　　　18. $\sin(-\frac{1}{4}\pi)$.　　　*19. $\sin \frac{7}{3}\pi$.　　　20. $\cos \frac{7}{3}\pi$.

*21. $\cos \frac{7}{4}\pi$.　　　　22. $\sin \frac{7}{4}\pi$.　　　　*23. $\sin \frac{5}{6}\pi$.　　　24. $\cos \frac{5}{6}\pi$.

*25. $\cos(-\frac{5}{2}\pi)$.　　　26. $\sin(-\frac{4}{3}\pi)$.　　　*27. $\sin \frac{2}{3}\pi$.　　　28. $\cos(-\frac{2}{3}\pi)$.

*29. Explain why for all positive integers n

$$\cos(t \pm 2n\pi) = \cos t \quad \text{and} \quad \sin(t \pm 2n\pi) = \sin t.$$

Calculate:

*30. $\cos 17\pi$.　　　　31. $\sin 17\pi$.　　　　*32. $\cos \frac{23}{4}\pi$.　　　33. $\sin \frac{23}{4}\pi$.

Determine all possible values for $\cos t$ from the given information:

*34. $\sin t = 0$.　　　　　　　　　　35. $\sin t = 1$.

*36. $\sin t = -\cos t$.　　　　　　　　37. $\sin t = 2 \cos t$.

*38. $3 \sin t + 4 \cos t = 0$.　　　　　　39. $2 \cos t - 7 \sin t = 0$.

5.3   The Addition Formulas

The values of the cosine and sine at numbers of the form $s \pm t$ are completely determined by their values at s and t.

The following *addition formulas* hold for all choices for s and t:

(5.3.1)

(i)	$\cos(s + t) = \cos s \cos t - \sin s \sin t.$
(ii)	$\cos(s - t) = \cos s \cos t + \sin s \sin t.$
(iii)	$\sin(s + t) = \sin s \cos t + \cos s \sin t.$
(iv)	$\sin(s - t) = \sin s \cos t - \cos s \sin t.$

We will begin by verifying formula (ii). To simplify matters, we assume that

$$0 \leq t \leq s \leq 2\pi$$

and refer to Figure 5.3.1.

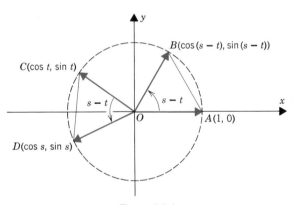

Figure 5.3.1

For reasons that you can figure out yourself, the two central angles marked in the figure both have radian measure $s - t$. The corresponding chords, $\overline{AB}$ and $\overline{CD}$, must therefore have the same length. Since

$$\text{length of } \overline{AB} = \sqrt{[\cos(s - t) - 1]^2 + [\sin(s - t) - 0]^2}$$

and

$$\text{length of } \overline{CD} = \sqrt{(\cos s - \cos t)^2 + (\sin s - \sin t)^2},$$

we must have

$$\sqrt{[\cos(s - t) - 1]^2 + \sin^2(s - t)} = \sqrt{(\cos s - \cos t)^2 + (\sin s - \sin t)^2}$$

and therefore

(∗) $[\cos(s-t) - 1]^2 + \sin^2(s-t) = (\cos s - \cos t)^2 + (\sin s - \sin t)^2.$

Carry out the squaring and simplify, remembering that $\cos^2 u + \sin^2 u = 1$, and you'll find that equation (∗) reduces to formula (ii).

To derive formula (i), we use formula (ii) together with the fact that the cosine function is even and the sine function is odd:

$$\cos(s+t) = \cos(s - (-t))$$

$$= \cos s \cos(-t) + \sin s \sin(-t) \qquad \text{(by formula (ii))}$$

$$= \cos s \cos t - \sin s \sin t.$$

The verification of formulas (iii) and (iv) is left to you in the exercises. □

Table 5.3.1

t	$\cos t$	$\sin t$
0	1	0
$\frac{1}{6}\pi$	$\frac{1}{2}\sqrt{3}$	$\frac{1}{2}$
$\frac{1}{4}\pi$	$\frac{1}{2}\sqrt{2}$	$\frac{1}{2}\sqrt{2}$
$\frac{1}{3}\pi$	$\frac{1}{2}$	$\frac{1}{2}\sqrt{3}$
$\frac{1}{2}\pi$	0	1

In Table 5.3.1 we have collected some values of the sine and cosine. We can calculate other values by using the addition formulas. Here are some examples:

1. $\sin\frac{1}{12}\pi = \sin(\frac{1}{3}\pi - \frac{1}{4}\pi)$

$= \sin\frac{1}{3}\pi \cos\frac{1}{4}\pi - \cos\frac{1}{3}\pi \sin\frac{1}{4}\pi$

$= (\frac{1}{2}\sqrt{3})(\frac{1}{2}\sqrt{2}) - (\frac{1}{2})(\frac{1}{2}\sqrt{2}) = \frac{1}{4}\sqrt{2}(\sqrt{3}-1).$

$\cos\frac{1}{12}\pi = \cos(\frac{1}{3}\pi - \frac{1}{4}\pi)$

$= \cos\frac{1}{3}\pi \cos\frac{1}{4}\pi + \sin\frac{1}{3}\pi \sin\frac{1}{4}\pi$

$= (\frac{1}{2})(\frac{1}{2}\sqrt{2}) + (\frac{1}{2}\sqrt{3})(\frac{1}{2}\sqrt{2}) = \frac{1}{4}\sqrt{2}(\sqrt{3}+1).$

2. $\sin\frac{5}{12}\pi = \sin(\frac{1}{4}\pi + \frac{1}{6}\pi)$

$= \sin\frac{1}{4}\pi \cos\frac{1}{6}\pi + \cos\frac{1}{4}\pi \sin\frac{1}{6}\pi$

$= (\frac{1}{2}\sqrt{2})(\frac{1}{2}\sqrt{3}) + (\frac{1}{2}\sqrt{2})(\frac{1}{2}) = \frac{1}{4}\sqrt{2}(\sqrt{3}+1).$

$\cos\frac{5}{12}\pi = \cos(\frac{1}{4}\pi + \frac{1}{6}\pi)$

$= \cos\frac{1}{4}\pi \cos\frac{1}{6}\pi - \sin\frac{1}{4}\pi \sin\frac{1}{6}\pi$

$= (\frac{1}{2}\sqrt{2})(\frac{1}{2}\sqrt{3}) - (\frac{1}{2}\sqrt{2})(\frac{1}{2}) = \frac{1}{4}\sqrt{2}(\sqrt{3}-1).$ □

If you look back to the examples we just worked out, you'll see that

$$\cos \tfrac{5}{12}\pi = \sin \tfrac{1}{12}\pi \quad \text{and} \quad \sin \tfrac{5}{12}\pi = \cos \tfrac{1}{12}\pi.$$

This is no accident. In general,

(5.3.2) $\boxed{\cos (\tfrac{1}{2}\pi - t) = \sin t \quad \text{and} \quad \sin (\tfrac{1}{2}\pi - t) = \cos t.}$

PROOF. By formula (ii),

$$\cos (\tfrac{1}{2}\pi - t) = \cos \tfrac{1}{2}\pi \cos t + \sin \tfrac{1}{2}\pi \sin t = 0 \cdot \cos t + 1 \cdot \sin t = \sin t.$$

We can use this result to prove the next one:

$$\sin (\tfrac{1}{2}\pi - t) = \cos [\tfrac{1}{2}\pi - (\tfrac{1}{2}\pi - t)] = \cos t. \quad \square$$

The addition formulas lead directly to the *double-angle formulas*

(5.3.3) $\boxed{\cos 2t = \cos^2 t - \sin^2 t, \quad \sin 2t = 2 \sin t \cos t.}$

PROOF

$$\cos 2t = \cos (t + t) = \cos t \cos t - \sin t \sin t = \cos^2 t - \sin^2 t,$$
$$\sin 2t = \sin (t + t) = \sin t \cos t + \cos t \sin t = 2 \sin t \cos t. \quad \square$$

Thus, for example,

$$\cos \tfrac{2}{3}\pi = \cos^2 \tfrac{1}{3}\pi - \sin^2 \tfrac{1}{3}\pi = (\tfrac{1}{2})^2 - (\tfrac{1}{2}\sqrt{3})^2 = \tfrac{1}{4} - \tfrac{3}{4} = -\tfrac{1}{2},$$
$$\underset{\text{Table 5.3.1}}{\nwarrow}$$

and

$$\sin \tfrac{2}{3}\pi = 2 \sin \tfrac{1}{3}\pi \cos \tfrac{1}{3}\pi = 2(\tfrac{1}{2}\sqrt{3})(\tfrac{1}{2}) = \tfrac{1}{2}\sqrt{3}. \quad \square$$
$$\underset{\text{Table 5.3.1}}{\nwarrow}$$

Finally we come to the *half-angle formulas*

(5.3.4) $\boxed{\cos^2 \tfrac{1}{2}t = \tfrac{1}{2}(1 + \cos t), \quad \sin^2 \tfrac{1}{2}t = \tfrac{1}{2}(1 - \cos t).}$

PROOF. We will derive the first formula and leave the second one to you. Since

$$\cos t = \cos [2(\tfrac{1}{2}t)] = \cos^2 \tfrac{1}{2}t - \sin^2 \tfrac{1}{2}t$$

and

$$\cos^2 \tfrac{1}{2}t + \sin^2 \tfrac{1}{2}t = 1,$$

we can write

$$\cos t = \cos^2 \tfrac{1}{2}t - (1 - \cos^2 \tfrac{1}{2}t) = 2 \cos^2 \tfrac{1}{2}t - 1$$

and thus

$$\cos^2 \tfrac{1}{2}t = \tfrac{1}{2}(1 + \cos t). \quad \square$$

Problem. Use a half-angle formula to calculate $\cos \frac{1}{12}\pi$.

SOLUTION. The half-angle formula for the cosine gives

$$\cos^2 \tfrac{1}{12}\pi = \cos^2 \left[\tfrac{1}{2}(\tfrac{1}{6}\pi)\right] = \tfrac{1}{2}(1 + \cos \tfrac{1}{6}\pi) = \tfrac{1}{2}(1 + \tfrac{1}{2}\sqrt{3}) = \tfrac{1}{4}(2 + \sqrt{3}).$$

Since the cosine function is positive in the first quadrant,

$$\cos \tfrac{1}{12}\pi = \sqrt{\tfrac{1}{4}(2 + \sqrt{3})} = \tfrac{1}{2}\sqrt{2 + \sqrt{3}}.$$

We can remove the outside radical by noting that

$$2 + \sqrt{3} = \frac{3 + 2\sqrt{3} + 1}{2} = \frac{(\sqrt{3} + 1)^2}{2}.$$

We can therefore write

$$\cos \tfrac{1}{12}\pi = \tfrac{1}{2}\sqrt{\frac{(\sqrt{3} + 1)^2}{2}} = \tfrac{1}{4}\sqrt{2}(\sqrt{3} + 1).$$

This is the value we obtained on page 188. ☐

Problem. Given that

$$\cos t = \tfrac{2}{3} \quad \text{and} \quad \tfrac{3}{2}\pi < t < 2\pi,$$

calculate   (a) $\cos 2t$,   (b) $\sin 2t$,   (c) $\cos \tfrac{1}{2}t$,   (d) $\sin \tfrac{1}{2}t$.

SOLUTION

(a) $\cos 2t = \cos^2 t - \sin^2 t = \cos^2 t - (1 - \cos^2 t) = 2\cos^2 t - 1.$

With $\cos t = \tfrac{2}{3}$, we have

$$\cos 2t = 2(\tfrac{2}{3})^2 - 1 = \tfrac{8}{9} - 1 = -\tfrac{1}{9}.$$

(b) $\sin^2 2t = 1 - \cos^2 2t.$

With t between $\tfrac{3}{2}\pi$ and 2π, $2t$ is between 3π and 4π and therefore $\sin 2t$ is negative. (The tip of the radius vector is below the x-axis.) It follows that

$$\sin 2t = -\sqrt{1 - \cos^2 2t}.$$

Since $\cos 2t = -\tfrac{1}{9}$, we have

$$\sin 2t = -\sqrt{1 - \tfrac{1}{81}} = -\tfrac{1}{9}\sqrt{80} = -\tfrac{4}{9}\sqrt{5}.$$

(c) $\cos^2 \tfrac{1}{2}t = \tfrac{1}{2}(1 + \cos t).$

With t between $\tfrac{3}{2}\pi$ and 2π, $\tfrac{1}{2}t$ is between $\tfrac{3}{4}\pi$ and π and therefore $\cos \tfrac{1}{2}t$ is negative. (The tip of the radius vector is to the left of the y-axis.) It follows that

$$\cos \tfrac{1}{2}t = -\sqrt{\tfrac{1}{2}(1 + \cos t)}.$$

Since $\cos \tfrac{1}{2}t = \tfrac{2}{3}$, we have

$$\cos \tfrac{1}{2}t = -\sqrt{\tfrac{1}{2}(1 + \tfrac{2}{3})} = -\sqrt{\tfrac{5}{6}} = -\tfrac{1}{6}\sqrt{30}.$$

(d) $$\sin^2 \tfrac{1}{2}t = \tfrac{1}{2}(1 - \cos t).$$

With $\tfrac{1}{2}t$ between $\tfrac{3}{4}\pi$ and π, $\sin \tfrac{1}{2}t$ is positive. (The tip of the radius vector is above the x-axis. Therefore we can write

$$\sin \tfrac{1}{2}t = \sqrt{\tfrac{1}{2}(1 - \cos t)},$$

which, with $\cos t = \tfrac{2}{3}$, gives

$$\sin \tfrac{1}{2}t = \sqrt{\tfrac{1}{2}(1 - \tfrac{2}{3})} = \sqrt{\tfrac{1}{6}} = \tfrac{1}{6}\sqrt{6}. \quad \square$$

Exercises

Use Table 5.3.1 and your ingenuity to calculate the following values:

*1. $\cos \tfrac{5}{6}\pi$.   2. $\sin \tfrac{5}{6}\pi$.   *3. $\cos \tfrac{7}{6}\pi$.

*4. $\sin \tfrac{7}{6}\pi$.   5. $\cos \tfrac{4}{3}\pi$.   *6. $\sin \tfrac{4}{3}\pi$.

*7. $\cos (-\tfrac{3}{4}\pi)$.   8. $\sin (-\tfrac{3}{4}\pi)$.   *9. $\cos \tfrac{7}{12}\pi$.

*10. $\sin \tfrac{7}{12}\pi$.   11. $\cos \tfrac{1}{8}\pi$.   *12. $\sin \tfrac{1}{8}\pi$.

*13. $\cos \tfrac{3}{8}\pi$.   14. $\sin \tfrac{3}{8}\pi$.   *15. $\cos (-\tfrac{7}{4}\pi)$.

*16. $\sin (-\tfrac{7}{4}\pi)$.   17. $\cos \tfrac{37}{12}\pi$.   *18. $\sin \tfrac{37}{12}\pi$.

19.   Derive the identity

$$\sin (s + t) = \sin s \cos t + \cos s \sin t$$

by noting that $\sin (s + t) = \cos [(\tfrac{1}{2}\pi - s) - t]$ and using a cosine expansion.

20.   Derive the identity

$$\sin (s - t) = \sin s \cos t - \cos s \sin t$$

by noting that $\sin (s - t) = \sin [s + (-t)]$ and expanding the right-hand side.

21.   Given that $\cos t = -\tfrac{4}{5}$ and $\tfrac{3}{4}\pi < t < \pi$, calculate
     *(a) $\cos 2t$.   (b) $\sin 2t$.   *(c) $\cos \tfrac{1}{2}t$.   (d) $\sin \tfrac{1}{2}t$.

22.   Given that $\sin t = -\tfrac{1}{5}$ and $\tfrac{7}{4}\pi < t < 2\pi$, calculate
     (a) $\cos 2t$.   *(b) $\sin 2t$.   (c) $\cos \tfrac{1}{2}t$.   *(d) $\sin \tfrac{1}{2}t$.

*23. Derive the half-angle formula: $\sin^2 \frac{1}{2}t = \frac{1}{2}(1 - \cos t)$.

24. Show that

(5.3.5) $\quad\boxed{\cos (t + \frac{1}{2}\pi) = -\sin t, \qquad \sin (t + \frac{1}{2}\pi) = \cos t.}$

5.4 Graphing Sines and Cosines

In Figure 5.4.1 we have drawn the unit circle and alongside it the graph of the sine function from 0 to π.

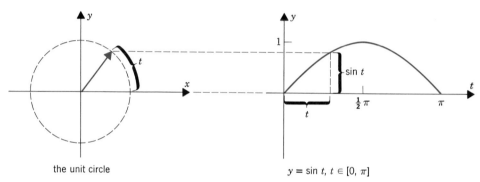

the unit circle $\qquad\qquad\qquad\qquad$ $y = \sin t, \, t \in [0, \pi]$

Figure 5.4.1

Since $\sin (t + \pi) = -\sin t$, the curve goes on in the same manner from π to 2π, this time below the horizontal axis. Figure 5.4.2 shows the graph from 0 to 2π with some points plotted.

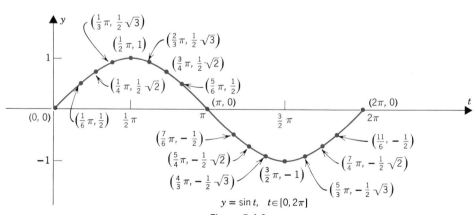

$y = \sin t, \quad t \in [0, 2\pi]$

Figure 5.4.2

Since the sine function has period 2π, the graph goes on with these same oscillations on every interval of length 2π. Figure 5.4.3, drawn to a smaller scale, exhibits this repetition.

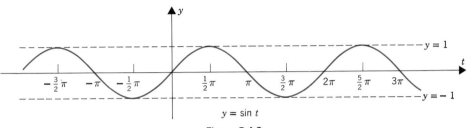

$$y = \sin t$$

Figure 5.4.3

As you saw before

$$\cos t = \sin (t + \tfrac{1}{2}\pi). \tag{5.3.5}$$

This identity tells us that the graph of the cosine function (Figure 5.4.4) is the graph of the sine function displaced $\tfrac{1}{2}\pi$ units to the left. [Recall (4.3.2).]

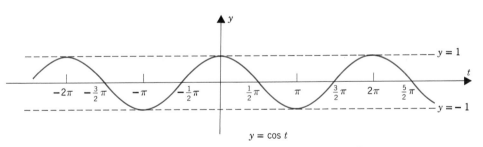

$$y = \cos t$$

Figure 5.4.4

Simple Harmonic Waves

Sines and cosines play a central role in the study of all wave motion. The simplest waves (called *simple harmonic waves*) have equations of the form

$$y = a \sin (bt + c)$$

with a and b positive and c in the interval $[0, 2\pi)$.

Problem. Write the equations

$$y = \sin(2t + 5\pi), \qquad y = 2\sin(-\tfrac{1}{3}t + \tfrac{1}{2}\pi), \qquad y = -2\sin(3t - \tfrac{1}{4}\pi)$$

in simple harmonic form.

SOLUTION

1. $y = \sin(2t + 5\pi) = \sin[(2t + \pi) + 4\pi] = \sin(2t + \pi)$.

$\qquad\qquad\qquad\qquad\qquad\qquad\qquad\quad$ $\underset{\displaystyle \sin(u + 4\pi) = \sin u}{}$

2. $y = 2\sin(-\tfrac{1}{3}t + \tfrac{1}{2}\pi) = 2[-\sin(\tfrac{1}{3}t - \tfrac{1}{2}\pi)] = 2\sin(\tfrac{1}{3}t + \tfrac{1}{2}\pi)$.

$\qquad\qquad\qquad\qquad\underset{\displaystyle \sin(-u) = -\sin u}{}\qquad\qquad \underset{\displaystyle -\sin u = \sin(u + \pi)}{}$

3. $y = -2\sin(3t - \tfrac{1}{4}\pi) = 2[-\sin(3t - \tfrac{1}{4}\pi)] = 2\sin(3t + \tfrac{3}{4}\pi)$. $\square$

$\qquad\qquad\qquad\qquad\qquad\qquad\qquad\underset{\displaystyle -\sin u = \sin(u + \pi)}{}$

Let's go back to

$$y = a\sin(bt + c)$$

and examine the significance of a, b, c. Since

$$-1 \le \sin u \le 1 \qquad \text{for all numbers } u,$$

you can see that

$$-1 \le \sin(bt + c) \le 1 \qquad \text{for all } t$$

and therefore

$$-a \le a\sin(bt + c) \le a.$$

This tells us that the curve $y = a\sin(bt + \pi)$ stays between the horizontal lines $y = -a$ and $y = a$. The number a, called the *amplitude* of the wave, measures the vertical magnitude of the oscillations. (See Figure 5.4.5.)

The significance of the constant b comes from the fact that the wave repeats itself on every interval of length $2\pi/b$. The y-coordinate at $t + 2\pi/b$ is the same as it is at t:

$$a\sin[b(t + 2\pi/b) + c] = a\sin(bt + 2\pi + c) = a\sin(bt + c).$$

The number $2\pi/b$ is called the *period* of the wave motion. (See Figure 5.4.5.) The smaller the period, the more frequently the oscillations occur. A curve with period $2\pi/b$ oscillates b times as frequently as the sine function (which has period 2π).

The number c, in conjunction with a, determines the y-intercept of the wave, the height or depth of the wave as it crosses the y-axis:

$$\text{at} \quad t = 0, \quad y = a\sin c. \qquad\qquad \text{(Figure 5.4.5)}$$

Since

$$y = a\sin(bt + c) = a\sin[b(t + c/b)],$$

the curve $y = a \sin (bt + c)$ is the curve $y = a \sin bt$ displaced c/b units to the left. [Recall (4.3.2).]

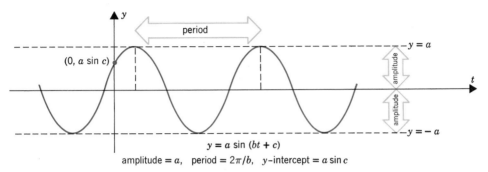

$$y = a \sin (bt + c)$$

amplitude $= a$,   period $= 2\pi/b$, y–intercept $= a \sin c$

Figure 5.4.5

Problem.   Find the amplitude, the period, and the y-intercept of

$$y = 2 \sin 3t$$

and then sketch the graph.

SOLUTION.   Here $a = 2, b = 3, c = 0$. The amplitude a is 2 (twice that of the sine curve); the period $2\pi/b$ is $\frac{2}{3}\pi$ (one third that of the sine curve); the y-intercept $a \sin c$ is 0 (like that of the sine curve).

  We can obtain the graph (Figure 5.4.6) by starting with the sine curve (Figure 5.4.3), doubling the magnitude of the oscillations and tripling their frequency.   □

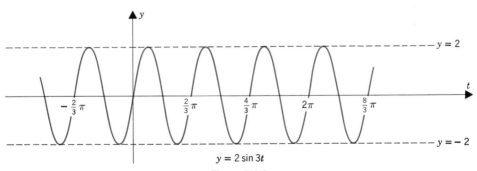

$$y = 2 \sin 3t$$

Figure 5.4.6

Problem.   Find the amplitude, the period, and the y-intercept of

$$y = 2 \sin (3t + \tfrac{1}{2}\pi)$$

and then sketch the graph.

SOLUTION. As before, the amplitude is 2 and the period is $\frac{2}{3}\pi$, but this time the y-intercept is $2 \sin \frac{1}{2}\pi = 2$. Since

$$2 \sin (3t + \tfrac{1}{2}\pi) = 2 \sin [3(t + \tfrac{1}{6}\pi)],$$

the graph (Figure 5.4.7) is the curve $y = 2 \sin 3t$ displaced $\frac{1}{6}\pi$ units to the left. It crosses the y-axis at the crest of a wave. ☐

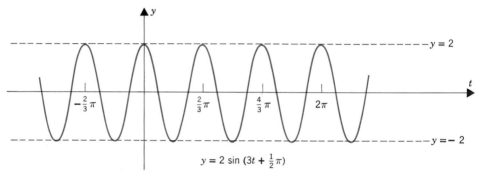

$$y = 2 \sin (3t + \tfrac{1}{2}\pi)$$

Figure 5.4.7

Problem. Find the amplitude, the period, and the y-intercept of

$$y = \tfrac{3}{4} \sin \tfrac{1}{2}t$$

and then sketch the graph.

SOLUTION. Here $a = \frac{3}{4}$, $b = \frac{1}{2}$, $c = 0$. The amplitude a is $\frac{3}{4}$ (three quarters that of the sine curve); the period $2\pi/b$ is 4π (twice that of the sine curve); the y-intercept $a \sin c$ is 0 (like that of the sine curve).

We can obtain the graph (Figure 5.4.8) by starting with the sine curve, reducing the magnitude of the oscillations by a factor of $\frac{3}{4}$ and halving their frequency. ☐

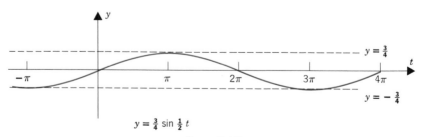

$$y = \tfrac{3}{4} \sin \tfrac{1}{2} t$$

Figure 5.4.8

Problem. Sketch the graph of
$$y = \tfrac{3}{4} \sin\left(\tfrac{1}{2}t + \tfrac{3}{2}\pi\right).$$

SOLUTION. The amplitude and the period are those of the last curve. The y-intercept $a \sin c$ is $\tfrac{3}{4} \sin \tfrac{3}{2}\pi = -\tfrac{3}{4}$. Since
$$\tfrac{3}{4} \sin\left(\tfrac{1}{2}t + \tfrac{3}{2}\pi\right) = \tfrac{3}{4} \sin\left[\tfrac{1}{2}(t + 3\pi)\right],$$
the graph (Figure 5.4.9) is the curve $y = \tfrac{3}{4} \sin \tfrac{1}{2}t$ displaced 3π units to the left.† It crosses the y-axis at the trough of a wave. $\square$

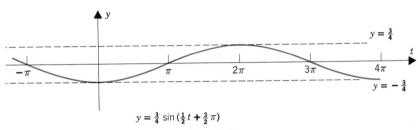

$$y = \tfrac{3}{4} \sin\left(\tfrac{1}{2}t + \tfrac{3}{2}\pi\right)$$

Figure 5.4.9

We have placed no emphasis on curves of the form
$$y = A \cos\left(Bt + C\right)$$
because all of them can be written in the form
$$y = a \sin\left(bt + c\right).$$
This follows directly from the observation that
$$\cos u = \sin\left(u + \tfrac{1}{2}\pi\right). \qquad\qquad (5.3.5)$$
Thus, for example, the curve
$$y = 2 \cos\left(\tfrac{1}{3}t - \tfrac{1}{4}\pi\right)$$
can be written
$$y = 2 \sin\left[\left(\tfrac{1}{3}t - \tfrac{1}{4}\pi\right) + \tfrac{1}{2}\pi\right] = 2 \sin\left(\tfrac{1}{3}t + \tfrac{1}{4}\pi\right). \quad \square$$

---

† The curve $y = \tfrac{3}{4} \sin\left(\tfrac{1}{2}t + \tfrac{3}{2}\pi\right)$ can also be viewed as the curve $y = \tfrac{3}{4} \sin \tfrac{1}{2}t$ displaced π units to the right:
$$\tfrac{3}{4} \sin\left(\tfrac{1}{2}t + \tfrac{3}{2}\pi\right) = \tfrac{3}{4} \sin\left[\left(\tfrac{1}{2}t + \tfrac{3}{2}\pi\right) - 2\pi\right] = \tfrac{3}{4} \sin\left(\tfrac{1}{2}t - \tfrac{1}{2}\pi\right) = \tfrac{3}{4} \sin\left[\tfrac{1}{2}(t - \pi)\right]. \quad \square$$

Exercises

Write each of the following curves in the form

$$y = a \sin (bt + c) \quad \text{with} \quad a > 0,\ b > 0,\ c \in [0, 2\pi).$$

Find the amplitude, the period, and the y-intercept:

*1. $y = -5 \sin (t + 6\pi)$.   2. $y = \frac{1}{5} \sin (\pi - 2t)$.

*3. $y = -\frac{1}{3} \sin (\frac{1}{2}\pi - t)$.   4. $y = 3 \sin (\frac{5}{2}\pi - 2t)$.

*5. $y = 2 \cos 3t$.   6. $y = -3 \cos 2t$.

*7. $y = -2 \cos (3t + \frac{1}{2}\pi)$.   8. $y = \frac{1}{2} \cos (\pi - t)$.

Each of the following curves is the curve $y = 4 \sin \frac{1}{3}t$ displaced left or right. Describe the displacement.

*9. $y = 4 \sin [\frac{1}{3}(t - \pi)]$.   10. $y = 4 \sin [\frac{1}{3}(t + \pi)]$.

*11. $y = 4 \sin (\frac{1}{3}t + \frac{1}{2}\pi)$.   12. $y = 4 \sin (\frac{1}{3}t - \frac{1}{2}\pi)$.

*13. $y = 4 \sin (\frac{1}{3}t - \frac{1}{4}\pi)$.   14. $y = 4 \cos \frac{1}{3}t$.

Find the amplitude, the period, and the y-intercept. Then sketch the graph.

*15. $y = \sin 2t$.   *16. $y = \frac{1}{2} \sin 2t$.

17. $y = 3 \sin 2t$.   18. $y = 3 \sin (2t + \frac{1}{2}\pi)$.

*19. $y = -3 \sin 2t$.   *20. $y = \cos 2t$.

21. $y = \cos (2t + \frac{1}{4}\pi)$.   22. $y = \sin \frac{3}{4}t$.

*23. $y = 2 \sin \frac{3}{4}t$.   *24. $y = 2 \sin (\frac{3}{4}t + \pi)$.

5.5   The Other Trigonometric Functions

There are four other trigonometric functions: the *tangent*, the *cotangent*, the *secant*, and the *cosecant*. They can all be defined in terms of sines and cosines:

(5.5.1)

$$\tan t = \frac{\sin t}{\cos t}, \qquad \cot t = \frac{\cos t}{\sin t},$$

$$\sec t = \frac{1}{\cos t}, \qquad \csc t = \frac{1}{\sin t}.$$

By far the most important of these is the tangent.

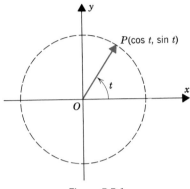

Figure 5.5.1

In Figure 5.5.1 we have drawn a unit circle and marked a point P with coordinates $(\cos t, \sin t)$;

(5.5.2) $\qquad \boxed{\tan t = \dfrac{\sin t}{\cos t} = \text{the slope of the radius vector } \overrightarrow{OP}.}$

At $t = 0$, the radius vector is horizontal and the slope is 0:

$$\tan 0 = \frac{\sin 0}{\cos 0} = \frac{0}{1} = 0.$$

At $t = \frac{1}{4}\pi$, the radius vector lies along the line $y = x$, and its slope is 1:

$$\tan \tfrac{1}{4}\pi = \frac{\sin \frac{1}{4}\pi}{\cos \frac{1}{4}\pi} = \frac{\frac{1}{2}\sqrt{2}}{\frac{1}{2}\sqrt{2}} = 1.$$

As t continues to increase between $\frac{1}{4}\pi$ and $\frac{1}{2}\pi$, the radius vector becomes steeper and steeper and its slope, $\tan t$, increases without bound. When t finallly reaches $\frac{1}{2}\pi$, the radius vector is perpendicular and its slope, $\tan t$, is not defined:

$$\tan \tfrac{1}{2}\pi = \frac{\sin \frac{1}{2}\pi}{\cos \frac{1}{2}\pi} = \frac{1}{0} \text{ is not defined.}$$

Figure 5.5.2 displays a short table of values and the graph of $y = \tan t$ for $t \in [0, \frac{1}{2}\pi)$. The points generated by the table are marked on the graph.

t	y
0	0
$\frac{1}{8}\pi$	$\sqrt{2}-1 \cong 0.41$
$\frac{1}{6}\pi$	$\frac{1}{3}\sqrt{3} \cong 0.58$
$\frac{1}{4}\pi$	1
$\frac{1}{3}\pi$	$\sqrt{3} \cong 1.73$
$\frac{3}{8}\pi$	$1+\sqrt{2} \cong 2.41$
$\frac{5}{12}\pi$	$2+\sqrt{3} \cong 3.73$
$\frac{1}{2}\pi$	undefined

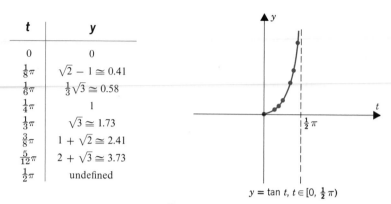

$$y = \tan t,\ t \in [0,\ \tfrac{1}{2}\pi)$$

Figure 5.5.2

Since

$$\tan(-t) = \frac{\sin(-t)}{\cos(-t)} = -\frac{\sin t}{\cos t} = -\tan t,$$

the tangent function is odd and its graph is symmetric about the origin. Hence Figure 5.5.3.

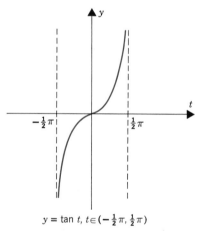

$$y = \tan t,\ t \in (-\tfrac{1}{2}\pi,\ \tfrac{1}{2}\pi)$$

Figure 5.5.3

To understand the complete graph of the tangent function, observe that

$$\tan(t + \pi) = \frac{\sin(t + \pi)}{\cos(t + \pi)} = \frac{-\sin t}{-\cos t} = \tan t.$$

The tangent function is periodic with period π. Its graph (Figure 5.5.4) repeats itself on every interval of length π. □

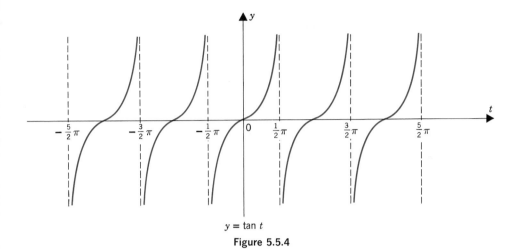

$$y = \tan t$$

Figure 5.5.4

We have shown that

(5.5.3)
$$\tan (-t) = -\tan t$$

and

(5.5.4)
$$\tan (t + \pi) = \tan t.$$

The addition formulas for the tangent function read

(5.5.5)
$$\tan (s + t) = \frac{\tan s + \tan t}{1 - \tan s \tan t}, \qquad \tan (s - t) = \frac{\tan s - \tan t}{1 + \tan s \tan t}.$$

You can derive the first formula as follows:

$$\tan (s + t) = \frac{\sin (s + t)}{\cos (s + t)} = \frac{\sin s \cos t + \cos s \sin t}{\cos s \cos t - \sin s \sin t}.$$

Divide both numerator and denominator by $\cos s \cos t$ and you have

$$\tan (s + t) = \frac{\tan s + \tan t}{1 - \tan s \tan t}.$$

The second formula follows from the first one together with the observation that $\tan (-t) = -\tan t$:

$$\tan (s - t) = \tan [s + (-t)] = \frac{\tan s + \tan (-t)}{1 - \tan s \tan (-t)} = \frac{\tan s - \tan t}{1 + \tan s \tan t}. \quad \square$$

Figure 5.5.5 shows the graph of the cotangent function. Like the tangent function the cotangent is odd and has period π.

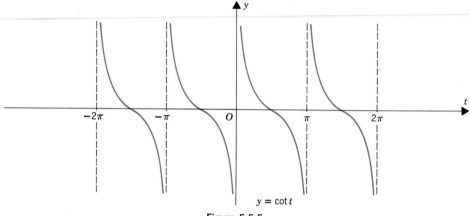

$$y = \cot t$$

Figure 5.5.5

The graphs of the secant and cosecant are shown from 0 to 2π in Figure 5.5.6. Since each function has period 2π, the patterns repeat themselves ad infinitum. As you can verify, the secant is even and the cosecant is odd.

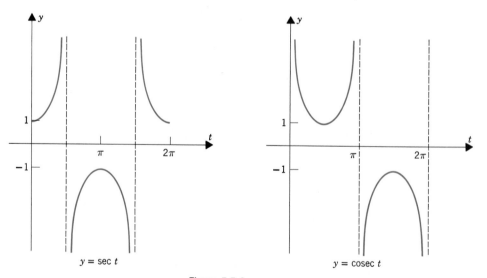

$$y = \sec t \qquad\qquad y = \csc t$$

Figure 5.5.6

More on the tangent, the secant, and the cosecant in the exercises.

Exercises

Calculate using Table 5.3.1 if necessary:

*1.　$\tan \frac{5}{4}\pi$.	2.　$\tan (-\frac{1}{4}\pi)$.	*3.　$\tan 3\pi$.
*4.　$\tan \frac{7}{6}\pi$.	5.　$\tan \frac{3}{4}\pi$.	*6.　$\tan (-\frac{3}{4}\pi)$.
*7.　$\tan \frac{5}{6}\pi$.	8.　$\tan \frac{8}{3}\pi$.	*9.　$\tan (-\frac{7}{3}\pi)$.
*10.　$\cot \frac{1}{6}\pi$.	11.　$\cot \frac{1}{4}\pi$.	*12.　$\cot \frac{1}{3}\pi$.
*13.　$\cot (-\frac{7}{6}\pi)$.	14.　$\cot \frac{3}{4}\pi$.	*15.　$\cot (-\frac{8}{3}\pi)$.
*16.　$\sec \frac{1}{6}\pi$.	17.　$\sec \frac{1}{3}\pi$.	*18.　$\sec \frac{1}{4}\pi$.
*19.　$\sec 3\pi$.	20.　$\sec (-\frac{8}{3}\pi)$.	*21.　$\sec \frac{5}{4}\pi$.
*22.　$\operatorname{cosec} \frac{1}{6}\pi$.	23.　$\operatorname{cosec} \frac{1}{3}\pi$.	*24.　$\operatorname{cosec} \frac{1}{4}\pi$.
*25.　$\operatorname{cosec} \frac{1}{2}\pi$.	26.　$\operatorname{cosec} (-\frac{3}{2}\pi)$.	*27.　$\operatorname{cosec} (-\frac{8}{3}\pi)$.

Determine the period:

*28.　$y = \tan 2t$.	29.　$y = \tan \frac{1}{2}t$.	*30.　$y = 2 \tan 3t$.
*31.　$y = 4 \tan \frac{1}{3}t$.	32.　$y = \tan \pi t$.	*33.　$y = \tan \frac{1}{4}\pi t$.

Find the approximate value given that $\tan A \cong 2.5$ and $\tan B \cong 1.2$:

*34.　$\tan (A + B)$.	35.　$\tan (A - B)$.	*36.　$\tan 2A$.
*37.　$\tan (2A - B)$.	38.　$\cot (A - B)$.	*39.　$\cot 2B$.

40.　Show that
　　*(a) the cotangent function is odd.
　　(b) the secant function is even.
　　*(c) the cosecant function is odd.

41.　Show that
　　(a) the cotangent function has period π.
　　*(b) the secant function has period 2π.
　　(c) the cosecant function has period 2π.

Sketch the graph.

*42.　$y = \tan 2t$,　$t \in (-\frac{1}{4}\pi, \frac{1}{4}\pi)$.	43.　$\tan \frac{1}{2}t$,　$t \in (-\pi, \pi)$.
*44.　$y = \tan (t + \frac{1}{2}\pi)$,　$t \in (0, \pi)$.	45.　$y = \tan (t - \frac{1}{2}\pi)$,　$t \in (\pi, \frac{3}{2}\pi]$.

46. Derive the following identities:
$$1 + \tan^2 t = \sec^2 t, \qquad 1 + \cot^2 t = \csc^2 t.$$

[HINT: $\cos^2 t + \sin^2 t = 1$.]

*47. Express $\tan 2t$ in terms of $\tan t$. HINT: $\tan 2t = \tan (t + t)$.

Verify the following identities.

*48. $(\tan t + \sec t)(1 - \sin t) = \cos t$.   49. $\csc t - \sin t = \cos t \cot t$.

*50. $\tan t + \cot t = \sec t \csc t$.   51. $\sec^2 t - 1 = \sin^2 t \sec^2 t$.

*52. $\dfrac{1}{\csc t + \cot t} = \csc t - \cot t$.   53. $(\sec t - \tan t)^2 = \dfrac{1 - \sin t}{1 + \sin t}$.

5.6 The Inverse Trigonometric Functions

The Arc Sine

We begin with a number b in the interval $[-1, 1]$, the range of the sine function. Although there are infinitely many numbers at which the sine function takes on the value b, only one of these numbers lies in the interval $[-\frac{1}{2}\pi, \frac{1}{2}\pi]$. This unique number is written *arc sin b* and is called the *arc sine of b*. In symbols,

(5.6.1) $\boxed{a = \arcsin b \qquad \text{iff} \qquad \sin a = b \quad \text{and} \quad a \in [-\tfrac{1}{2}\pi, \tfrac{1}{2}\pi].}$

Thus, for example,

$\arcsin 1 = \tfrac{1}{2}\pi,$ $(\sin \tfrac{1}{2}\pi = 1 \quad \text{and} \quad \tfrac{1}{2}\pi \in [-\tfrac{1}{2}\pi, \tfrac{1}{2}\pi])$

$\arcsin (-\tfrac{1}{2}\sqrt{2}) = -\tfrac{1}{4}\pi,$ $(\sin (-\tfrac{1}{4}\pi) = -\tfrac{1}{2}\sqrt{2} \quad \text{and} \quad -\tfrac{1}{4}\pi \in [-\tfrac{1}{2}\pi, \tfrac{1}{2}\pi])$

$\arcsin 0 = 0.$ $(\sin 0 = 0 \quad \text{and} \quad 0 \in [-\tfrac{1}{2}\pi, \tfrac{1}{2}\pi])$

Table 1 at the end of the book gives the values of the sine, cosine, and tangent from 0 to $1.58 \cong \tfrac{1}{2}\pi$. Reading backward from this table you can see, for instance, that

$\arcsin 0.315 \cong 0.32,$ $\arcsin 0.724 \cong 0.81,$ $\arcsin (-0.556) \cong -0.59.$

(5.6.2)

The function
$$g(t) = \arcsin t$$
is called the *inverse sine function*.

The inverse sine function is *not* the inverse of the entire sine function (the latter, not being one-to-one, has no inverse), but rather it is the inverse of the function

$$f(t) = \sin t, \qquad t \in [-\tfrac{1}{2}\pi, \tfrac{1}{2}\pi].$$

In other words, the inverse sine function is the inverse of the sine function restricted to the interval $[-\tfrac{1}{2}\pi, \tfrac{1}{2}\pi]$.

The graphs of f and g are pictured in Figure 5.6.1. The graph of g is the graph of f reflected in the line $y = t$.

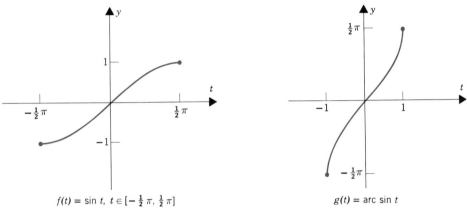

$$f(t) = \sin t,\ t \in [-\tfrac{1}{2}\pi, \tfrac{1}{2}\pi] \qquad\qquad g(t) = \text{arc sin } t$$

Figure 5.6.1

The Arc Tangent

This time we begin with an arbitrary real number b. (The range of the tangent function is the set of all real numbers.) Although there are infinitely many numbers at which the tangent function takes on the value b, only one of these numbers lies in the open interval $(-\tfrac{1}{2}\pi, \tfrac{1}{2}\pi)$. This unique number is written *arc tan b* and is called the *arc tangent of b*. In symbols,

(5.6.3) $\boxed{a = \text{arc tan } b \quad \text{iff} \quad \tan a = b \ \text{ and }\ a \in (-\tfrac{1}{2}\pi, \tfrac{1}{2}\pi).}$

Thus, for example,

$\text{arc tan } 1 = \tfrac{1}{4}\pi,$ $\qquad\qquad$ $[\tan \tfrac{1}{4}\pi = 1 \ \text{ and }\ \tfrac{1}{4}\pi \in (-\tfrac{1}{2}\pi, \tfrac{1}{2}\pi)]$

$\text{arc tan } (-\sqrt{3}) = -\tfrac{1}{3}\pi,$ $\qquad$ $[\tan(-\tfrac{1}{3}\pi) = -\sqrt{3} \ \text{ and }\ -\tfrac{1}{3}\pi \in (-\tfrac{1}{2}\pi, \tfrac{1}{2}\pi)]$

$\text{arc tan } 0 = 0.$ $\qquad\qquad$ $[\tan 0 = 0 \ \text{ and }\ 0 \in (-\tfrac{1}{2}\pi, \tfrac{1}{2}\pi)]$

Using Table 1 at the back of the book you can see that

$$\text{arc tan } 0.447 \cong 0.42 \qquad \text{and} \qquad \text{arc tan } (-5.798) \cong -1.40.$$

<table>
<tr><td>(5.6.4)</td><td>The function

$$g(t) = \text{arc tan } t$$

is called the *inverse tangent function.*</td></tr>
</table>

The inverse tangent function is *not* the inverse of the entire tangent function (the latter, not being one-to-one, has no inverse), but rather it is the inverse of the function

$$f(t) = \tan t, \qquad t \in (-\tfrac{1}{2}\pi, \tfrac{1}{2}\pi).$$

In other words, the inverse tangent function is the inverse of the tangent function restricted to the interval $(-\tfrac{1}{2}\pi, \tfrac{1}{2}\pi)$.

The graphs of f and g are pictured in Figure 5.6.2. The graph of g is the graph of f reflected in the line $y = t$.

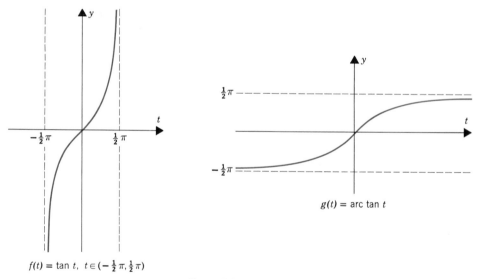

$$f(t) = \tan t, \ t \in (-\tfrac{1}{2}\pi, \tfrac{1}{2}\pi)$$

$$g(t) = \text{arc tan } t$$

Figure 5.6.2

While it is true that

<table>
<tr><td>(5.6.5)</td><td>sin (arc sin t) = t</td><td>for all t in the domain of the arc sine,</td></tr>
</table>

it is *not* true that

arc sin (sin t) = t for all t in the domain of the sine.

The relation

(5.6.6) $\boxed{\text{arc sin } (\sin t) = t \qquad \text{holds only for } t \in [-\tfrac{1}{2}\pi, \tfrac{1}{2}\pi].}$

For example,

$$\text{arc sin } (\sin \tfrac{1}{4}\pi) = \tfrac{1}{4}\pi \qquad \text{but} \qquad \text{arc sin } (\sin 2\pi) = \text{arc sin } (\sin 0) = 0.$$

Similarly, the relation

(5.6.7) $\boxed{\text{tan } (\text{arc tan } t) = t \qquad \text{is valid for all real } t,}$

but the relation

(5.6.8) $\boxed{\text{arc tan } (\tan t) = t \qquad \text{holds only for } t \in (-\tfrac{1}{2}\pi, \tfrac{1}{2}\pi).}$

Examples

1. $\sin (\text{arc sin } \tfrac{1}{2}) = \tfrac{1}{2}.$
              (5.6.5)

2. $\text{arc sin } (\sin \tfrac{3}{2}\pi) = \text{arc sin } [\sin (-\tfrac{1}{2}\pi)] = -\tfrac{1}{2}\pi.$
            $\sin t = \sin (t - 2\pi)$ (5.6.6)

3. $\tan [\text{arc tan } (-1)] = -1.$
             (5.6.7)

4. $\text{arc tan } (\tan \tfrac{5}{4}\pi) = \text{arc tan } (\tan \tfrac{1}{4}\pi) = \tfrac{1}{4}\pi.$ $\quad\square$
        $\tan t = \tan (t - \pi)$ (5.6.8)

Problem. Show that

(a) $\tan (\text{arc sin } t) = \dfrac{t}{\sqrt{1 - t^2}} \qquad$ for all $t \in (-1, 1).$

(b) $\sin (\text{arc tan } t) = \dfrac{t}{\sqrt{1 + t^2}} \qquad$ for all real $t.$

SOLUTION

(a) $\tan^2 (\text{arc sin } t) + 1 = \sec^2 (\text{arc sin } t)$
     $\tan^2 u + 1 = \sec^2 u$

$$= \frac{1}{\cos^2 (\text{arc sin } t)} = \frac{1}{1 - \sin^2 (\text{arc sin } t)} = \frac{1}{1 - t^2}.$$

This gives

$$\tan^2 (\arc \sin t) = \frac{1}{1 - t^2} - 1 = \frac{t^2}{1 - t^2}$$

and thus

(∗)
$$\tan (\arc \sin t) = \frac{\pm t}{\sqrt{1 - t^2}}.$$

Note that

if $\;0 \leq t < 1,\;$ then $\;0 \leq \arc \sin t < \frac{1}{2}\pi\;$ and $\;\tan (\arc \sin t) \geq 0;$

if $\;-1 < t < 0,\;$ then $\;-\frac{1}{2}\pi < \arc \sin t < 0\;$ and $\;\tan (\arc \sin t) < 0.$

We have shown that $\tan (\arc \sin t)$ and t have the same sign. We therefore reject the minus sign in (∗) and write

$$\tan (\arc \sin t) = \frac{t}{\sqrt{1 - t^2}}. \quad \square$$

(b) Left to you as an exercise. $\quad \square$

By part (a) of the last problem

$$\tan (\arc \sin \tfrac{1}{2}) = \frac{\frac{1}{2}}{\sqrt{1 - \frac{1}{4}}} = \frac{\frac{1}{2}}{\frac{1}{2}\sqrt{3}} = \tfrac{1}{3}\sqrt{3}.$$

By part (b) of the last problem

$$\sin (\arc \tan \tfrac{5}{12}) = \frac{\frac{5}{12}}{\sqrt{1 + \frac{25}{144}}} = \frac{\frac{5}{12}}{\frac{13}{12}} = \tfrac{5}{13}.$$

We can also carry out these computations directly:

$$\tan (\arc \sin \tfrac{1}{2}) = \tan \tfrac{1}{6}\pi = \tfrac{1}{3}\sqrt{3},$$

$$\sin (\arc \tan \tfrac{5}{12}) \cong \sin (\arc \tan 0.4167) \cong \sin 0.39 \cong 0.380 \cong \tfrac{5}{13}.$$
$$\underset{\text{Table 1}}{\smile}$$

The Other Trigonometric Inverses

There are four other trigonometric inverses: the arc cosine, the arc secant, the arc cosecant, and the arc cotangent. Of these only the arc cosine is much used. By definition

(5.6.9)
$$\boxed{a = \arc \cos b \quad \text{iff} \quad \cos a = b \;\;\text{and}\;\; a \in [0, \pi].}$$

The function
$$g(t) = \text{arc cos } t,$$
called the *inverse cosine function,* is the inverse of the function
$$f(t) = \cos t, \qquad t \in [0, \pi].$$

A word on notation: some textbook writers use $\sin^{-1} t$ for arc sin t, $\tan^{-1} t$ for arc tan t, $\cos^{-1} t$ for arc cos t, etc. We won't use this notation.

Exercises

Calculate the exact value:

*1.　arc sin $\frac{1}{2}\sqrt{3}$.

2.　arc sin $(-\frac{1}{2})$.

*3.　arc tan $(-\frac{1}{3}\sqrt{3})$.

4.　arc tan $\sqrt{3}$.

*5.　arc sin $\frac{1}{2}\sqrt{2}$.

6.　arc tan $\frac{1}{3}\sqrt{3}$.

*7.　arc cos 0.

8.　arc cos $\frac{1}{2}\sqrt{3}$.

*9.　arc cos $(-\frac{1}{2})$.

10.　arc cos $(-\frac{1}{2}\sqrt{3})$.

*11.　sin (arc sin $\frac{1}{2}$).

12.　cos (arc sin $\frac{1}{2}$).

*13.　sin (arc cos $\frac{1}{2}$).

14.　cos (arc tan 0).

*15.　arc tan (cos 0).

16.　arc sin (sin $\frac{7}{4}\pi$).

*17.　arc tan (tan $\frac{7}{4}\pi$).

18.　arc tan (tan $\frac{5}{6}\pi$).

*19.　arc sin [sin $(-\frac{7}{4}\pi)$].

20.　sin [arc sin (-1)].

Find the approximate value by using Table 1 at the back of the book.

*21.　arc sin 0.918.

22.　arc sin (-0.795).

*23.　arc tan (-0.493).

24.　arc tan 3.111.

*25.　arc cos (0.960).

26.　arc cos (-0.142).

*27.　Sketch the graph of the inverse cosine function.

28.　State whether the function is odd, even, or neither.
　　(a) The inverse sine function.
　　*(b) The inverse tangent function.
　　(c) The inverse cosine function.

29.　The sum arc cos t + arc sin t has a constant value. What is it?

Optional | *30. Calculate (a) sin (arc cos t) and (b) cos (arc sin t).

31. Show that

$$\sin (\text{arc tan } t) = \frac{t}{\sqrt{1 + t^2}} \quad \text{for all real } t.$$

HINT: Find a way to use the relation

$$\tan^2 u + 1 = \sec^2 u.$$

*32. By definition

$$a = \text{arc cot } b \quad \text{iff} \quad \cot a = b \quad \text{and} \quad a \in (0, \pi).$$

Sketch the graph of

$$g(t) = \text{arc cot } t.$$

HINT: This function is the inverse of

$$f(t) = \cot t, \quad t \in (0, \pi).$$

*33. By definition

$$a = \text{arc sec } b \quad \text{iff} \quad \sec a = b \quad \text{and} \quad a \in [0, \tfrac{1}{2}\pi) \cup (\tfrac{1}{2}\pi, \pi].$$

Sketch the graph of

$$g(t) = \text{arc sec } t.$$

HINT: This function is the inverse of

$$f(t) = \sec t, \quad t \in [0, \tfrac{1}{2}\pi) \cup (\tfrac{1}{2}\pi, \pi].$$

*34. By definition

$$a = \text{arc cosec } b \quad \text{iff} \quad \text{cosec } a = b \quad \text{and} \quad a \in [-\tfrac{1}{2}\pi, 0) \cup (0, \tfrac{1}{2}\pi].$$

Sketch the graph of

$$g(t) = \text{arc cosec } t.$$

HINT: This function is the inverse of

$$f(t) = \text{cosec } t, \quad t \in [-\tfrac{1}{2}\pi, 0) \cup (0, \tfrac{1}{2}\pi].$$

5.7 Right-Triangle Trigonometry

Here we apply sines, cosines, and tangents to right triangles. We will follow custom and measure angles in degrees. Remember the relation

$$360° = 2\pi \text{ radians}.$$

From this relation,

$$\sin 0° = \sin 0 = 0 = 0.0000,$$

$$\sin 30° = \sin \tfrac{1}{6}\pi = \tfrac{1}{2} = 0.5000,$$

$$\sin 45° = \sin \tfrac{1}{4}\pi = \tfrac{1}{2}\sqrt{2} \cong 0.7071,$$

$$\sin 60° = \sin \tfrac{1}{3}\pi = \tfrac{1}{2}\sqrt{3} \cong 0.8660,$$

and so on. Table 2 at the back of the book gives the sine, cosine, and tangent of angles from 0° to 90° in decimal form with enough accuracy for our present purposes.

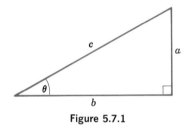

Figure 5.7.1

Let's begin. In Figure 5.7.1 you can see a right triangle with one of the acute angles marked θ, the hypotenuse marked c, and the other sides marked a and b. With respect to θ, a is called the *opposite side* and b the *adjacent side*. In our present notation,

$$0 < \theta < 90°.$$

What we want to do here is express $\sin \theta$, $\cos \theta$, and $\tan \theta$ in terms of a, b, c. To do this, we draw a set of coordinate axes as in Figure 5.7.2, and sketch in the unit circle.

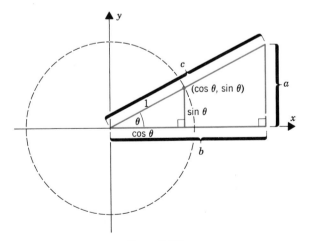

Figure 5.7.2

Since the two triangles marked in the figure are similar (they have equal angles), the corresponding sides are proportional:

$$\frac{\sin \theta}{a} = \frac{\cos \theta}{b} = \frac{1}{c}.$$

These relations give

(5.7.1)

$$
\begin{aligned}
\sin \theta &= \frac{a}{c} = \frac{\text{opposite side}}{\text{hypotenuse}}, \\[2mm]
\cos \theta &= \frac{b}{c} = \frac{\text{adjacent side}}{\text{hypotenuse}}, \\[2mm]
\tan \theta &= \frac{a}{b} = \frac{\text{opposite side}}{\text{adjacent side}}.
\end{aligned}
$$

By means of these formulas and the trigonometric tables, we can determine all the dimensions of a right triangle if we know one of the acute angles and one of the sides, or if we know two of the sides. Take as examples the triangles of Figure 5.7.3.

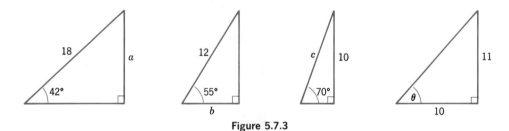

Figure 5.7.3

We can find a in the first triangle by using the sine function and Table 2 at the end of the book:

$$\sin 42° = \frac{\text{opposite side}}{\text{hypotenuse}} = \frac{a}{18},$$

$$a = 18 \sin 42° \cong 18(0.6691) \cong 12.04.$$

We have used $\cong$ to indicate approximate equality. The sine of 42° is not exactly 0.6691, but it is that rounded off to four decimal places.

The sides marked b and c and the angle marked θ can be found with similar ease:

$$\cos 55° = \frac{\text{adjacent side}}{\text{hypotenuse}} = \frac{b}{12},$$

$$b = 12 \cos 55° \cong 12(0.5736) \cong 6.88;$$

$$\sin 70° = \frac{\text{opposite side}}{\text{hypotenuse}} = \frac{10}{c},$$

$$c = \frac{10}{\sin 70°} \cong \frac{10}{0.9397} \cong 10.64;$$

$$\tan \theta = \frac{\text{opposite side}}{\text{adjacent side}} = \frac{11}{10} = 1.1,$$

$$\theta \cong 48°. \quad \square$$

Problem. If the shadow cast by a tree is 100 feet long when the angle of elevation of the sun is 20°, how tall is the tree?

SOLUTION. We begin by drawing a picture, Figure 5.7.4. With respect to the angle marked 20°, h is the opposite side and 100 is the adjacent side. The formula

$$\tan \theta = \frac{\text{opposite side}}{\text{adjacent side}}$$

gives

$$\tan 20° = \frac{h}{100},$$

$$0.3640 \cong \frac{h}{100},$$

$$h \cong 36.4.$$

The tree is approximately 36.4 feet tall. $\square$

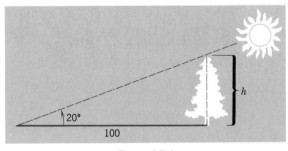

Figure 5.7.4

Exercises

The first eight exercises deal with right triangles labeled as in Figure 5.7.5.

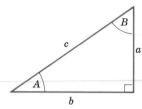

Figure 5.7.5

*1. Find a and b given that $A = 50°$ and $c = 2$.

2. Find b and c given that $B = 34°$ and $a = 2.5$.

*3. Find b and c given that $A = 60°$ and $a = 100$.

4. Find a and c given that $B = 10°$ and $b = 2.6$.

*5. Find A and b given that $a = 1.44$ and $c = 12$.

6. Find a and B given that $b = 7$ and $c = 10$.

*7. Find b and c given that $A = 19°$ and $a = 4$.

8. Find A and c given that $a = 3.2$ and $b = 6.5$.

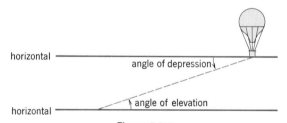

Figure 5.7.6

*9. An observer in a balloon 1000 feet high peers down at a fixed point below. If the angle of depression is $25°$ (see Figure 5.7.6), how far must he travel before he passes directly over that point?

10. From a point level with the base of a building 50 feet away, the angle of elevation of the top of the building is $33°$. How tall is the building?

*11.   A ship sails 2 miles 10° east of north. How far north has it gone? How far east?

12.   Lighthouse A is 3.41 miles east of lighthouse B. If a ship observes A due south and B at 43° west of south, how far is the ship from B?

*13.   A 7-foot tapestry is hung on a wall with its lower edge even with an observer's eye. As the observer stares at the lower edge of the tapestry, he must raise his gaze 20° to see the upper edge. How far from the wall is he standing?

14.   The angle between the mast of a sailboat and a stay that supports it is 15°. How tall is the mast if the stay is 20 feet long and reaches from the deck to a point two-thirds of the height of the mast?

*15.   A 10-foot ladder rests flat on the ground perpendicular to a wall, but 1 foot away from it. If the end closer to the wall is raised 65°, how far must the ladder be moved for it to reach the wall? How high will the ladder be at that point?

16.   Find the angles between the two diagonals of a rectangle if the sides measure 40 feet and 120 feet.

5.8   The Law of Sines and The Law of Cosines

In the last section we dealt only with right triangles. Here we take up two trigonometric relations that hold for all triangles: the *law of sines* and the *law of cosines*. Throughout we'll be making use of the following identities:

(5.8.1) $\boxed{\sin(180° - \theta) = \sin\theta \quad \text{and} \quad \cos(180° - \theta) = -\cos\theta.}$

The verification of these identities is left to you.

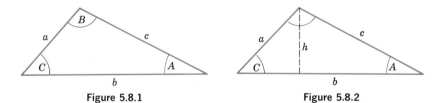

Figure 5.8.1   Figure 5.8.2

Theorem   5.8.2   The Law of Sines

In a triangle with sides a, b, c and opposite angles A, B, C (see Figure 5.8.1)

$$\frac{\sin A}{a} = \frac{\sin B}{b} = \frac{\sin C}{c}.$$

PROOF.   We begin by dropping a perpendicular as in Figure 5.8.2. From the figure,

$$\sin A = \frac{h}{c}, \qquad \sin C = \frac{h}{a},$$

so that

$$c \sin A = h = a \sin C$$

and therefore

(∗) $$\frac{\sin A}{a} = \frac{\sin C}{c}.$$

We are halfway there.

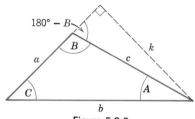

Figure 5.8.3

Now we draw a perpendicular as in Figure 5.8.3. Here

$$\sin B = \sin (180° - B) = \frac{k}{c}, \qquad \sin C = \frac{k}{b},$$

so that

$$c \sin B = k = b \sin C$$

and

(∗∗) $$\frac{\sin B}{b} = \frac{\sin C}{c}.$$

Combining (∗∗) with (∗), we have

$$\frac{\sin A}{a} = \frac{\sin B}{b} = \frac{\sin C}{c}. \quad \square$$

The law of sines enables us to find all the dimensions of a triangle if we know two of the angles and the included side.†

Problem.   Find the sides a, b and the angle C in the triangle of Figure 5.8.4.

---

†The so-called "ambiguous case," where we are given two sides and an opposite angle, is discussed in the supplement to this section.

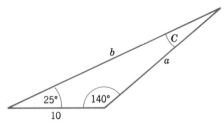

Figure 5.8.4

SOLUTION. Since the sum of the angles of a triangle is always 180°,

$$25° + 140° + C = 180°,$$

$$165° + C = 180°,$$

$$C = 15°.$$

By the law of sines,

$$\frac{\sin 25°}{a} = \frac{\sin 15°}{10} = \frac{\sin 140°}{b}.$$

The first equality gives

$$a = \frac{10 \sin 25°}{\sin 15°} \cong \frac{10(0.4226)}{0.2588} \cong 16.33.$$

The second equality gives

$$b = \frac{10 \sin 140°}{\sin 15°} \cong \frac{10 \sin 40°}{\sin 15°} \cong \frac{10(0.6428)}{0.2588} \cong 24.84. \quad \square$$

$$\sin \theta = \sin (180° - \theta)$$

A second important relationship between the angles and sides of a triangle is given by *the law of cosines*.

Theorem 5.8.3   The Law of Cosines

In a triangle with sides a, b, c

$$c^2 = a^2 + b^2 - 2ab \cos C,$$

where C is the angle opposite side c.

PROOF. We begin with the triangle of Figure 5.8.1 and coordinatize the vertices as in Figure 5.8.5.

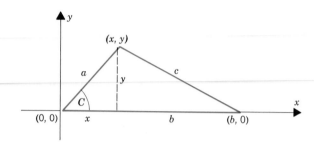

Figure 5.8.5

Observe first of all that

$$\cos C = \frac{x}{a}, \qquad \sin C = \frac{y}{a},$$

and therefore

$$x = a \cos C, \qquad y = a \sin C.$$

Now apply the distance formula:

$$c = \sqrt{(x - b)^2 + (y - 0)^2}$$
$$= \sqrt{(a \cos C - b)^2 + (a \sin C)^2}$$
$$= \sqrt{a^2 \cos^2 C - 2ab \cos C + b^2 + a^2 \sin^2 C}$$
$$= \sqrt{a^2(\cos^2 C + \sin^2 C) + b^2 - 2ab \cos C}$$
$$= \sqrt{a^2 + b^2 - 2ab \cos C}.$$

$$\underset{\cos^2 C + \sin^2 C = 1.}{}$$

By squaring,

$$c^2 = a^2 + b^2 - 2ab \cos C. \quad \square$$

The law of cosines is a generalization of the Pythagorean theorem: if C is a right angle, then $\cos C = 0$, and the equation

$$c^2 = a^2 + b^2 - 2ab \cos C$$

reduces to the familiar

$$c^2 = a^2 + b^2. \quad \square$$

Using the law of cosines, we can find the dimensions of any triangle if we know the three sides or if we know two of the sides and the included angle.

Problem. Find the angles A, B, C of a triangle given that $a = 4$, $b = 5$, $c = 6$. (Figure 5.8.6)

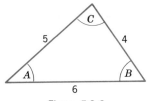

Figure 5.8.6

SOLUTION. Since we know the lengths of the three sides, we can use the law of cosines and Table 2 to find the angles:

$$16 = 25 + 36 - 2(5)(6) \cos A,$$

$$-45 = -60 \cos A,$$

$$\cos A = 0.7500,$$

$$A \cong 41°.$$

$$25 = 16 + 36 - 2(4)(6) \cos B,$$

$$-27 = -48 \cos B,$$

$$\cos B = 0.5625,$$

$$B \cong 56°.$$

We could use the law of cosines to find the third angle as well, but it is easier to use the fact that the sum of the angles of a triangle is 180°:

$$C \cong 180° - 41° - 56° = 83°. \quad \square$$

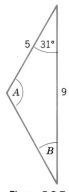

Figure 5.8.7

Problem. Find the other dimensions of the triangle given that $a = 9$, $b = 5$, and $C = 31°$. (Figure 5.8.7)

SOLUTION. Since we know two sides and the included angle, we can find the third side by the law of cosines:

$$c^2 = 81 + 25 - 2(9)(5) \cos 31°,$$

$$c^2 \cong 106 - 90(0.8572) = 28.852,$$

$$c \cong \sqrt{28.852} \cong 5.37.†$$

Now we can find either A or B, again by the law of cosines. To find B, we write the law of cosines as

$$b^2 = a^2 + c^2 - 2ac \cos B.$$

This gives

$$25 \cong 81 + 28.852 - 2(9)(5.37) \cos B,$$

$$-84.852 \cong -(96.66) \cos B,$$

$$\cos B \cong \frac{84.852}{96.66} \cong 0.8778,$$

$$B \cong 29°.$$

Finally,

$$A \cong 180° - 31° - 29° = 120°. \quad \square$$

Exercises

Find the other dimensions of each triangle. Take $a, b, c,$ as the sides and A, B, C as the opposite angles.

*1. $A = 40°, B = 80°, c = 6.$ 2. $B = 10°, C = 20°, a = 100.$

*3. $a = 7, b = 8, c = 9.$ 4. $a = 6, b = 11, c = 15.$

*5. $A = 16°, C = 56°, b = 24.$ 6. $a = 1.1, b = 2, c = 3.$

*7. $a = 10, b = 10, C = 40°.$ 8. $A = 60°, C = 30°, b = 4.$

*9. $a = 4, c = 2, B = 12°.$ 10. $B = 50°, C = 30°, a = 10.$

*11. $a = 1.2, b = 1.3, c = 2.$ 12. $b = 6.2, c = 7, A = 60°.$

*13. $A = 79°, B = 61°, c = 5.$ 14. $a = 20, b = 10, c = 15.$

*15. Show that

$$\sin(180° - \theta) = \sin \theta \quad \text{and} \quad \cos(180° - \theta) = -\cos \theta.$$

---

†There is a table of square roots at the back of the book.

*16. A ship receives a distress signal from a sailboat 100 miles due west. If the ship immediately heads for the point from which the signal was sent at a rate of 30 mph, while the sailboat drifts southeast at 5 mph, how far apart will they be after one hour?

17. Two lookouts, lookout A being 10 miles west of lookout B, sight a ship offshore. If A determines the ship's position to be 20° north of east, and B determines it to be exactly northeast, how far is the ship from each of the lookouts?

*18. A triangle has sides with lengths 4, 5, 7. What is the length of the perpendicular drawn to the longest side?

19. Two steamships start from the same place at the same time. One sails 12 miles 20° east of north and the other sails 8 miles due east. What is the distance between them?

20. (a) Show that for every triangle

$$ab \sin C = bc \sin A = ac \sin B.$$

(b) Verify the formula

$$\text{area} = \tfrac{1}{2}ab \sin C.$$

*(c) Find the area of a 2,3,4 triangle without referring to Trigonometric Tables.

(d) Find the area of a 3,8,10 triangle without referring to Trigonometric Tables.

Optional | **Supplement to Section 5.8.   The Law of Sines: The Ambiguous Case**

As you have seen, two angles and the included side uniquely determine a triangle. But if you are given *two sides and an opposite angle* and asked to find the remaining dimensions, there are three possibilities:

1. That no such triangle exists.

2. That two such triangles exist.

3. That only one such triangle exists.

For this reason we refer to this as "the ambiguous case." Below we look at each of the three possibilities. In each case sides a and b are given and also angle A.

1. No triangle exists if

$$a < b \sin A,$$

Optional   because then side a is just not long enough to form a triangle. (Figure 5.8.8)

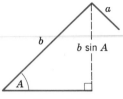

Figure 5.8.8

For example, if $A = 30°$, $a = 10$, and $b = 30$, we have by the law of sines

$$\frac{\sin 30°}{10} = \frac{\sin B}{30},$$

$$\frac{0.5000}{10} = \frac{\sin B}{30},$$

$$\sin B = 1.5.$$

This, of course, is impossible because there is no B such that $\sin B > 1$. ☐

2. Two triangles exist if

$$b \sin A < a < b.$$

This situation is illustrated in Figure 5.8.9.

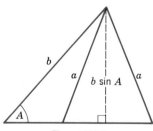

Figure 5.8.9

For example, if $A = 30°$, $a = 10$, and $b = 15$, then

$$\frac{\sin A}{10} = \frac{\sin B}{15},$$

$$\frac{0.5000}{10} = \frac{\sin B}{15},$$

$$\sin B = 0.75,$$

Optional

so that
$$B \cong 49° \quad \text{or} \quad B \cong 180° - 49° = 131°.$$

Taking $B \cong 49°$, we have a triangle with dimensions

$$A = 30°, \quad B \cong 49°, \quad C \cong 180° - (30° + 49°) = 101°,$$

$$a = 10, \quad b = 15, \quad c \cong \underset{\text{by the law of sines}}{\frac{(10)(0.9816)}{0.5}} = 19.632.$$

Taking $B \cong 131°$, we have a different triangle. Here

$$A = 30°, \quad B \cong 131°, \quad C \cong 180° - (30° + 131°) = 29°,$$

$$a = 10, \quad b = 15, \quad c \cong \underset{\text{by the law of sines}}{\frac{(10)(0.4848)}{0.5}} = 9.696. \quad \square$$

Figure 5.8.10

3. Only one triangle exists if
$$b \le a.$$

See Figure 5.8.10. For example, if $A = 30°$, $a = 10$, and $b = 5$, then

$$\frac{\sin 30°}{10} = \frac{\sin B}{5},$$

$$\frac{0.5000}{10} = \frac{\sin B}{5},$$

$$\sin B = 0.25,$$

so that
$$B \cong 14° \quad \text{or} \quad B \cong 166°.$$

But we know that $B \ne 166°$ because there are only 180 degrees in a triangle. Thus, the triangle is uniquely determined by the dimensions given:

$$A = 30°, \quad B \cong 14°, \quad C \cong 180° - (30° + 14°) = 148°,$$

$$a = 10, \quad b = 5, \quad c \cong \underset{\text{by the law of sines}}{\frac{(10)(0.5592)}{0.5}} = 11.184. \quad \square$$

5.9   The Inclination of a Line; The Angle Between Two Lines

We continue to measure angles in degrees. The *inclination* of a line *l* (Figure 5.9.1) is the angle that *l* makes with the *x*-axis as measured counterclockwise from the *x*-axis. The *inclination* of a horizontal line is by definition 0°. (Figure 5.9.2)

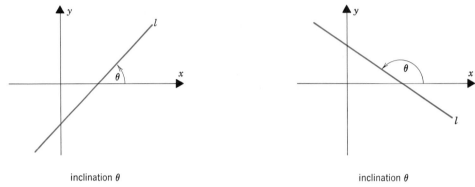

inclination *θ*   inclination *θ*

Figure 5.9.1

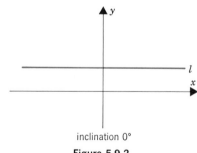

inclination 0°

Figure 5.9.2

There is a simple relation between the slope of a line and its inclination:

(5.9.1)

> if *m* is the slope of *l* and *θ* is the inclination, then
>
> $$m = \tan \theta.$$

PROOF.   If the line is horizontal, then the inclination and the slope are both zero, and the relation holds:

$$0 = \tan 0°.$$

If the line is vertical, then *m* does not exist, but, since *θ* is 90°, $\tan \theta$ does not exist either.

For the rest we refer to Figure 5.9.3:

if $0° < \theta < 90°$, then

$$\tan \theta = \frac{\text{opposite side}}{\text{adjacent side}} = \frac{y_1}{x_1 - c} = \frac{y_1 - 0}{x_1 - c} = \text{slope of } l;$$

if $90° < \theta < 180°$, then

$$\tan \theta = -\tan(180° - \theta)$$

$$= -\frac{\text{opposite side}}{\text{adjacent side}} = -\frac{y_1}{c - x_1} = \frac{y_1 - 0}{x_1 - c} = \text{slope of } l.$$

This takes care of all the possibilities. ☐

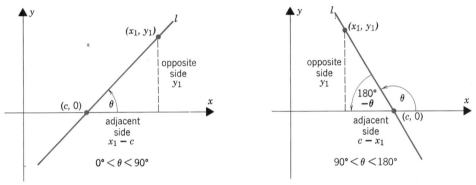

Figure 5.9.3

Problem. Find the inclination of the line

$$3x - 4y + 2 = 0.$$

SOLUTION. We first rewrite the equation in the slope-intercept form:

$$3x - 4y + 2 = 0,$$

$$4y = 3x + 2,$$

$$y = \tfrac{3}{4}x + \tfrac{1}{2},$$

so that $m = \tfrac{3}{4}$. Here

$$\tan \theta = \tfrac{3}{4} = 0.7500 \quad \text{and therefore} \quad \theta \cong 37°. ☐$$

Problem.   Find an equation for the line that intersects the *x*-axis at $x = 2$ with inclination 135°.

SOLUTION.   The equation of a line with *x*-intercept *a* and slope *m* can be written

$$y = m(x - a).$$

Here $a = 2$ and $m = \tan 135° = -\tan 45° = -1$.
   The equation we want is

$$y = -x + 2. \quad \square$$

The Angle Between Two Lines

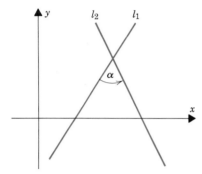

Figure 5.9.4

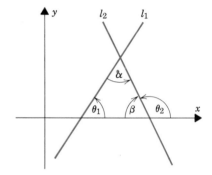

Figure 5.9.5

In Figure 5.9.4 you can see two intersecting lines l_1 and l_2 and the angle between them marked α. To find this angle α, we mark the inclinations θ_1 and θ_2 as in Figure 5.9.5 and also the angle β. Since

$$\alpha + \theta_1 + \beta = 180° \quad \text{and} \quad \theta_2 + \beta = 180°, \qquad \text{(explain)}$$

you can see that

$$\alpha + \theta_1 = \theta_2$$

and therefore

(5.9.2) $\boxed{\alpha = \theta_2 - \theta_1.}$

Problem.   Find the angle α between the lines

$$l: x - 2y + 7 = 0 \quad \text{and} \quad l': 2x - 7y + 31 = 0.$$

SOLUTION. We first rewrite the equations in slope-intercept form.

$$l: x - 2y + 7 = 0, \qquad\qquad l': 2x - 7y + 31 = 0,$$
$$2y = x + 7, \qquad\qquad 7y = 2x + 31,$$
$$y = \tfrac{1}{2}x + \tfrac{7}{2}, \qquad\qquad y = \tfrac{2}{7}x + \tfrac{31}{7},$$
$$\tan \theta = \tfrac{1}{2} = 0.5000, \qquad \tan \theta' = \tfrac{2}{7} \cong 0.2857,$$
$$\theta \cong 27°. \qquad\qquad \theta' \cong 16°.$$

To find α, we subtract the smaller inclination from the larger one:

$$\alpha = \theta - \theta' \cong 27° - 16° = 11°. \quad \square$$

Problem. Find the angle α between the lines

$$l: 2x - 5y + 4 = 0 \quad \text{and} \quad l': 3x + 4y + 7 = 0.$$

SOLUTION. From the first equation,

$$\tan \theta = \tfrac{2}{5} = 0.4000 \quad \text{and therefore} \quad \theta \cong 22°.$$

From the second equation,

$$\tan \theta' = -\tfrac{3}{4} = -0.7500$$

so that

$$\tan (180° - \theta') = 0.7500,$$
$$180° - \theta' \cong 37°,$$
$$\theta' \cong 143°.$$

To find α, we subtract the smaller inclination from the larger one:

$$\alpha = \theta' - \theta \cong 143° - 22° = 121°. \quad \square$$

Exercises

Determine the inclination of the line:

*1. $x - y + 2 = 0.$ 2. $6y + 5 = 0.$

*3. $2x - 3 = 0.$ 4. $2x - 3y + 6 = 0.$

*5. $4x + 3y - 7 = 0.$ 6. $x - 2y + 18 = 0.$

Write an equation $y = mx + b$ for the line:

*7.   Inclination 30°, y-intercept 2.

8.   Inclination 60°, x-intercept 2.

*9.   Inclination 45°, y-intercept -1.

10.   Inclination 120°, y-intercept 4.

*11.   Inclination 150°, x-intercept 3.

12.   Inclination 135°, y-intercept 3.

Find the angle α between l and l':

*13. l: $2x - 3y - 3 = 0$, l': $3x - 2y - 3 = 0$.

14. l: $3x + 10y + 4 = 0$, l': $7x - 10y - 1 = 0$.

*15. l: $4x - y + 2 = 0$, l': $19x + y = 0$.

16. l: $5x - 6y + 1 = 0$, l': $8x + 5y + 2 = 0$.

Find the angles of the triangle with the given vertices:

*17. $(-2, -5)$, $(0, 0)$, $(5, -2)$.   18. $(1, -1)$, $(4, 0)$, $(3, 2)$.

*19. $(-4, -2)$, $(0, -1)$, $(1, 1)$.   20. $(-4, 2)$, $(1, 0)$, $(1, 1)$.

21.   Show that if l_1 and l_2 are as in Figure 5.9.5, then

(5.9.3)
$$\tan \alpha = \frac{m_2 - m_1}{1 + m_2 m_1}.$$

5.10   Additional Exercises

*1.   Convert to radians:
(a) 135°.   (b) 150°.   (c) 270°.   (d) 330°.

*2.   Convert to degrees:
(a) $\frac{1}{6}\pi$ radians.   (b) $\frac{3}{16}\pi$ radians.
(c) 2 radians.   (d) $\frac{5}{8}\pi$ radians.

*3.   What is the length of the arc subtended by a central angle of $\frac{5}{4}\pi$ radians in a circle of area 16π? What is the area of the sector?

*4.   What central angle subtends an arc of length $\sqrt{2}\pi$ in a circle of area 4π? What is the area of the corresponding sector?

*5.　What are the possible values of $\tan t$ given that $\cos t = \frac{1}{4}$?

*6.　Evaluate:

(a) $\cos \frac{5}{4}\pi$.　　　(b) $\sin \frac{5}{4}\pi$.　　　(c) $\cos \frac{13}{6}\pi$.　　　(d) $\tan \frac{7}{6}\pi$.

(e) $\sec \frac{5}{4}\pi$.　　　(f) $\cot \frac{5}{4}\pi$.　　　(g) $\tan \frac{5}{3}\pi$.　　　(h) $\operatorname{cosec} \frac{1}{6}\pi$.

*7.　Determine the exact value:

(a) $\sin (\arctan \sqrt{3})$.　　　(b) $\arccos (-1)$.　　　(c) $\cos (\arcsin \frac{1}{2})$.

(d) $\arcsin (\sin \frac{14}{3}\pi)$.　　　(e) $\sin [\arcsin (-1)]$.　　　(f) $\tan (\arctan \frac{3}{5})$.

(g) $\arctan (\tan \frac{14}{3}\pi)$.　　　(h) $\cos (\arcsin \frac{3}{5})$.　　　(i) $\cot (\arctan \frac{3}{4})$.

*8.　Given that

$$\cos t = \tfrac{2}{5} \quad\text{and}\quad \tfrac{3}{2}\pi < t < 2\pi,$$

calculate the exact value of

(a) $\cos 2t$.　　　(b) $\sin 2t$.　　　(c) $\cos \frac{1}{2}t$.　　　(d) $\sin \frac{1}{2}t$.

*9.　Find the amplitude, the period, and the y-intercept. Then sketch the graph.

(a) $y = \frac{1}{3}\cos 2t$.　　　　　　　　(b) $y = 2 \cos (2t + \pi)$.

(c) $y = \frac{1}{2}\sin (4t + \frac{1}{2}\pi)$.　　　　　(d) $y = \frac{1}{4}\sin (\frac{1}{2}t - \frac{13}{4}\pi)$.

Find the other dimensions of each triangle. Take a, b, c as the sides and A, B, C as the opposite angles:

*10.　$C = 90°, \quad a = 3, \quad c = 5.$

*11.　$B = 80°, \quad C = 90°, \quad c = 4.$

*12.　$C = 90°, \quad b = 10, \quad c = 15.$

*13.　$A = 63°, \quad C = 90°, \quad a = 8.91.$

*14.　$A = 32°, \quad C = 90°, \quad b = 1.$

*15.　$C = 90°, \quad a = 15, \quad b = 25.$

*16.　$B = 80°, \quad C = 70°, \quad a = 5.$

*17.　$C = 60°, \quad a = 2, \quad b = 1.$

*18.　$A = 30°, \quad B = 75°, \quad c = 10.$

*19.　$B = 63°, \quad a = 5, \quad c = 1.$

*20.　$a = 10, \quad b = 11, \quad c = 12.$

*21.　$A = 40°, \quad C = 60°, \quad b = 12.31.$

*22.　$A = 40°, \quad b = 3, \quad c = 2.$

*23.　$a = 2, \quad b = 4, \quad c = 5.$

*24. A ship sails 4 miles 6° south of east. How far south has it gone? How far east?

*25. If a man cast a shadow 10 feet long when the angle of elevation of the sun was 32°, determine his height if he was (a) standing erect, (b) leaning 10° directly away from the sun.

*26. A triangle has sides with lengths 7, 13, 15. What is the altitude drawn to the smallest side?

*27. The angle of elevation of a balloon from a point on a level road is 22°. From a point on the road 1 mile closer, the angle of elevation is 47°. How high is the balloon?

*28. Find the angles of the triangle enclosed by the lines

$$l_1: 4x - 5y + 5 = 0,$$
$$l_2: 5x - y - 20 = 0,$$
$$l_3: x + 4y - 4 = 0.$$

*29. From a ship offshore, the angle of elevation of a mountain peak is 20°. When the ship advances 1 mile toward the mountain, the angle of elevation increases 10°. What is the height of the mountain?

*30. A compass with arms 4 inches long draws a circle of radius 2 inches. How much wider should the angle between the arms be if the compass is to draw a circle of radius 5 inches?

Optional | *31. Prove that

$$\text{arc cos } t + \text{arc sin } t = \tfrac{1}{2}\pi \qquad \text{for all real } t.$$

THE CONIC
SECTIONS

6

6.1 Introduction

If a double right circular cone is cut by a plane, the resulting intersection is called a *conic section* or, more briefly, a *conic*. In Figure 6.1.1 we depict three possibilities.

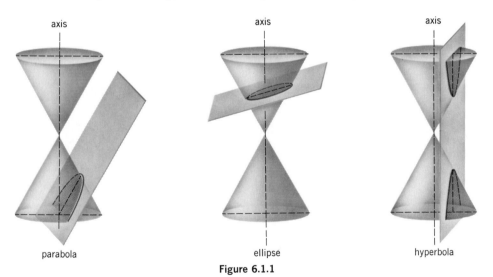

parabola   ellipse   hyperbola

Figure 6.1.1

By choosing a plane perpendicular to the axis of the cone, we can obtain a circle. The other possibilities are: a point, a line, or a pair of lines.

The study of conic sections from this point of view goes back to Apollonius of Perga, a Greek of the third century *B.C.* He wrote eight books on the subject.

For our purposes, it is useful to dispense with cones and three-dimensional geometry and, instead, define parabola, ellipse, and hyperbola entirely in terms of plane geometry.

Exercises

Indicate how a plane can be chosen so as to intersect the double right circular cone in

*1.   a point.   2.   a line.   *3.   a pair of lines.

6.2   The Parabola

We begin with a line *l* and a point *F* not on *l*. (Figure 6.2.1)

(6.2.1) | The set of points *P* equidistant from *F* and *l* is called a *parabola*. |

See Figure 6.2.2.

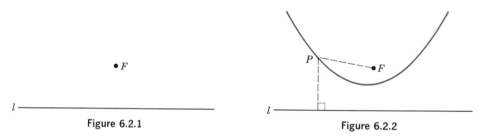

Figure 6.2.1   Figure 6.2.2

The line *l* is called the *directrix* of the parabola, and *F* is called the *focus*. The line through *F* that is perpendicular to *l* is called the *axis* of the parabola. The point at which the axis intersects the parabola is called the *vertex*. (See Figure 6.2.3.)

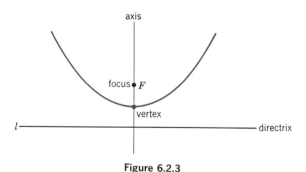

Figure 6.2.3

The equation of a parabola is particularly simple if we place the vertex at the origin and the focus on one of the coordinate axes. Suppose for a moment that the focus F is on the y-axis. Then F has coordinates of the form $(0, c)$. With the vertex at the origin, the directrix must have equation $y = -c$. (See Figure 6.2.4.)

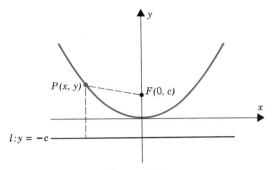

Figure 6.2.4

Every point $P(x, y)$ that lies on this parabola has the property that

$$d(P, F) = d(P, l).$$

Since

$$d(P, F) = \sqrt{x^2 + (y - c)^2} \quad \text{and} \quad d(P, l) = |y + c|,$$

check this out

you can see that

$$\sqrt{x^2 + (y - c)^2} = |y + c|.$$

Squaring both sides, we have

$$x^2 + (y - c)^2 = |y + c|^2 = (y + c)^2,$$
$$x^2 + y^2 - 2cy + c^2 = y^2 + 2cy + c^2.$$
$$x^2 = 4cy. \quad \square$$

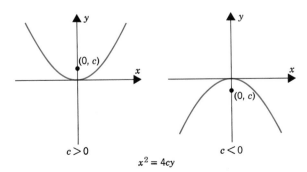

$$x^2 = 4cy$$

Figure 6.2.5

You have just seen that the equation

(6.2.2) $\boxed{x^2 = 4cy}$ (Figure 6.2.5)

represents a parabola with vertex at the origin and focus at $(0, c)$. By interchanging the roles of x and y, you can see that the equation

(6.2.3) $\boxed{y^2 = 4cx}$ (Figure 6.2.6)

represents a parabola with vertex at the origin and focus at $(c, 0)$.   □

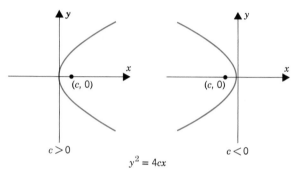

Figure 6.2.6

Problem.   Sketch the parabola specifying the focus, the directrix, and the axis:

$$(a) \ x^2 = -4y. \qquad (b) \ y^2 = 3x.$$

SOLUTION.   (a) The equation $x^2 = -4y$ has the form

$$x^2 = 4cy \qquad \text{with} \qquad c = -1.$$

The vertex is at the origin, and the focus is at $(0, -1)$; the directrix is the horizontal line $y = 1$; the axis of the parabola is the y-axis. The parabola is sketched in Figure 6.2.7.

(b) The equation $y^2 = 3x$ has the form

$$y^2 = 4cx \qquad \text{with} \qquad c = \tfrac{3}{4}.$$

The vertex is at the origin, and the focus is at $(\tfrac{3}{4}, 0)$; the directrix is the vertical line $x = -\tfrac{3}{4}$; the axis of the parabola is the x-axis. The parabola is sketched in Figure 6.2.7.   □

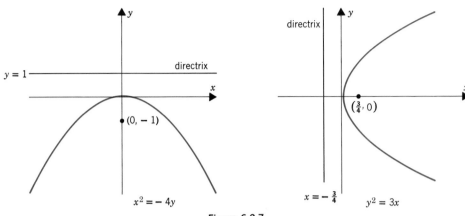

Figure 6.2.7

Problem. Find the points where the line $2x - 3y + 1 = 0$ intersects the parabola $y = x^2$.

SOLUTION. The idea is to find the pairs (x, y) that simultaneously satisfy both equations. One way to do this is to substitute $y = x^2$ into the first equation. Doing this, we get

$$2x - 3x^2 + 1 = 0,$$

which we can rewrite as

$$3x^2 - 2x - 1 = 0.$$

By the general quadratic formula,

$$x = \frac{2 \pm \sqrt{4 + 12}}{6} = \frac{2 \pm 4}{6} = \frac{1 \pm 2}{3},$$

so that x is $-\frac{1}{3}$ or 1. Substituting these values of x into $y = x^2$ (or into $2x - 3y + 1 = 0$), you can see that $x = -\frac{1}{3}$ corresponds to $y = \frac{1}{9}$ and that $x = 1$ corresponds to $y = 1$. The points of intersection are $(-\frac{1}{3}, \frac{1}{9})$ and $(1, 1)$.

Checking:

$$2(-\tfrac{1}{3}) - 3(\tfrac{1}{9}) + 1 = -\tfrac{2}{3} - \tfrac{1}{3} + 1 \overset{\checkmark}{=} 0, \Big\} \quad \text{(both points are on}$$
$$2(1) - 3(1) + 1 = 2 - 3 + 1 \overset{\checkmark}{=} 0, \Big\} \qquad 2x - 3y + 1 = 0)$$
$$\tfrac{1}{9} \overset{\checkmark}{=} (-\tfrac{1}{3})^2, \qquad 1 \overset{\checkmark}{=} 1^2. \qquad \text{(both points are on } y = x^2) \quad \square$$

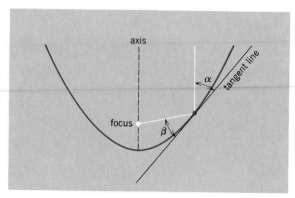

Figure 6.2.8

A Note on Parabolic Reflectors

Parabolic curves play an important role in the design of searchlights and telescopes. With elementary calculus it can be shown that the angles marked α and β in Figure 6.2.8 are equal. From physics we know that when light is reflected, the angle of incidence equals the angle of reflection. We can take either α or β as the angle of incidence; the other will then be the angle of reflection. The fact that $\alpha = \beta$ has important optical consequences. It means that *light from a source at the focus is reflected in a beam parallel to the axis.* It also means that *a beam of incoming light parallel to the axis is reflected by the parabola through the focus.* The parabolic mirror of a searchlight uses the first principle; the parabolic mirror of a reflecting telescope uses the second principle. (See Figure 6.2.9.)

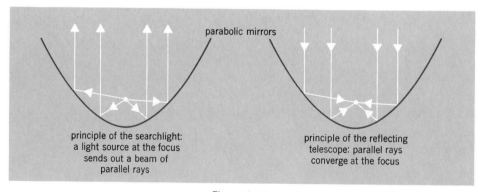

Figure 6.2.9

We'll return to parabolas in the next section.

Exercises

1. Give an equation for the parabola; then sketch the parabola displaying the directrix and the focus:
   *(a) vertex $(0, 0)$, focus $(2, 0)$. (b) vertex $(0, 0)$, focus $(-2, 0)$.
   *(c) vertex $(0, 0)$, focus $(0, 2)$. (d) vertex $(0, 0)$, focus $(0, -2)$.
   *(e) directrix $x = -\frac{1}{2}$, focus $(\frac{1}{2}, 0)$. (f) directrix $y = \frac{1}{2}$, focus $(0, -\frac{1}{2})$.
   *(g) directrix $y = -\frac{1}{2}$, focus $(0, \frac{1}{2})$. (h) directrix $x = \frac{1}{2}$, focus $(-\frac{1}{2}, 0)$.

2. Find the focus, the directrix, and the axis; then sketch the parabola:
   *(a) $y^2 = 2x$. (b) $x^2 = 5y$.
   *(c) $x^2 = -3y$. (d) $y^2 = -x$.

3. Find the points where the line $y = 4x$ intersects the parabola:
   *(a) $y^2 = 4x$. (b) $x^2 = 4y$.

4. Find the points where the parabola $x^2 = 3y$ intersects the circle:
   *(a) $x^2 + y^2 = 4$. (b) $x^2 + y^2 = 1$.

*5. Find the points where the parabolas $x^2 = 5y$ and $y^2 = -2x$ intersect.

6.3 Displacements and Reflections of Curves

If a curve has an equation in x and y, then that equation can be written in the form $f(x, y) = K$ with K a constant:

$$2y = 3x - 5 \qquad \text{can be written} \qquad 3x - 2y = 5,$$

$$y^2 = x^2 + x + 1 \qquad \text{can be written} \qquad x^2 - y^2 + x = -1,$$

$$y^2 = x^3 \qquad \text{can be written} \qquad x^3 - y^2 = 0, \text{ etc.}$$

Displacements

In Section 4.3 you studied displacements of graphs $y = f(x)$. Here we take up displacements of curves $f(x, y) = K$.

Let c be some positive number. Since $(a + c) - c = a$, it's obvious that

$$f(a, b) = K \qquad \text{iff} \qquad f[(a + c) - c, b] = K.$$

This tells us that the curve $f(x, y) = K$ contains the point (a, b) iff the curve $f(x - c, y) = K$ contains the point $(a + c, b)$. In other words, the curve $f(x - c, y) = K$ is the curve $f(x, y) = K$ displaced c units to the right.

In a similar manner you can see that the curve $f(x + c, y) = K$ is the curve $f(x, y) = K$ displaced c units to the left.

These results are important enough to be formally recorded:

> **Lateral Displacements**
> If c is a positive number, then the curve
>
> $$f(x - c, y) = K$$
>
> **(6.3.1)**   is the curve $f(x, y) = K$ displaced c units to the right and the curve
>
> $$f(x + c, y) = K$$
>
> is the curve $f(x, y) = K$ displaced c units to the left.

The addition of $\pm c$ to the y-term produces a vertical displacement:

> **Vertical Displacements**
> If c is a positive number, then the curve
>
> $$f(x, y - c) = K$$
>
> **(6.3.2)**   is the curve $f(x, y) = K$ raised c units and the curve
>
> $$f(x, y + c) = K$$
>
> is the curve $f(x, y) = K$ lowered c units.

Examples

1. We begin with the unit circle

$$x^2 + y^2 = 1.$$

(a) The equation $(x - 2)^2 + y^2 = 1$ represents the unit circle displaced 2 units to the right; the center is now at $(2, 0)$.

(b) The equation $(x + 2)^2 + y^2 = 1$ represents the unit circle displaced 2 units to the left; the center is at $(-2, 0)$.

(c) The equation $x^2 + (y - 2)^2 = 1$ represents the unit circle raised 2 units; the center is at $(0, 2)$.

(d) The equation $x^2 + (y + 2)^2 = 1$ represents the unit circle lowered 2 units; the center is at $(0, -2)$.

(e) The equation $(x - 2)^2 + (y + 3)^2 = 1$ represents the unit circle displaced 2 units to the right and lowered 3 units; the center is at $(2, -3)$.

(f) The equation $(x + 2)^2 + (y - 3)^2 = 1$ represents the unit circle displaced 2 units to the left and raised 3 units; the center is at $(-2, 3)$.

2. The line through the origin with slope m has equation

$$y = mx. \qquad\qquad\qquad [y - mx = 0]$$

This same line displaced 3 units to the left and raised 4 units has equation

$$y - 4 = m(x + 3). \qquad [y - 4 - m(x + 3) = 0]$$

3. In Figure 6.2.7 we sketched the parabola

$$x^2 = -4y. \qquad\qquad\qquad [x^2 + 4y = 0]$$

The vertex is at the origin and the focus is at $(0, -1)$. This same parabola displaced 5 units to the right and lowered 3 units has equation

$$(x - 5)^2 = -4(y + 3). \qquad [(x - 5)^2 + 4(y + 3) = 0]$$

The vertex is now at $(5, -3)$, and the focus is at $(5, -4)$. ☐

Problem. Identify the curve

$$(x - 4)^2 = 8(y + 3).$$

SOLUTION. The curve is a parabola. It is the parabola

$$x^2 = 8y \qquad\qquad\qquad (c = 2)$$

displaced 4 units to the right and lowered 3 units. The parabola $x^2 = 8y$ has vertex at the origin; the focus is at $(0, 2)$, and the directrix is the line $y = 2$. The new parabola has vertex at $(4, -3)$; the focus is at $(4, -1)$, and the directrix is the line $y = -5$. See Figure 6.3.1. ☐

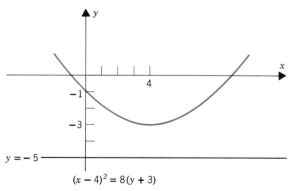

$$(x - 4)^2 = 8(y + 3)$$

Figure 6.3.1

Problem. Identify the curve

$$y = x^2 + 2x - 1.$$

SOLUTION. First complete the square on the right by adding 2 to both sides of the equation. This gives

$$y + 2 = x^2 + 2x + 1 = (x + 1)^2,$$

which you can rewrite as

$$(x + 1)^2 = y + 2.$$

This is the parabola

$$x^2 = y \qquad\qquad (c = \tfrac{1}{4})$$

displaced 1 unit to the left and lowered 2 units. The parabola $x^2 = y$ has vertex at the origin; the focus is at $(0, \tfrac{1}{4})$, and the directrix is the line $y = -\tfrac{1}{4}$. The new parabola has vertex at $(-1, -2)$; the focus is at $(-1, -\tfrac{7}{4})$, and the directrix is the line $y = -\tfrac{9}{4}$. ☐

By the method of the last problem you can show that every quadratic

$$y = Ax^2 + Bx + C$$

represents a parabola with vertical axis. It looks like ∪ if A is positive, and like ∩ if A is negative.

Reflections

First a brief review, all of which is illustrated in Figure 6.3.2.

1. The point $(a, -b)$ is the point (a, b) reflected in the x-axis.

2. The point $(-a, b)$ is the point (a, b) reflected in the y-axis.

3. The point $(-a, -b)$ is the point (a, b) reflected in the origin.

4. The point (b, a) is the point (a, b) reflected in the line $y = x$.

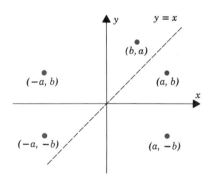

Figure 6.3.2

These observations have important implications for curves.

(6.3.3)

> **Reflections of Curves**
> 1. The curve $f(x, -y) = K$ is the curve $f(x, y) = K$ reflected in the x-axis.
>
> 2. The curve $f(-x, y) = K$ is the curve $f(x, y) = K$ reflected in the y-axis.
>
> 3. The curve $f(-x, -y) = K$ is the curve $f(x, y) = K$ reflected in the origin.
>
> 4. The curve $f(y, x) = K$ is the curve $f(x, y) = K$ reflected in in the line $y = x$.

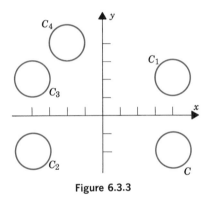

Figure 6.3.3

Examples

1. Figure 6.3.3 shows the circle

$$C: (x - 4)^2 + (y + 2)^2 = 1$$

and four other circles C_1, C_2, C_3, C_4.

(a) C_1 is C reflected in the x-axis. The equation for C_1 can be written

$$(x - 4)^2 + (-y + 2)^2 = 1, \quad \text{or more simply,} \quad (x - 4)^2 + (y - 2)^2 = 1.$$

(b) C_2 is C reflected in the y-axis. The equation for C_2 can be written

$$(-x - 4)^2 + (y + 2)^2 = 1, \quad \text{or more simply,} \quad (x + 4)^2 + (y + 2)^2 = 1.$$

(c) C_3 is C reflected in the origin. The equation for C_3 can be written

$$(-x - 4)^2 + (-y + 2)^2 = 1, \quad \text{or more simply,} \quad (x + 4)^2 + (y - 2)^2 = 1.$$

(d) C_4 is C reflected in the line $y = x$. The equation for C_4 can be written

$$(y - 4)^2 + (x + 2)^2 = 1, \quad \text{or} \quad (x + 2)^2 + (y - 4)^2 = 1.$$

2. Figure 6.3.4 shows the parabola

$$x^2 = 4y.$$

The parabola $y^2 = 4x$ is the parabola $x^2 = 4y$ reflected in the line $y = x$.   □

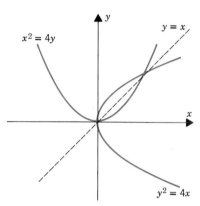

Figure 6.3.4

Exercises

1. Write an equation for the circle

$$x^2 + y^2 = 4$$

*(a) displaced 3 units to the right.
 (b) displaced 3 units to the right and lowered 5 units.
 (c) displaced 3 units to the left.
*(d) displaced 3 units to the left and raised 5 units.

2. Write an equation for the circle

$$(x - 5)^2 + (y - 10)^2 = 4$$

 (a) reflected in the x-axis.
*(b) reflected in the y-axis.
 (c) reflected in the origin.
*(d) reflected in the line $y = x$.

3. Write an equation for the parabola

$$(y + 1)^2 = -6(x + 2)$$

*(a) displaced 2 units left and raised 7 units.
 (b) displaced 7 units right and lowered 2 units.
*(c) reflected in the origin.
 (d) reflected in the line $y = x$.

4. Find the vertex, the focus, the axis, and the directrix; then sketch the parabola:
    *(a) $y^2 = 2(x - 1)$.   (b) $2y = 4x^2 - 1$.
    *(c) $(x + 2)^2 = 8y - 12$.   (d) $y - 3 = 2(x - 1)^2$.
    *(e) $y = x^2 + x + 1$.   (f) $x = y^2 + y + 1$.

5. Sketch the parabola and give an equation for it:
    (a) vertex $(1, 2)$,   focus $(1, 3)$.
    *(b) focus $(1, 1)$,   directrix $y = -1$.
    (c) focus $(1, 1)$,   directrix $x = 2$.
    *(d) focus $(2, -2)$,   directrix $x = 5$.

*6. Find the points where the parabolas $x = y^2$ and $x = -2y^2 + 3$ intersect.

7. Find the points where the line $2x - 3y - 4 = 0$ intersects the parabola $y^2 = x - 1$.

*8. Find the points where the parabola $y = x^2 - 5$ intersects the circle $x^2 + y^2 = 25$.

Optional | *9. Find the vertex, the focus, and the directrix of the parabola

$$y = Ax^2 + Bx + C, \qquad A \neq 0.$$

10. The equation of every parabola with vertical axis can be written in the form

$$y = Ax^2 + Bx + C, \qquad A \neq 0.$$

What is the corresponding general equation for a parabola with horizontal axis?

11. Find an equation for the parabola that has directrix $y = 1$, axis $x = 2$, and passes through the point $(5, 6)$.

*12. Find an equation for the parabola that has horizontal axis, vertex $(-1, 1)$, and passes through the point $(-6, 13)$.

6.4 The Ellipse

Start with two points F_1, F_2 and a number k greater than the distance between them.

(6.4.1)

> The set of all points P such that
> $$d(P, F_1) + d(P, F_2) = k$$
> is called an *ellipse*. F_1 and F_2 are called the *foci*.

The idea is illustrated in Figure 6.4.1. A string is looped over tacks placed at the foci. The pencil placed in the loop traces out an ellipse.

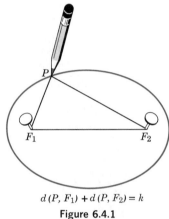

$$d(P, F_1) + d(P, F_2) = k$$

Figure 6.4.1

The *standard position* of an ellipse is with the foci along the x-axis at equal distances from the origin. (Figure 6.4.2)

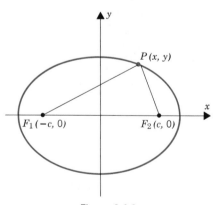

Figure 6.4.2

To find an equation for such an ellipse we set $k = 2a$. A point $P(x, y)$ will lie on the ellipse iff

$$d(P, F_1) + d(P, F_2) = 2a.$$

With F_1 at $(-c, 0)$ and F_2 at $(c, 0)$ we must have

$$\sqrt{(x + c)^2 + y^2} + \sqrt{(x - c)^2 + y^2} = 2a.$$

By transferring the second term to the right-hand side and squaring both sides, we obtain

$$(x + c)^2 + y^2 = 4a^2 + (x - c)^2 + y^2 - 4a\sqrt{(x - c)^2 + y^2}.$$

This reduces to

$$4a\sqrt{(x - c)^2 + y^2} = 4(a^2 - cx).$$

Cancelling the factor 4 and squaring again, we obtain

$$a^2(x^2 - 2cx + c^2 + y^2) = a^4 - 2a^2cx + c^2x^2.$$

This in turn reduces to

$$(a^2 - c^2)x^2 + a^2y^2 = a^2(a^2 - c^2),$$

and thus to

$$\frac{x^2}{a^2} + \frac{y^2}{a^2 - c^2} = 1. \quad \square$$

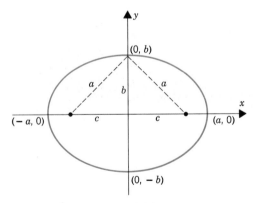

Figure 6.4.3

It is customary to set

$$b = \sqrt{a^2 - c^2}.$$

The equation for an ellipse in standard position then takes the form

(6.4.2)

$$\frac{x^2}{a^2} + \frac{y^2}{b^2} = 1, \qquad \text{with } a > b.$$

The relationship between a, b, and c is illustrated in Figure 6.4.3.

Every ellipse has four *vertices*. In Figure 6.4.3 these are marked $(a, 0)$, $(-a, 0)$, $(0, b)$, $(0, -b)$. The line segments that join opposite vertices are called the *axes* of the ellipse. The axis that contains the foci is called the *major axis,* the other the *minor axis.* In standard position the major axis is horizontal and has length $2a$; the minor axis is vertical and has length $2b$. The point at which the axes intersect is called the *center* of the ellipse. In standard position the center is at the origin.

Example 1. The equation

$$16x^2 + 25y^2 = 400$$

can be written

$$\frac{x^2}{25} + \frac{y^2}{16} = 1. \qquad \text{(divide by 400)}$$

Here $a = 5$, $b = 4$, and $c = \sqrt{a^2 - b^2} = \sqrt{9} = 3$. The equation has the form of (6.4.2). It is an ellipse in standard position with foci at $(-3, 0)$ and $(3, 0)$. The major axis has length $2a = 10$, and the minor axis has length $2b = 8$. The center is at the origin. The ellipse is sketched in Figure 6.4.4. □

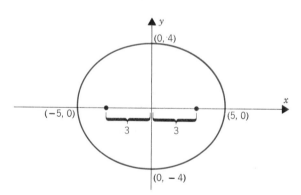

Figure 6.4.4

Example 2. The equation

$$\frac{x^2}{16} + \frac{y^2}{25} = 1$$

does not represent an ellipse in standard position because $25 > 16$. This equation is the equation of Example 1 with x and y interchanged. It represents the ellipse of

Example 1 reflected in the line $y = x$. (See Figure 6.4.5.) The foci are now on the y-axis, at $(0, -3)$ and $(0, 3)$. The major axis, now vertical, has length 10, and the minor axis has length 8. The center remains at the origin.   □

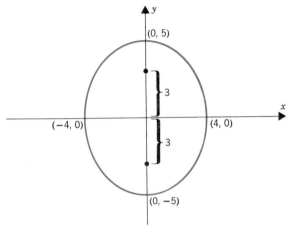

Figure 6.4.5

Example 3.   Figure 6.4.6 shows two ellipses:

$$\frac{x^2}{25} + \frac{y^2}{9} = 1 \quad \text{and} \quad \frac{(x-1)^2}{25} + \frac{(y+4)^2}{9} = 1.$$

The first ellipse is in standard position (6.4.2). Here $a = 5$, $b = 3$, $c = \sqrt{a^2 - b^2} = \sqrt{16} = 4$. Its foci are at $(-4, 0)$ and $(4, 0)$. The major axis has length 10, and the minor axis has length 6.

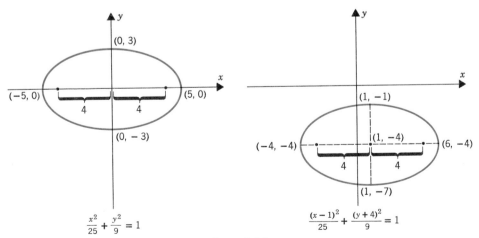

Figure 6.4.6

The second ellipse is the first ellipse displaced 1 unit to the right and lowered 4 units. The center now falls at the point $(1, -4)$. The foci are at $(-3, -4)$ and $(5, -4)$. ☐

Example 4.   To identify the curve

$$4x^2 - 8x + y^2 + 4y - 8 = 0,$$

we write

$$4(x^2 - 2x + \ \ ) + (y^2 + 4y + \ \ ) = 8.$$

By completing the squares within the parentheses, we get

$$4(x^2 - 2x + 1) + (y^2 + 4y + 4) = 16,$$

$$4(x - 1)^2 + (y + 2)^2 = 16,$$

(∗)
$$\frac{(x - 1)^2}{4} + \frac{(y + 2)^2}{16} = 1.$$

This is the ellipse

(∗∗)
$$\frac{x^2}{16} + \frac{y^2}{4} = 1 \qquad (a = 4, b = 2, c = \sqrt{16 - 4} = 2\sqrt{3})$$

reflected in the line $y = x$ and then displaced 1 unit to the right and lowered 2 units. Since the foci of (∗∗) are at $(-2\sqrt{3}, 0)$ and $(2\sqrt{3}, 0)$, the foci of (∗) are at $(1, -2 - 2\sqrt{3})$ and $(1, -2 + 2\sqrt{3})$. The major axis, now vertical, has length 8; the minor axis has length 4. ☐

The Reflecting Property of the Ellipse

Like the parabola, the ellipse has an interesting reflecting property. With calculus it can be shown that at each point P of the ellipse (see Figure 6.4.7) the focal radii $\overline{F_1 P}$

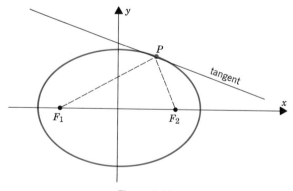

Figure 6.4.7

and $\overline{F_2P}$ make equal angles with the tangent. This has the following physical consequence: *an elliptical mirror takes light or sound originating at one focus and converges it at the other focus.* In elliptical rooms called "whispering chambers" a whisper at one focus is easily heard at the other focus.

Exercises

For each of the following ellipses find (a) the center, (b) the foci, (c) the length of the major axis, and (d) the length of the minor axis. Then sketch the figure.

*1. $\dfrac{x^2}{9} + \dfrac{y^2}{4} = 1.$
                                        2. $\dfrac{x^2}{4} + \dfrac{y^2}{9} = 1.$

*3. $3x^2 + 2y^2 = 12.$
                                        4. $(x - 1)^2 + 4y^2 = 64.$

*5. $3x^2 + 4y^2 - 12 = 0.$
                                        6. $4x^2 + y^2 - 6y + 5 = 0.$

*7. $4(x - 1)^2 + y^2 = 64.$
                                        8. $16(x - 2)^2 + 25(y - 3)^2 = 400.$

Find an equation for the ellipse that satisfies the given conditions.

*9. Foci at $(-1, 0)$, $(1, 0)$; major axis 6.

10. Foci at $(0, -1)$, $(0, 1)$; major axis 6.

*11. Foci at $(3, 1)$, $(9, 1)$; major axis 10.

12. Foci at $(1, 3)$, $(1, 9)$; major axis 8.

*13. Focus at $(1, 1)$; center at $(1, 3)$; major axis 10.

14. Center at $(2, 1)$; vertices at $(2, 6)$ and $(1, 1)$.

*15. Major axis 10; vertices at $(3, 2)$ and $(3, -4)$.

16. Find the points where the ellipse $3x^2 + 2y^2 = 11$ intersects the circle $x^2 + y^2 = 5.$

*17. Find the points where the ellipse $x^2 + 4y^2 = 24$ intersects the parabola $x^2 = 4y.$

18. Where does the line $4x + 3y + 6 = 0$ intersect the ellipse

$$\frac{x^2}{18} + \frac{y^2}{8} = 1?$$

6.5 The Hyperbola

Start with two points F_1, F_2 and a positive number k less than the distance between them.

(6.5.1)

> The set of all points P such that
>
> $$|d(P, F_1) - d(P, F_2)| = k$$
>
> is called a *hyperbola*. F_1 and F_2 are called the *foci*.

The *standard position* of a hyperbola is with the foci along the x-axis at equal distances from the origin. (Figure 6.5.1)

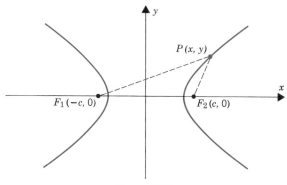

Figure 6.5.1

To find an equation for such a hyperbola we set $k = 2a$. A point $P(x, y)$ will lie on the hyperbola iff

$$|d(P, F_1) - d(P, F_2)| = 2a.$$

With F_1 at $(-c, 0)$ and F_2 at $(c, 0)$ we must have

$$\sqrt{(x + c)^2 + y^2} - \sqrt{(x - c)^2 + y^2} = \pm 2a. \qquad \text{(explain)}$$

Transferring the second term to the right and squaring both sides, we obtain

$$(x + c)^2 + y^2 = 4a^2 \pm 4a\sqrt{(x - c)^2 + y^2} + (x - c)^2 + y^2.$$

This equation reduces to

$$xc - a^2 = \pm a\sqrt{(x - c)^2 + y^2}.$$

Squaring once more, we obtain

$$x^2c^2 - 2a^2xc + a^4 = a^2(x^2 - 2xc + c^2 + y^2),$$

which reduces to

$$(c^2 - a^2)x^2 - a^2y^2 = a^2(c^2 - a^2),$$

and thus to

$$\frac{x^2}{a^2} - \frac{y^2}{c^2 - a^2} = 1. \quad \square$$

In this setting it is customary to set

$$b = \sqrt{c^2 - a^2}.$$

The equation for a hyperbola in standard position then takes the form

(6.5.2)
$$\boxed{\frac{x^2}{a^2} - \frac{y^2}{b^2} = 1.}$$

The hyperbola together with the relationship between a, b, and c is illustrated in Figure 6.5.2.

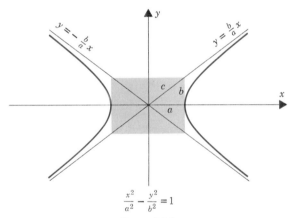

Figure 6.5.2

As the figure indicates, the hyperbola remains between the lines

$$y = \frac{b}{a}x \quad \text{and} \quad y = -\frac{b}{a}x.$$

These lines are called the *asymptotes* of the hyperbola. They can be obtained from Equation 6.5.2 by replacing the 1 on the right-hand side by 0:

$$\frac{x^2}{a^2} - \frac{y^2}{b^2} = 0 \quad \text{gives} \quad y = \pm\frac{b}{a}x.$$

As |x| increases without bound, the separation between the hyperbola and the asymptotes tends to zero. To see this, solve the equation

$$\frac{x^2}{a^2} - \frac{y^2}{b^2} = 1$$

for y. This gives

$$y = \pm \sqrt{\frac{b^2}{a^2}x^2 - b^2}.$$

For large $|x|$, b^2 is negligible compared to $(b^2/a^2)x^2$, and y is therefore approximately

$$\pm \sqrt{\frac{b^2}{a^2}x^2} = \pm \frac{b}{a}x. \quad \square$$

The line determined by the foci of a hyperbola intersects the hyperbola at two points called the *vertices*. The line segment that joins the vertices is called the *transverse* axis. The midpoint of the transverse axis is called the *center* of the hyperbola.

In standard position (Figure 6.5.2), the vertices are $(\pm a, 0)$, the transverse axis has length $2a$, and the center is at the origin.

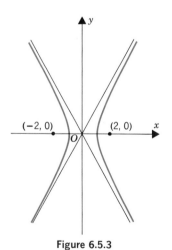

Figure 6.5.3

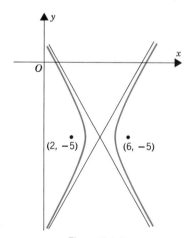

Figure 6.5.4

Example 1. The equation

$$\frac{x^2}{1} - \frac{y^2}{3} = 1, \qquad \text{(Figure 6.5.3)}$$

represents a hyperbola in standard position. Here $a = 1, b = \sqrt{3}, c = \sqrt{1 + 3} = 2$. The center is at the origin. The foci are at $(-2, 0)$ and $(2, 0)$. The transverse axis has length 2. The vertices are at $(-1, 0)$ and $(1, 0)$. We can obtain the asymptotes by setting

$$\frac{x^2}{1} - \frac{y^2}{3} = 0.$$

The asymptotes are the lines

$$y = \pm \sqrt{3}x. \quad \square$$

Example 2. The hyperbola

$$\frac{(x - 4)^2}{1} - \frac{(y + 5)^2}{3} = 1 \qquad \text{(Figure 6.5.4)}$$

is the hyperbola

$$\frac{x^2}{1} - \frac{y^2}{3} = 1$$

of Example 1 displaced 4 units to the right and lowered 5 units. The center of the hyperbola now falls at the point $(4, -5)$. The foci are at $(2, -5)$ and $(6, -5)$. The vertices are at $(3, -5)$ and $(5, -5)$. The new asymptotes are the lines

$$y + 5 = \pm \sqrt{3}(x - 4). \quad \square$$

Example 3. The hyperbola

$$\frac{y^2}{1} - \frac{x^2}{3} = 1 \qquad \text{(Figure 6.5.5)}$$

is the hyperbola

$$\frac{x^2}{1} - \frac{y^2}{3} = 1$$

of Example 1 reflected in the line $y = x$. The center is still at the origin. The foci are now at $(0, -2)$ and $(0, 2)$. The vertices are at $(0, -1)$ and $(0, 1)$. The asymptotes are now the lines

$$x = \pm \sqrt{3}y. \quad \square$$

Example 4. The hyperbola

$$\frac{(y - 5)^2}{1} - \frac{(x - 2)^2}{3} = 1 \qquad \text{(Figure 6.5.6)}$$

is the hyperbola of Example 3 displaced 2 units to the right and raised 5 units. $\square$

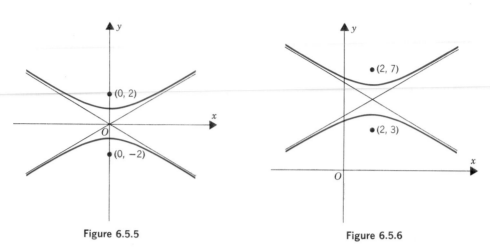

Figure 6.5.5   Figure 6.5.6

An Application to Range Finding

There is a simple application of the hyperbola to range-finding. If observers, located at two listening posts at a known distance apart, time the firing of a gun, the time difference multiplied by the velocity of sound gives the value of $2a$ and hence determines a hyperbola on which the gun must be located. A third listening post gives two more hyperbolas. The gun is found where the hyperbolas intersect.

Exercises

Find an equation for the indicated hyperbola:

*1. Foci at $(-5, 0)$, $(5, 0)$; transverse axis 6.

2. Foci at $(-13, 0)$, $(13, 0)$; transverse axis 10.

*3. Foci at $(0, -13)$, $(0, 13)$; transverse axis 10.

4. Foci at $(0, -13)$, $(0, 13)$; transverse axis 24.

*5. Foci at $(-5, 1)$, $(5, 1)$; transverse axis 6.

6. Foci at $(-3, 1)$, $(7, 1)$; transverse axis 6.

*7. Foci at $(-1, -1)$, $(-1, 1)$; transverse axis $\frac{1}{2}$.

For each of the following hyperbolas, find the length of the transverse axis, the vertices, the foci, and the asymptotes. Then sketch the figure.

*8. $x^2 - y^2 = 1$.   9. $y^2 - x^2 = 1$.

*10. $\dfrac{x^2}{9} - \dfrac{y^2}{16} = 1$.   11. $\dfrac{x^2}{16} - \dfrac{y^2}{9} = 1$.

*12. $\dfrac{y^2}{16} - \dfrac{x^2}{9} = 1.$ 13. $\dfrac{y^2}{9} - \dfrac{x^2}{16} = 1.$

*14. $\dfrac{(x-1)^2}{9} - \dfrac{(y-3)^2}{16} = 1.$ 15. $\dfrac{(x-1)^2}{16} - \dfrac{(y-3)^2}{9} = 1.$

*16. $-3x^2 + y^2 - 6x = 0.$ 17. $4x^2 - 8x - y^2 + 6y - 1 = 0.$

6.6   Additional Exercises

Write an equation for the given curve.

*1. The parabola with focus $F(4, 0)$ and directrix $y = -2$.

*2. The parabola with focus $F(4, 0)$ and vertex $V(4, 6)$.

*3. The ellipse centered at the origin with horizontal major axis 18 and minor axis 4.

*4. The ellipse centered at $P(2, 1)$ with vertical major axis 18 and minor axis 4.

*5. The hyperbola with foci $F_1(-1, 0)$, $F_2(1, 0)$ and transverse axis 1.

*6. The hyperbola with foci $F_1(0, -1)$, $F_2(0, 1)$ and transverse axis 1.

Describe the following curves in detail.

*7. $x^2 - 4y - 4 = 0.$ *8. $3x^2 + 2y^2 - 6 = 0.$

*9. $x^2 - 4y^2 - 10x + 41 = 0.$ *10. $9x^2 - 4y^2 - 18x - 8y - 31 = 0.$

*11. $x^2 + 3y^2 + 6x + 8 = 0.$ *12. $x^2 - 10x - 8y + 41 = 0.$

*13. $y^2 + 4y + 2x + 1 = 0.$ *14. $9x^2 + 4y^2 - 18x - 8y - 23 = 0.$

*15. $9x^2 + 25y^2 + 100y + 99 = 0.$ *16. $7x^2 - y^2 + 42x + 14y + 21 = 0.$

*17. $7x^2 - 5y^2 + 14x - 40y = 118.$ *18. $2x^2 - 3y^2 + 4\sqrt{3}x - 6\sqrt{3}y = 9.$

*19. $(x^2 - 4y)(4x^2 + 9y^2 - 36) = 0.$ *20. $(x^2 - 4y)(x^2 - 4y^2 - 1) = 0.$

SOME ADDITIONAL TOPICS

7

This chapter is a collection of odds and ends. Sections 7.1–7.4 deal with *polar coordinates,* used in calculus in the study of spiral-like curves. Section 7.5 is devoted to *induction,* an important technique of mathematical proof. In Section 7.6 we discuss the *least upper bound axiom.* This is the axiom by which we can make sense out of infinite decimals, prove the existence of $\sqrt{2}$, define the length of a curve, and give precise meaning to expressions such as $3^{\sqrt{2}}$ and 10^{π}. Section 7.7 gives Cramer's rule for sytems of linear equations, $n = 2$ and $n = 3$. In the last section we introduce the *complex numbers.*

7.1 Polar Coordinates

The purpose of coordinates is to fix position with respect to a given frame of reference. When we use rectangular coordinates, our frame of reference is a pair of lines that intersect at right angles. For a *polar coordinate system,* the frame of reference is a point O that we call the *pole* and a ray that emanates from it that we call the *polar axis.* (Figure 7.1.1)

O polar axis

Figure 7.1.1

In Figure 7.1.2 we have drawn two more rays from the pole. One lies at an angle of θ radians from the polar axis. We call it *ray θ*. The opposite ray lies at an angle of $\theta + \pi$ radians. We call it *ray $\theta + \pi$*.

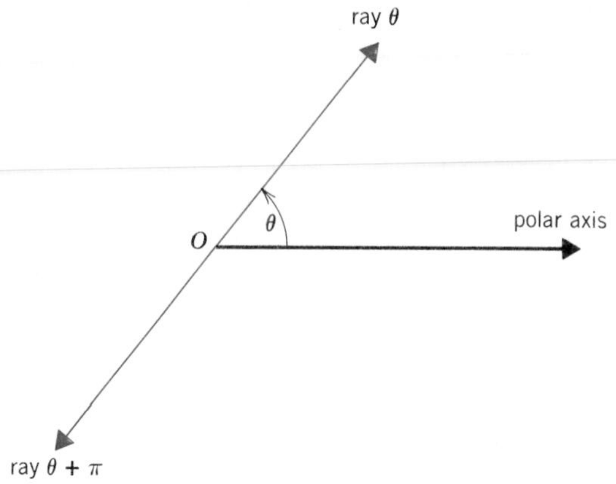

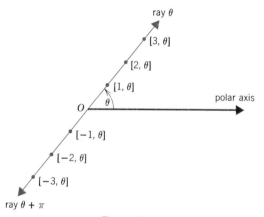

ray θ

polar axis

O

ray θ + π

Figure 7.1.2

Figure 7.1.3 shows some points along these same rays, labeled with *polar coordinates.*

ray θ

[3, θ]

[2, θ]

[1, θ]

polar axis

O

[−1, θ]

[−2, θ]

[−3, θ]

ray θ + π

Figure 7.1.3

	In general, a point is given *polar coordinates* $[r, \theta]$ iff it lies at a distance $\lvert r \rvert$ from the pole
(7.1.1)	
	along the ray θ, if $r \geq 0$, and along the ray $\theta + \pi$, if $r < 0$.

Figure 7.1.4 shows the point $[2, \frac{2}{3}\pi]$ at a distance of 2 units from the pole along the ray $\frac{2}{3}\pi$. The point $[-2, \frac{2}{3}\pi]$ also lies 2 units from the pole, not along the ray $\frac{2}{3}\pi$, but along the opposite ray.

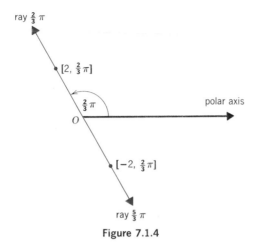

Figure 7.1.4

Unfortunately, polar coordinates are not unique. Many pairs $[r, \theta]$ can represent the same point.

1. If $r = 0$, it does not matter how we choose θ. The resulting point is still the pole:

(7.1.2)
$$\boxed{O = [0, \theta] \qquad \text{for all } \theta.}$$

2. Geometrically there is no distinction between angles that differ by an integral multiple of 2π. Consequently, as suggested in Figure 7.1.5,

(7.1.3)
$$\boxed{[r, \theta] = [r, \theta + 2n\pi] \qquad \text{for all integers } n.}$$

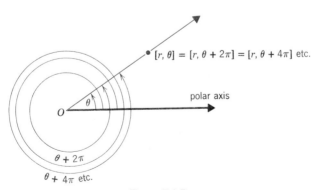

Figure 7.1.5

3. Adding π to the second coordinate is equivalent to changing the sign of the first coordinate:

(7.1.4) $$[r, \theta + \pi] = [-r, \theta].$$ (Figure 7.1.6)

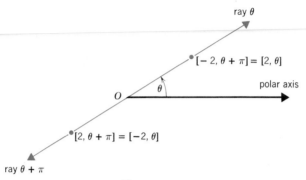

Figure 7.1.6

Relation to Rectangular Coordinates

In Figure 7.1.7 we have superimposed a polar coordinate system on a rectangular coordinate system. We have placed the pole at the origin and the polar axis along the positive x-axis.

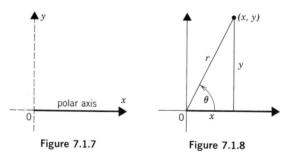

Figure 7.1.7   Figure 7.1.8

The relation between polar coordinates $[r, \theta]$ and rectangular coordinates (x, y) can now be written

(7.1.5) $$x = r \cos \theta, \qquad y = r \sin \theta.$$

PROOF. If $r = 0$, the formulas hold, since the point $[r, \theta]$ is then the origin and both x and y are 0:

$$0 = 0 \cos \theta, \qquad 0 = 0 \sin \theta.$$

For $r > 0$, we refer to Figure 7.1.8. From the figure,

$$\cos \theta = \frac{x}{r}, \qquad \sin \theta = \frac{y}{r}$$

and therefore
$$x = r \cos \theta, \qquad y = r \sin \theta.$$

Suppose now that $r < 0$. Since $[r, \theta] = [-r, \theta + \pi]$ and $-r > 0$, we know from the previous case that
$$x = -r \cos (\theta + \pi) \qquad \text{and} \qquad y = -r \sin (\theta + \pi).$$

By (5.2.4)
$$\cos (\theta + \pi) = -\cos \theta \qquad \text{and} \qquad \sin (\theta + \pi) = -\sin \theta,$$

so that once again we have
$$x = r \cos \theta, \qquad y = r \sin \theta. \quad \square$$

From the relations we just proved you can see that, subject to the obvious exclusions,

(7.1.6)
$$\boxed{\tan \theta = \frac{y}{x}}$$

and, under all circumstances,

(7.1.7)
$$\boxed{x^2 + y^2 = r^2.} \qquad \text{(check this out)}$$

Problem. Find the rectangular coordinates of the point with polar coordinates $[-2, \frac{1}{3}\pi]$.

SOLUTION. The relations
$$x = r \cos \theta, \qquad y = r \sin \theta$$

give
$$x = -2 \cos \tfrac{1}{3}\pi = -2(\tfrac{1}{2}) = -1$$

and
$$y = -2 \sin \tfrac{1}{3}\pi = -2(\tfrac{1}{2}\sqrt{3}) = -\sqrt{3}. \quad \square$$

Problem. Find all possible polar coordinates for the point with rectangular coordinates $(-2, 2\sqrt{3})$.

SOLUTION. Set
$$-2 = r \cos \theta, \qquad 2\sqrt{3} = r \sin \theta.$$

Since in general

$$r^2 = x^2 + y^2 = r^2 \cos^2 \theta + r^2 \sin^2 \theta,$$

$$\underset{(7.1.7)}{\qquad} \quad \underset{(7.1.5)}{\qquad}$$

here

$$r^2 = (-2)^2 + (2\sqrt{3})^2 = 4 + 12 = 16.$$

If we set $r = 4$, then we have

$$-2 = 4 \cos \theta, \qquad 2\sqrt{3} = 4 \sin \theta$$

and thus

$$-\tfrac{1}{2} = \cos \theta, \qquad \tfrac{1}{2}\sqrt{3} = \sin \theta.$$

This means that we can take

$$\theta = \tfrac{2}{3}\pi \quad \text{or more generally} \quad \theta = \tfrac{2}{3}\pi + 2n\pi.$$

If we set $r = -4$, then we can take

$$\theta = \tfrac{2}{3}\pi + \pi = \tfrac{5}{3}\pi \quad \text{or more generally} \quad \theta = \tfrac{5}{3}\pi + 2n\pi.$$

The possible polar coordinates are all pairs of the form

$$[4, \tfrac{2}{3}\pi + 2n\pi], \qquad [-4, \tfrac{5}{3}\pi + 2n\pi] \qquad \text{with } n \text{ an integer.} \quad \square$$

Exercises

Plot the following points:

1. $[1, \tfrac{1}{3}\pi]$.   2. $[1, \tfrac{1}{2}\pi]$.   3. $[-1, \tfrac{1}{3}\pi]$.   4. $[-1, -\tfrac{1}{3}\pi]$.

5. $[4, \tfrac{5}{4}\pi]$.   6. $[-2, 0]$.   7. $[-\tfrac{1}{2}, \pi]$.   8. $[\tfrac{1}{3}, \tfrac{2}{3}\pi]$.

Find the rectangular coordinates of each of the following points:

*9. $[3, \tfrac{1}{2}\pi]$.   10. $[4, \tfrac{1}{6}\pi]$.   *11. $[-1, -\pi]$.   12. $[-1, \tfrac{1}{4}\pi]$.

*13. $[-3, -\tfrac{1}{3}\pi]$.   14. $[2, 0]$.   *15. $[2, 3\pi]$.   16. $[3, -\tfrac{1}{2}\pi]$.

The following points are given in rectangular coordinates. Find all possible polar coordinates for each point.

*17. $(0, 1)$.   18. $(1, 0)$.   *19. $(-3, 0)$.

20. $(4, 4)$.   *21. $(2, -2)$.   22. $(3, -3\sqrt{3})$.

*23. $(4\sqrt{3}, 4)$.   24. $(\sqrt{3}, -1)$.   *25. $(-1, \sqrt{3})$.

Optional   26.   (a) Show that the distance between $[r_1, \theta_1]$ and $[r_2, \theta_2]$ is

$$\sqrt{r_1^2 + r_2^2 - 2r_1 r_2 \cos(\theta_1 - \theta_2)}.$$

[HINT: Change to rectangular coordinates.]

      (b) Show that for $r_1 > 0$, $r_2 > 0$, $|\theta_1 - \theta_2| < \pi$ the distance formula in part (a) is just the law of cosines.

7.2   Some Sets in Polar Coordinates; Symmetry

Let's specify some simple sets in polar coordinates.

1.   In rectangular coordinates the circle of radius a centered at the origin has equation

$$x^2 + y^2 = a^2.$$

The equation for this circle in polar coordinates is simply

$$r = a.$$

The interior of the circle is given by

$$0 \le r < a$$

and the exterior by

$$r > a.$$

2.   The line through the origin with inclination α is given by the polar equation

$$\theta = \alpha.$$

3.   The vertical line $x = a$ becomes

$$r \cos \theta = a$$

and the horizontal line $y = b$ becomes

$$r \sin \theta = b.$$

4.   The line $Ax + By + C = 0$ can be written

$$r(A \cos \theta + B \sin \theta) + C = 0. \quad \square$$

Problem.   Find an equation in polar coordinates for the equilateral hyperbola

$$x^2 - y^2 = a^2.$$

SOLUTION.   Setting

$$x = r \cos \theta, \qquad y = r \sin \theta$$

we have

$$r^2 \cos^2 \theta - r^2 \sin^2 \theta = a^2,$$

$$r^2(\cos^2 \theta - \sin^2 \theta) = a^2,$$

$$r^2 \cos 2\theta = a^2. \quad \square$$

Problem.   Show that the polar equation

$$r = 2a \cos \theta$$

represents a circle.

SOLUTION.   Multiplication by r gives

$$r^2 = 2ar \cos \theta,$$

$$x^2 + y^2 = 2ax,$$

$$x^2 - 2ax + y^2 = 0,$$

$$x^2 - 2ax + a^2 + y^2 = a^2,$$

$$(x - a)^2 + y^2 = a^2.$$

This is a circle of radius a centered at the point with rectangular coordinates $(a, 0)$. $\square$

Symmetry

Symmetry with respect to each of the coordinate axes and with respect to the origin is illustrated in Figures 7.2.1, 7.2.2, and 7.2.3. The coordinates marked are, of course, not the only ones possible.

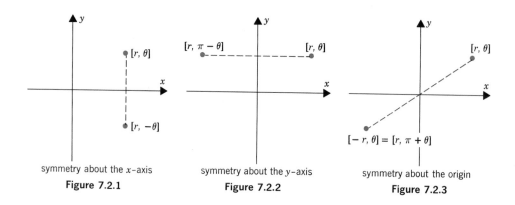

symmetry about the x–axis
Figure 7.2.1

symmetry about the y–axis
Figure 7.2.2

symmetry about the origin
Figure 7.2.3

Problem. Test the lemniscate

$$r^2 = \cos 2\theta$$

for symmetry.

SOLUTION. Since

$$\cos [2(-\theta)] = \cos (-2\theta) = \cos 2\theta,$$

you can see that, if $[r, \theta]$ is on the curve, then so is $[r, -\theta]$. This says that the curve is symmetric about the x-axis.
     Since

$$\cos [2(\pi - \theta)] = \cos (2\pi - 2\theta) = \cos (-2\theta) = \cos 2\theta,$$

you can see that, if $[r, \theta]$ is on the curve, then so is $[r, \pi - \theta]$. The curve is therefore symmetric about the y-axis.
     Being symmetric about both axes, the curve must also be symmetric about the origin. You can verify this directly by noting that

$$\cos [2(\pi + \theta)] = \cos (2\pi + 2\theta) = \cos 2\theta,$$

so that, if $[r, \theta]$ lies on the curve, then so does $[r, \pi + \theta]$. A sketch of the lemniscate appears in Figure 7.2.4. $\square$

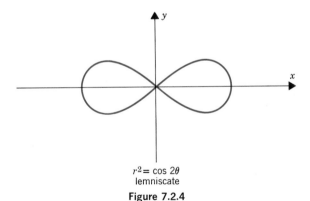

$r^2 = \cos 2\theta$
lemniscate
Figure 7.2.4

Exercises

Find the point $[r, \theta]$ symmetric to the given point
(a) about the x-axis.   (b) about the y-axis.   (c) about the origin.
Express your answer with $r > 0$ and $\theta \in [0, 2\pi]$.

*1. $[\frac{1}{2}, \frac{1}{6}\pi]$.   2. $[3, -\frac{5}{4}\pi]$.   *3. $[-2, \frac{1}{3}\pi]$.   4. $[-3, -\frac{7}{4}\pi]$.

Test these curves for symmetry about the coordinate axes and the origin.

*5. $r = 2 + \cos\theta$.   6. $r = \cos 2\theta$.   *7. $r(\sin\theta + \cos\theta) = 1$.

*8. $r^2 = \sin\theta$.   9. $r^2 \sin 2\theta = 1$.   *10. $r^2 \cos 2\theta = 1$.

Express the following equations in terms of polar coordinates.

*11. $2xy = 1$.   12. $y = mx$.

*13. $x^2 + y^2 = 4$.   14. $x^2 + (y - b)^2 = b^2$.

*15. $x^2 + y^2 + ax = a\sqrt{x^2 + y^2}$.   16. $(x^2 + y^2)^2 = a^2(x^2 - y^2)$.

Identify the following curves. Change to rectangular coordinates.

*17. $r \sin\theta = 4$.   18. $r \cos\theta = 4$.   *19. $\theta = \frac{1}{3}\pi$.

*20. $\theta^2 = \frac{1}{9}\pi^2$.   21. $r = 2(1 - \cos\theta)^{-1}$.   *22. $r = 4 \sin(\theta + \pi)$.

7.3   Graphing in Polar Coordinates

In rectangular coordinates the simplest equations to graph take the form

$$y = y_0 \quad \text{and} \quad x = x_0.$$

The first of these equations represents a horizontal line and the second a vertical line. The polar counterparts of these equations take the form

$$r = r_0 \quad \text{and} \quad \theta = \theta_0.$$

The first of these equations represents a circle. (Figure 7.3.1) The center is at the pole, and the radius is $|r_0|$. The second equation represents a straight line. (Figure 7.3.2) The line passes through the pole at an angle θ_0 from the polar axis.

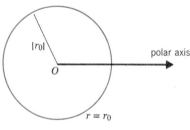

Figure 7.3.1

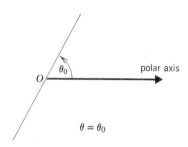

Figure 7.3.2

In earlier chapters we graphed equations of the form $y = f(x)$ in rectangular coordinates. Here we graph some equations of the form $r = f(\theta)$ in polar coordinates.

We begin with the equation

$$r = \theta, \qquad \theta \geq 0.$$

Its graph is a nonending spiral, the famous *spiral of Archimedes*. It is shown in detail from $\theta = 0$ to $\theta = 2\pi$ in Figure 7.3.3. At $\theta = 0, r = 0$; at $\theta = \frac{1}{4}\pi, r = \frac{1}{4}\pi$; at $\theta = \frac{1}{2}\pi$, $r = \frac{1}{2}\pi$; etc.

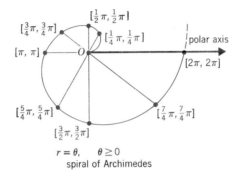

$r = \theta, \qquad \theta \geq 0$
spiral of Archimedes

Figure 7.3.3

Problem. Sketch the graph of the equation

$$r = a(1 + \cos \theta). \qquad\qquad (a > 0)$$

SOLUTION. Since the cosine has period 2π, we will concern ourselves only with θ between $-\pi$ and π. Outside of that interval the curve repeats itself.

Since the cosine function is even, that is, $\cos(-\theta) = \cos \theta$, the value of r corresponding to $-\theta$ is the same as the value corresponding to θ. This tells us that the curve is symmetric about the x-axis. All we have to do, then, is sketch the curve from 0 to π (that is, in the upper half plane). The rest we can fill in by symmetry.

Table 7.3.1

θ	r
0	$2a$
$\frac{1}{6}\pi$	$(1 + \frac{1}{2}\sqrt{3})a$
$\frac{1}{3}\pi$	$\frac{3}{2}a$
$\frac{1}{2}\pi$	a
$\frac{2}{3}\pi$	$\frac{1}{2}a$
π	0

Table 7.3.1 lists some representative values of θ from 0 to π and the corresponding values of r obtained from the equation $r = a(1 + \cos \theta)$. It is not hard to see that as θ increases from 0 to π, $\cos \theta$ decreases from 1 to 0, and r itself decreases from $2a$ to 0. From 0 to π the graph must look as in Figure 7.3.4.

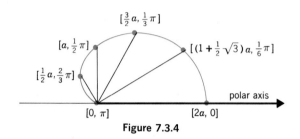

Figure 7.3.4

Filling in the lower part by symmetry (remember, the curve is symmetric about the x-axis), we have the complete curve as shown in Figure 7.3.5. This heart-shaped curve is called a *cardioid*.   □

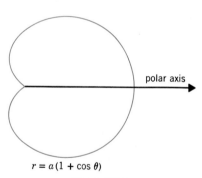

$$r = a(1 + \cos \theta)$$

Figure 7.3.5

Problem.   Sketch the three-leaved rose

$$r = a \sin 3\theta. \tag{$a > 0$}$$

SOLUTION.   As θ increases from 0 to $\frac{1}{6}\pi$, r increases from 0 to a:

$$a \sin 3(\tfrac{1}{6}\pi) = a \sin \tfrac{1}{2}\pi = a.$$

As θ increases from $\frac{1}{6}\pi$ to $\frac{1}{3}\pi$, r decreases back to 0:

$$a \sin 3(\tfrac{1}{3}\pi) = a \sin \pi = 0. \tag{Figure 7.3.6}$$

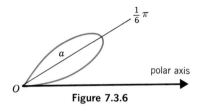

Figure 7.3.6

As θ increases from $\frac{1}{3}\pi$ to $\frac{1}{3}\pi + \frac{1}{6}\pi = \frac{1}{2}\pi$, r decreases from 0 to $-a$:

$$a \sin 3(\tfrac{1}{2}\pi) = -a.$$

As θ increases from $\frac{1}{2}\pi$ to $\frac{1}{2}\pi + \frac{1}{6}\pi = \frac{2}{3}\pi$, r increases back to 0:

$$a \sin 3(\tfrac{2}{3}\pi) = a \sin 2\pi = 0.\dagger \qquad\qquad\text{(Figure 7.3.7)}$$

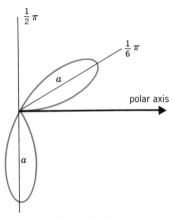

Figure 7.3.7

As θ increases from $\frac{2}{3}\pi$ to $\frac{2}{3}\pi + \frac{1}{6}\pi = \frac{5}{6}\pi$, r increases from 0 to a:

$$a \sin 3(\tfrac{5}{6}\pi) = a \sin \tfrac{5}{2}\pi = a \sin \tfrac{1}{2}\pi = a.$$

As θ increases from $\frac{5}{6}\pi$ to π, r decreases back to 0:

$$a \sin 3\pi = a \sin \pi = 0. \qquad\qquad\text{(Figure 7.3.8)}$$

From here on, the curve repeats itself. The value $\theta + \pi$ gives rise to the same point on the curve as did θ:

$$[a \sin 3(\theta + \pi), \theta + \pi] = [-a \sin 3\theta, \theta + \pi] = [a \sin 3\theta, \theta].$$

The last equality follows from the fact that in general $[r, \theta + \pi] = [-r, \theta]$. $\square$

$\dagger$ For θ between $\frac{1}{3}\pi$ and $\frac{2}{3}\pi$, $r = a \sin 3\theta$ is negative and the points appear on the opposite ray.

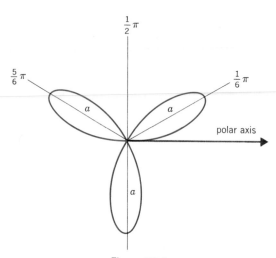

Figure 7.3.8

Exercises

Graph the following polar equations.

*1. $r = 2$.   2. $r = -3$.   *3. $\theta = \frac{1}{6}\pi$.   4. $\theta = -\frac{1}{6}\pi$.

*5. $r = 2\theta, \quad 0 \leq \theta \leq 2\pi$.   6. $r = \frac{1}{2}\theta, \quad 0 \leq \theta \leq 2\pi$.

*7. $r = \cos \theta$.   8. $r = \sin \theta$.   *9. $r = 2\cos \theta$.   10. $r = 2\sin \theta$.

Graph the following, taking $a > 0$.

*11. The three-leaved rose $r = a \cos 3\theta$.

12. The parabola $r = a \sec^2 \frac{1}{2}\theta$.

*13. The four-leaved rose $r = a \sin 2\theta$.

14. The four-leaved rose $r = a \cos 2\theta$.

*15. The two-leaved lemniscate $r^2 = a^2 \sin 2\theta$.

16. The eight-leaved rose $r = a \sin 4\theta$.

Optional | **7.4   The Intersection of Polar Curves**

The fact that a single point has many pairs of polar coordinates can cause complications. In particular, it means that a point $[r_1, \theta_1]$ can lie on a curve given by a polar equation although its coordinates r_1 and θ_1 do not

Optional | satisfy the equation. For example, the coordinates of $[2, \pi]$ do not satisfy the equation

$$r^2 = 4 \cos \theta.$$

$$(r^2 = 2^2 = 4 \quad \text{but} \quad 4 \cos \theta = 4 \cos \pi = -4.)$$

Nevertheless the point $[2, \pi]$ does lie on the curve $r^2 = 4 \cos \theta$. It lies on the curve because

$$[2, \pi] = [-2, 0]$$

and the coordinates of $[-2, 0]$ satisfy the equation:

$$r^2 = (-2)^2 = 4, \qquad 4 \cos \theta = 4 \cos 0 = 4. \quad \square$$

The difficulties are compounded when we deal with two or more curves. Here is an example.

Problem. Find the points where the cardioids

$$r = a(1 - \cos \theta) \quad \text{and} \quad r = a(1 + \cos \theta)$$

intersect.

SOLUTION. We begin by solving the two equations simultaneously. Adding these equations, we get $2r = 2a$ and thus $r = a$. This tells us that $\cos \theta = 0$ and therefore $\theta = \frac{1}{2}\pi + n\pi$. The points $[a, \frac{1}{2}\pi + n\pi]$ all lie on both curves. Not all of these points are distinct, however:

$$\text{for } n \text{ even, } [a, \tfrac{1}{2}\pi + n\pi] = [a, \tfrac{1}{2}\pi]$$

$$\text{for } n \text{ odd, } [a, \tfrac{1}{2}\pi + n\pi] = [a, \tfrac{3}{2}\pi].$$

In short, by solving the two equations simultaneously we have arrived at two common points:

$$[a, \tfrac{1}{2}\pi] = (0, a) \quad \text{and} \quad [a, \tfrac{3}{2}\pi] = (0, -a).$$

There is, however, a third point at which the curves intersect, and that is the origin O. (Figure 7.4.1 on page 272) The origin clearly lies on both curves:

$$\text{for } r = a(1 - \cos \theta) \quad \text{take } \theta = 0, 2\pi, \text{ etc.}$$

$$\text{for } r = a(1 + \cos \theta) \quad \text{take } \theta = \pi, 3\pi, \text{ etc.}$$

The reason that the origin does not appear when we solve the two equations simultaneously is that the curves do not pass through the origin "simultaneously"; that is, they do not pass through the origin for the same values of θ. Think of each of the equations

$$r = a(1 - \cos \theta) \quad \text{and} \quad r = a(1 + \cos \theta)$$

Optional | as giving the position of an object at time θ. At the points we found by solving the two equations simultaneously, the objects collide. (They both arrive there at the same time.) At the origin the situation is different. Both objects pass through the origin, but no collision takes place because the objects pass through the origin at *different* times. ☐

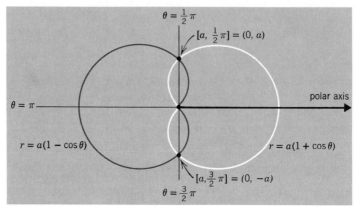

Figure 7.4.1

Exercises

Determine whether the point lies on the curve.

*1. $r^2 \cos \theta = 1$; $[1, \pi]$.   2. $r^2 = \cos 2\theta$; $[1, \frac{1}{4}\pi]$.

*3. $r = \sin \frac{1}{3}\theta$; $[\frac{1}{2}, \frac{1}{2}\pi]$.   4. $r^2 = \sin 3\theta$; $[1, -\frac{5}{6}\pi]$.

5. Show that the point $[2, \pi]$ lies on both $r^2 = 4 \cos \theta$ and $r = 3 + \cos \theta$.

Find the points at which the curves intersect. Express your answers in rectangular coordinates.

*6. $r = \cos^2 \theta$, $r = -1$.   7. $r = 2 \sin \theta$, $r = 2 \cos \theta$.

*8. $r = \dfrac{1}{1 - \cos \theta}$, $r \sin \theta = b$.   9. $r^2 = \sin \theta$, $r = 2 - \sin \theta$.

*10. $r = 1 - \cos \theta$, $r = \cos \theta$.   11. $r = 1 - \cos \theta$, $r = \sin \theta$.

7.5 Induction

Suppose that you were asked to show that a certain set S contains the set of positive integers. You could start by verifying that $1 \in S$, and $2 \in S$, and $3 \in S$, and so on, but even if each such step took you only one-hundredth of a second, you would still never finish.

To avoid such a bind, mathematicians use a special procedure called *induction*. That induction works is an *assumption* that we make.

Axiom 7.5.1   Axiom of Induction

Let S be a set of integers. If

$$\text{(A)} \ 1 \in S \quad \text{and} \quad \text{(B)} \ k \in S \text{ implies } k + 1 \in S,$$

then all the positive integers are in S.

You can think of the axiom of induction as a kind of "domino theory." If the first domino falls (Figure 7.5.1), and if each domino that falls causes the next one to fall, then, according to the axiom of induction, all the dominoes will fall.

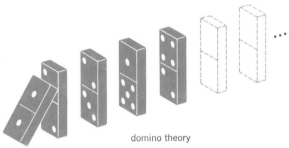

domino theory

Figure 7.5.1

While we cannot prove that this axiom is valid (axioms are by their very nature assumptions and therefore not subject to proof), we can argue that it is *plausible*.

Let's assume that we have a set S that satisfies conditions (A) and (B). Now let's choose a positive integer m and "argue" that $m \in S$.

From (A) we know that $1 \in S$. Since $1 \in S$, we know from (B) that $1 + 1 \in S$, and thus that $(1 + 1) + 1 \in S$, and so on. Since m can be obtained from 1 by adding 1 successively $(m - 1)$ times, it *seems clear* that $m \in S$. $\quad\square$

As an example of this procedure, note that

$$5 = \{[(1 + 1) + 1] + 1\} + 1$$

and thus $5 \in S$.

Problem.   Show that

$$\text{if } 0 \le a < b \quad \text{then} \quad a^n < b^n \quad \text{for all positive integers } n.$$

SOLUTION.   Suppose that $0 \le a < b$, and let S be the set of positive integers n for which $a^n < b^n$.

Obviously $1 \in S$. Let's assume now that $k \in S$. This assures us that $a^k < b^k$. It follows that

$$a^{k+1} = a \cdot a^k < a \cdot b^k < b \cdot b^k = b^{k+1}$$

and thus that $k + 1 \in S$.

We have shown that

$$1 \in S \quad \text{and that} \quad k \in S \quad \text{implies} \quad k + 1 \in S.$$

By the axiom of induction, we can conclude that all the positive integers are in S. $\quad\square$

Problem. Show that, if $x \geq -1$, then

$$(1 + x)^n \geq 1 + nx \qquad \text{for all positive integers } n.$$

SOLUTION. Take $x \geq -1$ and let S be the set of positive integers n for which

$$(1 + x)^n \geq 1 + nx.$$

Since

$$(1 + x)^1 \geq 1 + 1 \cdot x,$$

you can see that $1 \in S$.

Assume now that $k \in S$. By the definition of S,

$$(1 + x)^k \geq 1 + kx.$$

Since

$$(1 + x)^{k+1} = (1 + x)^k(1 + x) \geq (1 + kx)(1 + x) \qquad \text{(explain)}$$

and

$$(1 + kx)(1 + x) = 1 + (k + 1)x + kx^2 \geq 1 + (k + 1)x,$$

it follows that

$$(1 + x)^{k+1} \geq 1 + (k + 1)x$$

and thus that $k + 1 \in S$.

We have shown that

$$1 \in S \quad \text{and that} \quad k \in S \quad \text{implies} \quad k + 1 \in S.$$

By the axiom of induction, all the positive integers are in S. $\quad\square$

Exercises

Show that the following statements hold for all positive integers n:

*1. $n(n + 1)$ is divisible by 2. [HINT: $(k + 1)(k + 2) = k(k + 1) + 2(k + 1)$.]

2. $n(n + 1)(n + 2)$ is divisible by 6.

3. $1 + 2 + 3 + \cdots + n = \frac{1}{2}n(n + 1)$.

*4. $1 + 3 + 5 + \cdots + (2n - 1) = n^2$.

5. $1^2 + 2^2 + 3^2 + \cdots + n^2 = \frac{1}{6}n(n + 1)(2n + 1)$.

6. $1^3 + 2^3 + 3^3 + \cdots + n^3 = (1 + 2 + 3 + \cdots + n)^2$ [HINT: Use Exercise 3.]

*7. $3^{2n+1} + 2^{n+2}$ is divisible by 7.

8.   For what integers n is $9^n - 8n - 1$ divisible by 64? Prove that your answer is correct.

9.   Prove that all sets with n elements have 2^n subsets. Count the empty set $\varnothing$ as a subset.

10.   Show that for all positive integers n

$$1^3 + 2^3 + \cdots + (n - 1)^3 < \tfrac{1}{4}n^4 < 1^3 + 2^3 + \cdots + n^3.$$

11.   Find a simplifying expression for the product

$$\left(1 - \frac{1}{2}\right)\left(1 - \frac{1}{3}\right) \cdots \left(1 - \frac{1}{n}\right)$$

and verify its validity for all integers $n \geq 2$.

7.6   The Least Upper Bound Axiom

Let S be a set of real numbers. As we indicated before (Section 2.6), a number M is called an *upper bound* for S iff

$$x \leq M \quad \text{for all } x \in S.$$

Not all sets of real numbers have upper bounds. Those that do are said to be *bounded above*.

It is obvious that every set that has a largest element has an upper bound; if b is the largest element of S, then

$$x \leq b \quad \text{for all } x \in S$$

and therefore b is an upper bound for S. The converse is false: the sets

$$(-\infty, 0) \quad \text{and} \quad \left\{\frac{1}{2}, \frac{2}{3}, \frac{3}{4}, \cdots, \frac{n}{n + 1}, \cdots\right\}$$

both have upper bounds (2 for instance) but neither has a largest element.

Let's return to the set $(-\infty, 0)$. While $(-\infty, 0)$ does not have a largest element, the set of its upper bounds, $[0, \infty)$, does have a least element, namely 0. A similar remark can be made about the second set. While the set of quotients

$$\frac{n}{n + 1} = 1 - \frac{1}{n + 1}$$

does not have a greatest element, the set of its upper bounds, $[1, \infty)$, does have a least element, namely 1. Along these lines there is a key *assumption* that we make about the real number system. It is called the *least upper bound axiom*.

Axiom 7.6.1   The Least Upper Bound Axiom

Every nonempty set of real numbers that has an upper bound has a *least* upper bound.

To indicate the least upper bound of a set S, we will write lub S. Here are some examples:

1. lub $(-\infty, 0) = 0$, lub $(-\infty, 0] = 0$.

2. lub $(-4, -1) = -1$, lub $(-4, -1] = -1$.

3. lub $\left\{ \dfrac{1}{2}, \dfrac{2}{3}, \dfrac{3}{4}, \ldots, \dfrac{n}{n+1}, \ldots \right\} = 1$.

4. lub $\left\{ -\dfrac{1}{2}, -\dfrac{1}{8}, -\dfrac{1}{27}, \ldots, -\dfrac{1}{n^3}, \ldots \right\} = 0$.

The least upper bound of a set has a property that deserves particular attention. The idea is this: the fact that M is the least upper bound of the set S does not tell us that M is in S (it need not be), but it does tell us that we can approximate M as closely as we wish by members of S. We state this as a theorem.

Theorem   7.6.2

If M is the least upper bound of the set S and ϵ† is positive, then there is a number s in S such that

$$M - \epsilon < s \leq M.$$

Proof. Since all numbers in S are less than or equal to M, the right-hand side of the inequality causes no difficulty. All we have to show therefore is that there exists some number s in S such that

$$M - \epsilon < s.$$

Suppose on the contrary that no such number exists. We then have

$$x \leq M - \epsilon \quad \text{for all } x \in S,$$

---

†The Greek letter *epsilon*, used almost universally in calculus to denote an arbitrary positive real number.

so that $M - \epsilon$ becomes an upper bound for S. But this cannot happen, for it makes $M - \epsilon$ an upper bound that is less than M, and M is by assumption the *least* upper bound.   □

The theorem we just proved is illustrated in Figure 7.6.1. If $M = \text{lub } S$, then S has an element in every half-open interval of the form $(M - \epsilon, M]$.

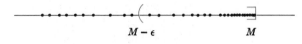

$$M - \epsilon \qquad\qquad M$$

Figure 7.6.1

Examples

1. Let

$$S = \left\{ \frac{1}{2}, \frac{2}{3}, \frac{3}{4}, \cdots, \frac{n}{n+1}, \cdots \right\}$$

and take

$$\epsilon = 0.0001.$$

Since 1 is the least upper bound of S, there must be a number s in S such that

$$1 - 0.0001 < s \leq 1.$$

There is: take, for example, $s = \frac{99999}{100000}$.   □

2. Let

$$S = \{1, 2, 3\}$$

and take

$$\epsilon = 0.000001.$$

Since 3 is the least upper bound of S, there must be a number s in S such that

$$3 - 0.000001 < s \leq 3.$$

There is: $s = 3$.   □

The least upper bound axiom is a powerful tool, and it has an immense number of applications. Here are four of them.

1. Decimal Representations
Terminating decimals such as

$$0.147, \quad 0.2165, \quad 0.32946$$

are easily defined in terms of arithmetic:

$$0.147 = \tfrac{147}{1000}, \qquad 0.2165 = \tfrac{2165}{10000}, \qquad 0.32946 = \tfrac{32946}{100000}.$$

More generally,

$$0.a_1a_2 \cdots a_n = \frac{a_1a_2 \cdots a_n}{10^n}.$$

But what about nonterminating decimals? How is

$$0.a_1a_2 \cdots a_n \cdots$$

to be defined? The easiest way to handle this problem is to note that the set

$$\{0.a_1, 0.a_1a_2, 0.a_1a_2a_3, \cdots\}$$

has an upper bound (1 for instance) and therefore, by the least upper bound axiom, it
has a least upper bound. We set

(7.6.3) $\boxed{0.a_1a_2 \cdots a_n \cdots = \text{lub } \{0.a_1, 0.a_1a_2, 0.a_1a_2a_3, \cdots\}.}$

Every real number is thus the least upper bound of a set of terminating decimals:

$\tfrac{5}{4} = 1.25 = \text{lub } \{1, 1.2, 1.25\}.$

$\tfrac{1}{3} = 0.333 \cdots = \text{lub } \{0.3, 0.33, 0.333, \cdots\}.$

$\sqrt{2} = 1.14139 \cdots = \text{lub } \{1, 1.1, 1.14, 1.141, 1.1413, 1.14139, \cdots\}.$

$\pi = 3.14159 \cdots = \text{lub } \{3, 3.1, 3.14, 3.141, 3.1415, 3.14159, \cdots\}.$ $\square$

2. Arc Length

In Figure 7.6.2 we have sketched a curve C and a polygonal approximation to it.

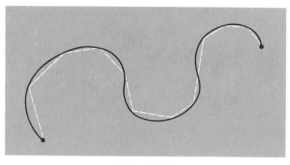

Figure 7.6.2

Question: What is the length of this curve C?

Answer: The length of C is by definition the least upper bound of the set of lengths of
all such polygonal approximations.

To say that 2π is the circumference of the unit circle is to say that 2π is the least upper bound of the set of lengths of all polygons inscribed in the unit circle. See Figure 7.6.3. $\square$

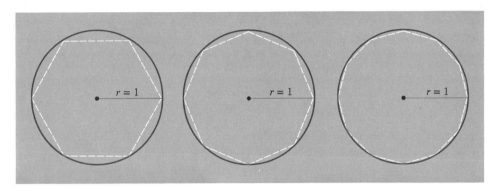

Figure 7.6.3

3. *n*th Roots

We have talked about square roots, cube roots, etc. without proving that they actually exist. One way to prove the existence of such roots is to use the least upper bound axiom.

We will illustrate the idea by showing that the square root of 2 exists; namely, we will show that there exists a real number x_0 such that $x_0^2 = 2$.

PROOF. We begin by setting
$$S = \{x : x^2 < 2\}.$$

This set S is nonempty ($1 \in S$) and it is bounded above (by 5 for instance). It follows from the least upper bound axiom that S has a least upper bound. Call it x_0.

We will show that
$$x_0^2 = 2.$$

We do this by showing that x_0^2 cannot be less than 2 and cannot be greater than 2.

Suppose for a moment that
$$x_0^2 < 2.$$

Observe that $2 - x_0^2 > 0$ and choose a positive integer
$$n > \frac{2x_0 + 1}{2 - x_0^2} \cdot †$$

---

† We choose such an n because (as you'll see in a moment) for such an n we can prove that

$$\left(x_0 + \frac{1}{n}\right)^2 < 2.$$

This in turn will enable us to generate a contradiction.

From this inequality,

$$n(2 - x_0^2) > 2x_0 + 1,$$

$$\frac{1}{n}(2x_0 + 1) < 2 - x_0^2,$$

$$\frac{2x_0}{n} + \frac{1}{n} < 2 - x_0^2,$$

$$\frac{2x_0}{n} + \frac{1}{n^2} < 2 - x_0^2, \qquad\qquad \left(\frac{1}{n^2} \le \frac{1}{n}\right)$$

$$x_0^2 + \frac{2x_0}{n} + \frac{1}{n^2} < 2,$$

$$\left(x_0 + \frac{1}{n}\right)^2 < 2.$$

This last inequality says that

$$x_0 + \frac{1}{n} \in S,$$

which, since

$$x_0 < x_0 + \frac{1}{n},$$

means that the number x_0 cannot be an upper bound for S. But by assumption x_0 is an upper bound for S. This contradiction shows that x_0^2 cannot be less than 2.

In a similar manner, you can show that x_0^2 cannot be greater than 2. For if

$$x_0^2 > 2,$$

then you can find an integer k such that

$$\left(x_0 - \frac{1}{k}\right)^2 > 2. \qquad\qquad \left(\text{take } k > \frac{2x_0}{x_0^2 - 2}\right)$$

But this makes

$$x_0 - \frac{1}{k}$$

an upper bound for S, and that cannot be, because

$$x_0 - \frac{1}{k} < x_0$$

and x_0 is, by definition, the *least* upper bound for S.

This last contradiction shows that x_0^2 cannot be greater than 2. The only possibility that remains is that $x_0^2 = 2$. □

4. Irrational Exponents

In Section 4.11 we gave meaning to irrational powers by invoking a betweenness condition. We can arrive at the same result with more dispatch and seemingly with more precision by means of the least upper bound axiom.

Assume, for instance, that $b > 1$ and that z is irrational. To define b^z, we form the set of all rational powers of b where the exponent p/q is less than z:

$$\{b^{p/q} : p/q < z\}.$$

This set is nonempty and bounded above. It therefore has a least upper bound. This least upper bound is what we mean by b^z: for $b > 1$

(7.6.4)
$$b^z = \operatorname{lub} \{b^{p/q} : p/q < z\}. \qquad \square$$

Lower Bounds

We come now to lower bounds. In the first place, a number m is called a *lower bound* for S iff

$$m \leq x \qquad \text{for all } x \in S.$$

Sets that have lower bounds are said to be *bounded below*. Not all sets have lower bounds, but those that do have *greatest lower bounds*. This need not be taken as an axiom. Using the least upper bound axiom, we can prove it as a theorem.

Theorem   7.6.5

Every nonempty set of real numbers that has a lower bound has a *greatest* lower bound.

PROOF. Suppose that S is nonempty and that it has a lower bound x. Then
$$x \leq s \qquad \text{for all } s \in S.$$
It follows that
$$-s \leq -x \qquad \text{for all } s \in S;$$
that is,
$$\{-s : s \in S\} \quad \text{has an upper bound } -x.$$
From the least upper bound axiom we conclude that $\{-s : s \in S\}$ has a least upper bound. Call it x_0. Since
$$-s \leq x_0 \qquad \text{for all } s \in S,$$
you can see that
$$-x_0 \leq s \qquad \text{for all } s \in S,$$

and thus $-x_0$ is a lower bound for S. We now assert that $-x_0$ is the greatest lower bound of the set S. To see this, note that if there existed x_1 satisfying

$$-x_0 < x_1 \leq s \qquad \text{for all } s \in S,$$

then we would have

$$-s \leq -x_1 < x_0 \qquad \text{for all } s \in S,$$

and thus x_0 would not be the *least* upper bound of $\{-s : s \in S\}$.   □†

As in the case of the least upper bound, the greatest lower bound of a set need not be in the set, but it can be approximated as closely as we wish by members of the set. In short, we have the following theorem, the proof of which is left as an exercise.

Theorem 7.6.6

If m is the greatest lower bound of the set S and ϵ is positive, then there is a number s in S such that

$$m \leq s < m + \epsilon.$$

The theorem is illustrated in Figure 7.6.4. If $m = \text{glb } S$ (that is, if m is the greatest lower bound of the set S), then S has an element in every half-open interval of the form $[m, m + \epsilon)$.

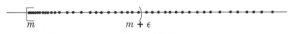
$$m \qquad\qquad\qquad m + \epsilon$$

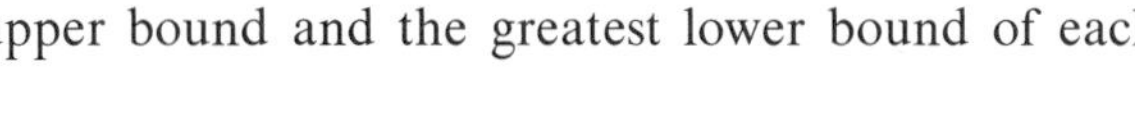

Figure 7.6.4

Exercises

Find the least upper bound and the greatest lower bound of each of the following sets:

*1. $(0, 2)$.   2. $[0, 2]$.

*3. $(0, \infty)$.   4. $(-\infty, 1)$.

*5. $\{x : x^2 < 4\}$.   6. $\{x : x^2 \leq 10\}$.

*7. $\{\frac{1}{2}, \frac{3}{4}, \frac{7}{8}, 1\}$.   8. $\{2\frac{1}{2}, 2\frac{1}{3}, 2\frac{1}{4}, \cdots\}$.

---

† We proved Theorem 7.6.5 by assuming the least upper bound axiom. We could have proceeded the other way. We could have set Theorem 7.6.5 as an axiom and then proved the least upper bound axiom as a theorem.

*9. $\{\frac{1}{2}, (\frac{1}{2})^2, (\frac{1}{2})^3, \cdots\}$.

10. $\{-1, -\frac{1}{2}, -\frac{1}{3}, -\frac{1}{4}, \cdots\}$.

*11. $\{10, 10^{1/2}, 10^{1/3}, 10^{1/4}, \cdots\}$.

12. $\{0.1, 0.11, 0.111, 0.1111, \cdots\}$.

*13. $\{0.6, 0.66, 0.666, 0.6666, \cdots\}$.

14. $\{8^{0.3}, 8^{0.33}, 8^{0.333}, \cdots\}$.

*15. $\{\frac{3}{4}, (\frac{3}{4})^2, (\frac{3}{4})^3, \cdots\}$.

16. $\{\frac{1}{2}, -\frac{2}{3}, \frac{3}{4}, -\frac{4}{5}, \cdots\}$.

17. Let $0 < b < 1$. Use the notion of a greatest lower bound to define b^z for z irrational.

18. Illustrate the validity of Theorem 7.6.6
     *(a) taking $S = \{\frac{1}{11}, (\frac{1}{11})^2, (\frac{1}{11})^3, \cdots, (\frac{1}{11})^n, \cdots\}$, $\epsilon = 0.001$.
     (b) taking $S = \{1, 2, 3, 4\}$, $\epsilon = 0.0001$.

Optional | 19. Prove Theorem 7.6.6 by imitating the proof of Theorem 7.6.2.

| 20. Prove that there is a real number x_0 such that $x_0^2 = 3$.

7.7 Cramer's Rule

In Section 3.6 you saw how to solve linear equations simultaneously. Here we discuss the subject more systematically.

For convenience we write the equations in the form

(7.7.1)
$$ax + by = e,$$
$$cx + dy = f.$$

We can solve for x by multiplying the first equation by d, the second equation by b, and then subtracting:

$$d(ax + by) = de,$$
$$b(cx + dy) = bf;$$

$$adx + bdy = de$$
$$bcx + bdy = bf;$$

$$(ad - bc)x = de - bf,$$

(*)   thus   if $ad - bc \neq 0$,   then $x = \dfrac{de - bf}{ad - bc}$.

In a similar manner you can verify that

(**)   if $ad - bc \neq 0$,   then $y = \dfrac{af - ce}{ad - bc}$.

In their present form formulas (∗) and (∗∗) are difficult to remember. The corresponding formulas for three linear equations in three unknowns are even more troublesome. The entire subject can be simplified enormously by the introduction of *matrices* and *determinants*. We begin with the 2 by 2 case.

A square array of numbers

$$\begin{pmatrix} a & b \\ c & d \end{pmatrix}$$

is called a *2 by 2 matrix*. The *determinant* of such a matrix, indicated by replacing the parentheses by vertical bars, is the number $ad - bc$. In symbols we write

(7.7.2)
$$\begin{vmatrix} a & b \\ c & d \end{vmatrix} = ad - bc.$$

Thus, for example,

$$\begin{vmatrix} 2 & 3 \\ 4 & 5 \end{vmatrix} = (2)(5) - (3)(4) = 10 - 12 = -2$$

and

$$\begin{vmatrix} 2 & 5 \\ -4 & 3 \end{vmatrix} = (2)(3) - (5)(-4) = 6 + 20 = 26. \quad \square$$

The explicit solution of Equations 7.7.1 written in determinant notation is called Cramer's Rule.

Theorem 7.7.3   Cramer's Rule (2 by 2 case)

If

$$\begin{vmatrix} a & b \\ c & d \end{vmatrix} \neq 0,$$

then the equations

$$ax + by = e$$
$$cx + dy = f$$

have the unique simultaneous solution

$$x = \frac{\begin{vmatrix} e & b \\ f & d \end{vmatrix}}{\begin{vmatrix} a & b \\ c & d \end{vmatrix}}, \qquad y = \frac{\begin{vmatrix} a & e \\ c & f \end{vmatrix}}{\begin{vmatrix} a & b \\ c & d \end{vmatrix}}.$$

In each case the denominator is just the determinant of the matrix of coefficients:

$$ax + by = e,$$
$$cx + dy = f.$$

To obtain the numerator for x replace the x-coefficient column $\begin{pmatrix} a \\ c \end{pmatrix}$ by the column of constants $\begin{pmatrix} e \\ f \end{pmatrix}$. To obtain the numerator for y replace the y-coefficient column $\begin{pmatrix} b \\ d \end{pmatrix}$ by the column of constants $\begin{pmatrix} e \\ f \end{pmatrix}$.

Problem. Solve the equations

$$2x - 5y = 1,$$
$$x + 4y = -2$$

simultaneously by Cramer's rule.

SOLUTION

$$x = \frac{\begin{vmatrix} 1 & -5 \\ -2 & 4 \end{vmatrix}}{\begin{vmatrix} 2 & -5 \\ 1 & 4 \end{vmatrix}} = \frac{(1)(4) - (-5)(-2)}{(2)(4) - (-5)(1)} = \frac{4 - 10}{13} = -\frac{6}{13},$$

$$y = \frac{\begin{vmatrix} 2 & 1 \\ 1 & -2 \end{vmatrix}}{\begin{vmatrix} 2 & -5 \\ 1 & 4 \end{vmatrix}} = \frac{(2)(-2) - (1)(1)}{(2)(4) - (-5)(1)} = \frac{-4 - 1}{13} = -\frac{5}{13}.$$

You can check these results by substitution.   □

A 3 by 3 square array of numbers

$$\begin{pmatrix} a & b & c \\ d & e & f \\ g & h & i \end{pmatrix}$$

is called a *3 by 3 matrix*. The determinant of such a matrix can be defined by the equation

(7.7.4)
$$\begin{vmatrix} a & b & c \\ d & e & f \\ g & h & i \end{vmatrix} = a \begin{vmatrix} e & f \\ h & i \end{vmatrix} - b \begin{vmatrix} d & f \\ g & i \end{vmatrix} + c \begin{vmatrix} d & e \\ g & h \end{vmatrix}.$$

Note that

$$\begin{vmatrix} e & f \\ h & i \end{vmatrix} = \begin{vmatrix} a & b & c \\ d & e & f \\ g & h & i \end{vmatrix}, \qquad \begin{vmatrix} d & f \\ g & i \end{vmatrix} = \begin{vmatrix} a & b & c \\ d & e & f \\ g & h & i \end{vmatrix}, \qquad \begin{vmatrix} d & e \\ g & h \end{vmatrix} = \begin{vmatrix} a & b & c \\ d & e & f \\ g & h & i \end{vmatrix}.$$

Example

$$\begin{vmatrix} 3 & 2 & 4 \\ -2 & 3 & 2 \\ 5 & 1 & -1 \end{vmatrix} = 3\begin{vmatrix} 3 & 2 \\ 1 & -1 \end{vmatrix} - 2\begin{vmatrix} -2 & 2 \\ 5 & -1 \end{vmatrix} + 4\begin{vmatrix} -2 & 3 \\ 5 & 1 \end{vmatrix}$$

$$= 3[(3)(-1) - (2)(1)] - 2[(-2)(-1) - (2)(5)] + 4[(-2)(1) - (3)(5)]$$

$$= 3(-3 - 2) - 2(2 - 10) + 4(-2 - 15)$$

$$= -67. \quad \square$$

The simultaneous solution of 3 linear equations in 3 unknowns

$$ax + by + cz = j$$

(7.7.5) $$dx + ey + fz = k$$

$$gx + hy + iz = l$$

can be obtained by manipulating the equations as we did in the case of 2 unknowns. As before, the result is much easier to state in terms of determinants.

Theorem   7.7.6   Cramer's Rule (3 by 3 case)

If

$$\begin{vmatrix} a & b & c \\ d & e & f \\ g & h & i \end{vmatrix} \neq 0,$$

Equations 7.7.5 have the unique simultaneous solution

$$x = \frac{\begin{vmatrix} j & b & c \\ k & e & f \\ l & h & i \end{vmatrix}}{\begin{vmatrix} a & b & c \\ d & e & f \\ g & h & i \end{vmatrix}}, \qquad y = \frac{\begin{vmatrix} a & j & c \\ d & k & f \\ g & l & i \end{vmatrix}}{\begin{vmatrix} a & b & c \\ d & e & f \\ g & h & i \end{vmatrix}}, \qquad z = \frac{\begin{vmatrix} a & b & j \\ d & e & k \\ g & h & l \end{vmatrix}}{\begin{vmatrix} a & b & c \\ d & e & f \\ g & h & i \end{vmatrix}}.$$

Problem. Solve the equations

$$3x - 2y + z = 7$$
$$x + y - z = -2$$
$$2x + 3y + 5z = 9$$

simultaneously by Cramer's Rule.

SOLUTION. First we compute the determinant of the matrix of coefficients:

$$\begin{vmatrix} 3 & -2 & 1 \\ 1 & 1 & -1 \\ 2 & 3 & 5 \end{vmatrix} = 3\begin{vmatrix} 1 & -1 \\ 3 & 5 \end{vmatrix} + 2\begin{vmatrix} 1 & -1 \\ 2 & 5 \end{vmatrix} + \begin{vmatrix} 1 & 1 \\ 2 & 3 \end{vmatrix}$$

$$= 3(5 + 3) + 2(5 + 2) + (3 - 2) = 24 + 14 + 1 = 39.$$

Cramer's rule now gives

$$x = \tfrac{1}{39}\begin{vmatrix} 7 & -2 & 1 \\ -2 & 1 & -1 \\ 9 & 3 & 5 \end{vmatrix} = \tfrac{1}{39}\left[7\begin{vmatrix} 1 & -1 \\ 3 & 5 \end{vmatrix} + 2\begin{vmatrix} -2 & -1 \\ 9 & 5 \end{vmatrix} + \begin{vmatrix} -2 & 1 \\ 9 & 3 \end{vmatrix}\right]$$

$$= \tfrac{1}{39}[7(5 + 3) + 2(-10 + 9) + (-6 - 9)]$$

$$= \tfrac{1}{39}(56 - 2 - 15) = \tfrac{39}{39} = 1,$$

$$y = \tfrac{1}{39}\begin{vmatrix} 3 & 7 & 1 \\ 1 & -2 & -1 \\ 2 & 9 & 5 \end{vmatrix} = \tfrac{1}{39}\left[3\begin{vmatrix} -2 & -1 \\ 9 & 5 \end{vmatrix} - 7\begin{vmatrix} 1 & -1 \\ 2 & 5 \end{vmatrix} + \begin{vmatrix} 1 & -2 \\ 2 & 9 \end{vmatrix}\right]$$

$$= \tfrac{1}{39}[3(-10 + 9) - 7(5 + 2) + (9 + 4)]$$

$$= \tfrac{1}{39}(-3 - 49 + 13) = -\tfrac{39}{39} = -1,$$

$$z = \tfrac{1}{39}\begin{vmatrix} 3 & -2 & 7 \\ 1 & 1 & -2 \\ 2 & 3 & 9 \end{vmatrix} = \tfrac{1}{39}\left[3\begin{vmatrix} 1 & -2 \\ 3 & 9 \end{vmatrix} + 2\begin{vmatrix} 1 & -2 \\ 2 & 9 \end{vmatrix} + 7\begin{vmatrix} 1 & 1 \\ 2 & 3 \end{vmatrix}\right]$$

$$= \tfrac{1}{39}[3(9 + 6) + 2(9 + 4) + 7(3 - 2)]$$

$$= \tfrac{1}{39}(45 + 26 + 7) = \tfrac{78}{39} = 2. \quad \square$$

Determinant Zero

Cramer's rule presupposes that the matrix of coefficients has a nonzero determinant. If the matrix of coefficients has determinant zero, then there are two possibilities: *Possibility 1.* The equations are incompatible and have no simultaneous solution. As an example take the equations

$$3x + 4y = 0,$$
$$3x + 4y = 1.$$

The second equation contradicts the first one.

Possibility 2. The equations are redundant and there is an infinite number of solutions. As an example take the equations

$$3x + 4y = 1,$$
$$6x + 8y = 2.$$

The second equation is just the first equation in disguise. The infinite number of pairs (x, y) that satisfy the first equation also satisfy the second equation.

Exercises

Solve the equations simultaneously by Cramer's rule.

*1. $3x - 2y = 7$
     $x + 3y = 6.$

2. $x + 6y = 24$
     $2x - 5y = -20.$

*3. $2x + 7y = 25$
     $5x - 3y = 1.$

*4. $6x - 5y = 3$
     $8x + 7y = 4.$

5. $\frac{1}{2}x + 6y = 21$
     $\frac{1}{3}x + \frac{1}{2}y = 0.$

*6. $4x - 5y = 7$
     $3x + 4y = 1.$

*7. $x - 2y + 3z = 4$
     $4x - 5y + 3z = 7$
     $2x + 4y - 5z = -3.$

8. $6x - \frac{1}{2}y + 3z = 1$
     $x + 4y - 5z = 3$
     $5x - 3y + 2z = 7.$

*9. $7x + 4y - 6z = 0$
     $2x - 3y + 3z = 1$
     $x + 2y - 2z = 2.$

10. $12x - 5y + 3z = 1$
     $6x + 2y - 3z = 0$
     $4x + 5y - 6z = 14.$

7.8 Complex Roots of Quadratic Equations

While there are real numbers x with the property that

$$x^2 = 1,$$

there are no real numbers x such that

$$x^2 = -1.$$

More generally, while a quadratic

$$ax^2 + bx + c = 0$$

has real roots if $b^2 - 4ac \geq 0$, it has no real roots if $b^2 - 4ac < 0$. We can overcome this deficiency by using *complex numbers*.

What are the Complex Numbers?

Just as the real numbers can be thought of as points of the number line, the *complex numbers* can be thought of as points of the coordinate plane. (See Figure 7.8.1.) For *x*-axis write *real axis*; for *y*-axis write *imaginary axis*; for the point (a, b) write $a + bi$.

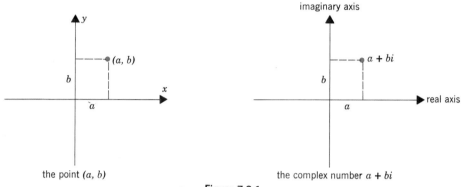

Figure 7.8.1

Some examples are given in Figure 7.8.2.

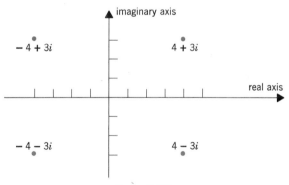

Figure 7.8.2

If $b = 0$, then $a + bi$ reduces to $a + 0i$, which we write simply as a. By this device it is clear that the set of real numbers forms a subset of the set of complex numbers:

$$\text{if} \quad x \text{ is real}, \quad \text{then} \quad x = x + 0i.$$

If $a = 0$, then $a + bi$ reduces to $0 + bi$, which we write simply as bi. Complex numbers of the form bi are called *purely imaginary*.

Thus a complex number $a + bi$ consists of two parts: a *real part a* and a *purely imaginary part bi*.

Some Arithmetic with Complex Numbers

To add or subtract complex numbers, we add the real and imaginary parts separately. By definition,

(7.8.1)

$$(a + bi) + (c + di) = (a + c) + (b + d)i,$$
$$(a + bi) - (c + di) = (a - c) + (b - d)i.$$

Thus, for example,

$$(2 + 3i) + (6 - 5i) = 8 - 2i,$$
$$(2 + 3i) - (6 - 5i) = -4 + 8i,$$
$$(2 + 3i) - (2 + 3i) = 0 + 0i = 0. \quad \square$$

Multiplication (the heart of the matter for our purposes) is defined by setting

(7.8.2)

$$(a + bi)(c + di) = (ac - bd) + (ad + bc)i.$$

Here are some routine multiplications carried out:

$$(2 + 3i)(6 + 5i) = [(2)(6) - (3)(5)] + [(2)(5) + (3)(6)]i$$
$$= -3 + 28i.$$
$$(2 + 3i)(6 - 5i) = [(2)(6) - (3)(-5)] + [(2)(-5) + (3)(6)]i$$
$$= 27 + 8i.$$
$$(2 - 3i)(6 - 5i) = [(2)(6) - (-3)(-5)] + [(2)(-5) + (-3)(6)]i$$
$$= -3 - 28i. \quad \square$$

Square Roots of Negative Numbers

The special merit of multiplication according to the rule

$$(a + bi)(c + di) = (ac - bd) + (ad + bc)i$$

is that it gives us a square root of -1. In fact it gives us two of them: $\pm i$.

Theorem 7.8.3   Square Roots of -1

$$i^2 = -1 \quad \text{and} \quad (-i)^2 = -1.$$

PROOF

$$i^2 = (0 + i)(0 + i) = [(0)(0) - (1)(1)] + [(0)(1) + (1)(0)]i$$
$$= -1 + 0i = -1,$$
$$(-i)^2 = (0 - i)(0 - i) = [(0)(0) - (-1)(-1)] + [(0)(-1) + (-1)(0)]i$$
$$= -1 + 0i = -1. \quad \square$$

Not only can we take the square root of -1, but we can take the square root of any negative number $-p$.

Theorem 7.8.4   Square Roots of Negative Numbers

If $p > 0$, the equation $x^2 = -p$ has two purely imaginary roots

$$x = \pm \sqrt{p}\,i.$$

PROOF

$$(\sqrt{p}\,i)^2 = (0 + \sqrt{p}\,i)(0 + \sqrt{p}\,i)$$
$$= [(0)(0) - \sqrt{p}\sqrt{p}] + [(0)(\sqrt{p}) + (\sqrt{p})(0)]i$$
$$= -p + 0i = -p.$$
$$(-\sqrt{p}\,i)^2 = (0 - \sqrt{p}\,i)(0 - \sqrt{p}\,i)$$
$$= [(0)(0) - (-\sqrt{p})(-\sqrt{p})] + [(0)(-\sqrt{p}) + (-\sqrt{p})(0)]i$$
$$= -p + 0i = -p. \quad \square$$

Complex Numbers and Quadratic Equations

With complex numbers at our disposal, we can take the square root of every real number—positive, zero, or negative. We can therefore apply the general quadratic formula

$$x = \frac{-b \pm \sqrt{b^2 - 4ac}}{2a}$$

to all quadratic equations

$$ax^2 + bx + c = 0.$$

We have solved equations with $b^2 - 4ac \geq 0$ before. Here we solve some where $b^2 - 4ac < 0$.

Problem. Solve the equation

$$x^2 + x + 1 = 0.$$

SOLUTION. According to the general quadratic formula,

$$x = \frac{-1 \pm \sqrt{1 - 4}}{2} = -\tfrac{1}{2} \pm \tfrac{1}{2}\sqrt{-3} = -\tfrac{1}{2} \pm \tfrac{1}{2}\sqrt{3}\,i.$$

We can check these roots by substituting them into the initial equation. Taking the root with the $+$ sign, we have

$$(-\tfrac{1}{2} + \tfrac{1}{2}\sqrt{3}\,i)^2 = (-\tfrac{1}{2} + \tfrac{1}{2}\sqrt{3}\,i)(-\tfrac{1}{2} + \tfrac{1}{2}\sqrt{3}\,i)$$

$$= [(-\tfrac{1}{2})(-\tfrac{1}{2}) - (\tfrac{1}{2}\sqrt{3})(\tfrac{1}{2}\sqrt{3})] + [(-\tfrac{1}{2})(\tfrac{1}{2}\sqrt{3}) + (\tfrac{1}{2}\sqrt{3})(-\tfrac{1}{2})]i$$

$$= (\tfrac{1}{4} - \tfrac{3}{4}) + (-\tfrac{1}{4}\sqrt{3} - \tfrac{1}{4}\sqrt{3})i$$

$$= -\tfrac{1}{2} - \tfrac{1}{2}\sqrt{3}\,i,$$

so that

$$(-\tfrac{1}{2} + \tfrac{1}{2}\sqrt{3}\,i)^2 + (-\tfrac{1}{2} + \tfrac{1}{2}\sqrt{3}\,i) + 1 = (-\tfrac{1}{2} - \tfrac{1}{2}\sqrt{3}\,i) + (-\tfrac{1}{2} + \tfrac{1}{2}\sqrt{3}\,i) + (1 + 0i)$$

$$= (-1 + 1) + 0i = 0 + 0i \overset{\checkmark}{=} 0.$$

As you can verify, the other root checks out also. □

Problem. Solve the equation

$$2x^2 + 3x + 7 = 0.$$

SOLUTION. The general formula gives

$$x = \frac{-3 \pm \sqrt{9 - 56}}{4} = -\tfrac{3}{4} \pm \tfrac{1}{4}\sqrt{-47} = -\tfrac{3}{4} \pm \tfrac{1}{4}\sqrt{47}\,i.$$

Once again you can check these roots by substitution. □

Exercises

*1. Plot these complex numbers:
    (a) $1 + 2i$.　　(b) $1 - 2i$.　　(c) $-1 + 2i$.　　(d) $-1 - 2i$.
    (e) $3i$.　　　　(f) $-3i$.　　　(g) 1.　　　　(h) -1.

Calculate:

*2. $(5 + i) + (3 - 2i)$.　　　　　　3. $(5 + i) - (3 - 2i)$.

*4. $(6 - 4i) - (5 - 3i) + i.$ 5. $(6 - 4i) + (5 - 3i) - i.$

*6. $(5 + i)(3 - 2i).$ 7. $(5 + i)(3 + 2i).$

*8. $(6 - 4i)(5 - 3i).$ 9. $(6 - 4i)(-5 - 3i).$

*10. $i(5 + i).$ 11. $i(5 - i).$

*12. $i^3.$ 13. $i^4.$

*14. $(1 + i)^3.$ 15. $(1 - i)^3.$

*16. $(\frac{1}{2}\sqrt{2} + \frac{1}{2}\sqrt{2}i)^2.$ 17. $(\frac{1}{2}\sqrt{2} + \frac{1}{2}\sqrt{2}i)^4.$

Solve the following quadratic equations:

*18. $x^2 + 9 = 0.$ 19. $x^2 + 2x + 9 = 0.$

*20. $3x^2 + 2x + 1 = 0.$ 21. $3x^2 + 2x + 10 = 0.$

*22. $2x^2 + 2x + 1 = 0.$ 23. $5x^2 - x + 7 = 0.$

24. Show that the addition of complex numbers is commutative:

$$(a + bi) + (c + di) = (c + di) + (a + bi).$$

25. Show that the multiplication of complex numbers is commutative:

$$(a + bi)(c + di) = (c + di)(a + bi).$$

*26. Interpret the addition of complex numbers geometrically.

27. The *absolute value* of a complex number $z = a + bi$ is defined by setting

(7.8.5) $$\boxed{|z| = \sqrt{a^2 + b^2}.}$$

(a) Interpret $|z|$ geometrically.

(b) Show that the absolute value of the product of two complex numbers is the product of their absolute values:

(7.8.6) $$\boxed{|z_1 z_2| = |z_1||z_2|.}$$

[HINT: Set $z_1 = a + bi$, $z_2 = c + di$ and carry out the calculations.]

(c) The *complex conjugate* $\bar{z}$ of a complex number $z = a + bi$ is defined by setting

$$\bar{z} = a - bi.$$

Show that

(7.8.7) $$\boxed{|\bar{z}| = |z| \quad \text{and} \quad z\bar{z} = |z|^2.}$$

28. The quotient s/t of two real numbers ($t \neq 0$) is the real number x that satisfies the equation

$$tx = s.$$

The quotient z_1/z_2 of two complex numbers ($z \neq 0$) is the complex number w that satisfies the equation

$$z_2 w = z_1.$$

(a) Show that if $c + di \neq 0$, then

(7.8.8)
$$\boxed{\frac{a + bi}{c + di} = \frac{ac + bd}{c^2 + d^2} + \frac{bc - ad}{c^2 + d^2}i.}$$

(b) Show that the division formula given in part (a) can be obtained by multiplying numerator and denominator of

$$\frac{a + bi}{c + di}$$

by $c - di$.

Compute the following quotients by multiplying numerator and denominator by the complex conjugate of the denominator:

*29. $\dfrac{1}{1 + i}$.

30. $\dfrac{1}{1 - i}$.

*31. $\dfrac{3 + 2i}{2 + i}$.

*32. $\dfrac{i}{3 - 5i}$.

33. $\dfrac{2 + 5i}{5 - 2i}$.

*34. $\dfrac{6 - 7i}{4 + 5i}$.

TABLES

Table 1. Sines, cosines, tangents (radian measure).

t	sin t	cos t	tan t	t	sin t	cos t	tan t
0.00	0.000	1.000	0.000	0.34	0.333	0.943	0.354
0.01	0.010	1.000	0.010	0.35	0.343	0.939	0.365
0.02	0.020	1.000	0.020	0.36	0.352	0.936	0.376
0.03	0.030	1.000	0.030	0.37	0.362	0.932	0.388
0.04	0.040	0.999	0.040	0.38	0.371	0.929	0.399
0.05	0.050	0.999	0.050	0.39	0.380	0.925	0.411
0.06	0.060	0.998	0.060	0.40	0.389	0.921	0.423
0.07	0.070	0.998	0.070	0.41	0.399	0.917	0.435
0.08	0.080	0.997	0.080	0.42	0.408	0.913	0.447
0.09	0.090	0.996	0.090	0.43	0.417	0.909	0.459
0.10	0.100	0.995	0.100	0.44	0.426	0.905	0.471
0.11	0.110	0.994	0.110	0.45	0.435	0.900	0.483
0.12	0.120	0.993	0.121	0.46	0.444	0.896	0.495
0.13	0.130	0.992	0.131	0.47	0.453	0.892	0.508
0.14	0.140	0.990	0.141	0.48	0.462	0.887	0.521
0.15	0.149	0.989	0.151	0.49	0.471	0.882	0.533
0.16	0.159	0.987	0.161	0.50	0.479	0.878	0.546
0.17	0.169	0.986	0.172	0.51	0.488	0.873	0.559
0.18	0.179	0.984	0.182	0.52	0.497	0.868	0.573
0.19	0.189	0.982	0.192	0.53	0.506	0.863	0.586
0.20	0.199	0.980	0.203	0.54	0.514	0.858	0.599
0.21	0.208	0.978	0.213	0.55	0.523	0.853	0.613
0.22	0.218	0.976	0.224	0.56	0.531	0.847	0.627
0.23	0.228	0.974	0.234	0.57	0.540	0.842	0.641
0.24	0.238	0.971	0.245	0.58	0.548	0.836	0.655
0.25	0.247	0.969	0.255	0.59	0.556	0.831	0.670
0.26	0.257	0.966	0.266	0.60	0.565	0.825	0.684
0.27	0.267	0.964	0.277	0.61	0.573	0.820	0.699
0.28	0.276	0.961	0.288	0.62	0.581	0.814	0.714
0.29	0.286	0.958	0.298	0.63	0.589	0.808	0.729
0.30	0.296	0.955	0.309	0.64	0.597	0.802	0.745
0.31	0.305	0.952	0.320	0.65	0.605	0.796	0.760
0.32	0.315	0.949	0.331	0.66	0.613	0.790	0.776
0.33	0.324	0.946	0.343	0.67	0.621	0.784	0.792

Table 1. (continued)

t	sin t	cos t	tan t	t	sin t	cos t	tan t
0.68	0.629	0.778	0.809	1.14	0.909	0.418	2.176
0.69	0.637	0.771	0.825	1.15	0.913	0.408	2.234
0.70	0.644	0.765	0.842	1.16	0.917	0.399	2.296
0.71	0.652	0.758	0.860	1.17	0.921	0.390	2.360
0.72	0.659	0.752	0.877	1.18	0.925	0.381	2.427
0.73	0.667	0.745	0.895	1.19	0.928	0.372	2.498
0.74	0.674	0.738	0.913	1.20	0.932	0.362	2.572
0.75	0.682	0.732	0.932	1.21	0.936	0.353	2.650
0.76	0.689	0.725	0.950	1.22	0.939	0.344	2.733
0.77	0.696	0.718	0.970	1.23	0.942	0.334	2.820
0.78	0.703	0.711	0.989	1.24	0.946	0.325	2.912
0.79	0.710	0.704	1.009	1.25	0.949	0.315	3.010
0.80	0.717	0.697	1.030	1.26	0.952	0.306	3.113
0.81	0.724	0.689	1.050	1.27	0.955	0.296	3.224
0.82	0.731	0.682	1.072	1.28	0.958	0.287	3.341
0.83	0.738	0.675	1.093	1.29	0.961	0.277	3.467
0.84	0.745	0.667	1.116	1.30	0.964	0.267	3.602
0.85	0.751	0.660	1.138	1.31	0.966	0.258	3.747
0.86	0.758	0.652	1.162	1.32	0.969	0.248	3.903
0.87	0.764	0.645	1.185	1.33	0.971	0.238	4.072
0.88	0.771	0.637	1.210	1.34	0.973	0.229	4.256
0.89	0.777	0.629	1.235	1.35	0.976	0.219	4.455
0.90	0.783	0.622	1.260	1.36	0.978	0.209	4.673
0.91	0.790	0.614	1.286	1.37	0.980	0.199	4.913
0.92	0.796	0.606	1.313	1.38	0.982	0.190	5.177
0.93	0.802	0.598	1.341	1.39	0.984	0.180	5.471
0.94	0.808	0.590	1.369	1.40	0.985	0.170	5.798
0.95	0.813	0.582	1.398	1.41	0.987	0.160	6.165
0.96	0.819	0.574	1.428	1.42	0.989	0.150	6.581
0.97	0.825	0.565	1.459	1.43	0.990	0.140	7.055
0.98	0.830	0.557	1.491	1.44	0.991	0.130	7.602
0.99	0.836	0.549	1.524	1.45	0.993	0.121	8.238
1.00	0.841	0.540	1.557	1.46	0.994	0.111	8.989
1.01	0.847	0.532	1.592	1.47	0.995	0.101	9.887
1.02	0.852	0.523	1.628	1.48	0.996	0.091	10.983
1.03	0.857	0.515	1.665	1.49	0.997	0.081	12.350
1.04	0.862	0.506	1.704	1.50	0.997	0.071	14.101
1.05	0.867	0.498	1.743	1.51	0.998	0.061	16.428
1.06	0.872	0.489	1.784	1.52	0.999	0.051	19.670
1.07	0.877	0.480	1.827	1.53	0.999	0.041	24.498
1.08	0.882	0.471	1.871	1.54	1.000	0.031	32.461
1.09	0.887	0.462	1.917	1.55	1.000	0.021	48.078
1.10	0.891	0.454	1.965	1.56	1.000	0.011	92.620
1.11	0.896	0.445	2.014	1.57	1.000	0.001	1255.770
1.12	0.900	0.436	2.066	1.58	1.000	-0.009	-108.649
1.13	0.904	0.427	2.120				

Table 2.   Sines, cosines, tangents (degree measure).

θ	sin θ	cos θ	tan θ	θ	sin θ	cos θ	tan θ
0°	0.0000	1.0000	0.0000	45°	0.7071	0.7071	1.000
1	0.0175	0.9998	0.0175	46	0.7193	0.6947	1.036
2	0.0349	0.9994	0.0349	47	0.7314	0.6820	1.072
3	0.0523	0.9986	0.0524	48	0.7431	0.6691	1.111
4	0.0698	0.9976	0.0699	49	0.7547	0.6561	1.150
5	0.0872	0.9962	0.0875	50	0.7660	0.6428	1.192
6	0.1045	0.9945	0.1051	51	0.7771	0.6293	1.235
7	0.1219	0.9925	0.1228	52	0.7880	0.6157	1.280
8	0.1392	0.9903	0.1405	53	0.7986	0.6018	1.327
9	0.1564	0.9877	0.1584	54	0.8090	0.5878	1.376
10	0.1736	0.9848	0.1763	55	0.8192	0.5736	1.428
11	0.1908	0.9816	0.1944	56	0.8290	0.5592	1.483
12	0.2079	0.9781	0.2126	57	0.8387	0.5446	1.540
13	0.2250	0.9744	0.2309	58	0.8480	0.5299	1.600
14	0.2419	0.9703	0.2493	59	0.8572	0.5150	1.664
15	0.2588	0.9659	0.2679	60	0.8660	0.5000	1.732
16	0.2756	0.9613	0.2867	61	0.8746	0.4848	1.804
17	0.2924	0.9563	0.3057	62	0.8829	0.4695	1.881
18	0.3090	0.9511	0.3249	63	0.8910	0.4540	1.963
19	0.3256	0.9455	0.3443	64	0.8988	0.4384	2.050
20	0.3420	0.9397	0.3640	65	0.9063	0.4226	2.145
21	0.3584	0.9336	0.3839	66	0.9135	0.4067	2.246
22	0.3746	0.9272	0.4040	67	0.9205	0.3907	2.356
23	0.3907	0.9205	0.4245	68	0.9272	0.3746	2.475
24	0.4067	0.9135	0.4452	69	0.9336	0.3584	2.605
25	0.4226	0.9063	0.4663	70	0.9397	0.3420	2.747
26	0.4384	0.8988	0.4877	71	0.9455	0.3256	2.904
27	0.4540	0.8910	0.5095	72	0.9511	0.3090	3.078
28	0.4695	0.8829	0.5317	73	0.9563	0.2924	3.271
29	0.4848	0.8746	0.5543	74	0.9613	0.2756	3.487
30	0.5000	0.8660	0.5774	75	0.9659	0.2588	3.732
31	0.5150	0.8572	0.6009	76	0.9703	0.2419	4.011
32	0.5299	0.8480	0.6249	77	0.9744	0.2250	4.331
33	0.5446	0.8387	0.6494	78	0.9781	0.2079	4.705
34	0.5592	0.8290	0.6745	79	0.9816	0.1908	5.145
35	0.5736	0.8192	0.7002	80	0.9848	0.1736	5.671
36	0.5878	0.8090	0.7265	81	0.9877	0.1564	6.314
37	0.6018	0.7986	0.7536	82	0.9903	0.1392	7.115
38	0.6157	0.7880	0.7813	83	0.9925	0.1219	8.144
39	0.6293	0.7771	0.8098	84	0.9945	0.1045	9.514
40	0.6428	0.7660	0.8391	85	0.9962	0.0872	11.43
41	0.6561	0.7547	0.8693	86	0.9976	0.0698	14.30
42	0.6691	0.7431	0.9004	87	0.9986	0.0523	19.08
43	0.6820	0.7314	0.9325	88	0.9994	0.0349	28.64
44	0.6947	0.7193	0.9657	89	0.9998	0.0175	57.29
45	0.7071	0.7071	1.0000	90	1.0000	0.0000	—

Table 3.   Four-place logarithms (base 10).   1.00–5.49

x	0	1	2	3	4	5	6	7	8	9
1.0	0.0000	0.0043	0.0086	0.0128	0.0170	0.0212	0.0253	0.0294	0.0334	0.0374
1.1	0.0414	0.0453	0.0492	0.0531	0.0569	0.0607	0.0645	0.0682	0.0719	0.0755
1.2	0.0792	0.0828	0.0864	0.0899	0.0934	0.0969	0.1004	0.1038	0.1072	0.1106
1.3	0.1139	0.1173	0.1206	0.1239	0.1271	0.1303	0.1335	0.1367	0.1399	0.1430
1.4	0.1461	0.1492	0.1523	0.1553	0.1584	0.1614	0.1644	0.1673	0.1703	0.1732
1.5	0.1761	0.1790	0.1818	0.1847	0.1875	0.1903	0.1931	0.1959	0.1987	0.2014
1.6	0.2041	0.2068	0.2095	0.2122	0.2148	0.2175	0.2201	0.2227	0.2253	0.2279
1.7	0.2304	0.2330	0.2355	0.2380	0.2405	0.2430	0.2455	0.2480	0.2504	0.2529
1.8	0.2553	0.2577	0.2601	0.2625	0.2648	0.2672	0.2695	0.2718	0.2742	0.2765
1.9	0.2788	0.2810	0.2833	0.2856	0.2878	0.2900	0.2923	0.2945	0.2967	0.2989
2.0	0.3010	0.3032	0.3054	0.3075	0.3096	0.3118	0.3139	0.3160	0.3181	0.3201
2.1	0.3222	0.3243	0.3263	0.3284	0.3304	0.3324	0.3345	0.3365	0.3385	0.3404
2.2	0.3424	0.3444	0.3464	0.3483	0.3502	0.3522	0.3541	0.3560	0.3579	0.3598
2.3	0.3617	0.3636	0.3655	0.3674	0.3692	0.3711	0.3729	0.3747	0.3766	0.3784
2.4	0.3802	0.3820	0.3838	0.3856	0.3874	0.3892	0.3909	0.3927	0.3945	0.3962
2.5	0.3979	0.3997	0.4014	0.4031	0.4048	0.4065	0.4082	0.4099	0.4116	0.4133
2.6	0.4150	0.4166	0.4183	0.4200	0.4216	0.4232	0.4249	0.4265	0.4281	0.4298
2.7	0.4314	0.4330	0.4346	0.4362	0.4378	0.4393	0.4409	0.4425	0.4440	0.4456
2.8	0.4472	0.4487	0.4502	0.4518	0.4533	0.4548	0.4564	0.4579	0.4594	0.4609
2.9	0.4624	0.4639	0.4654	0.4669	0.4683	0.4698	0.4713	0.4728	0.4742	0.4757
3.0	0.4771	0.4786	0.4800	0.4814	0.4829	0.4843	0.4857	0.4871	0.4886	0.4900
3.1	0.4914	0.4928	0.4942	0.4955	0.4969	0.4983	0.4997	0.5011	0.5024	0.5038
3.2	0.5051	0.5065	0.5079	0.5092	0.5105	0.5119	0.5132	0.5145	0.5159	0.5172
3.3	0.5185	0.5198	0.5211	0.5224	0.5237	0.5250	0.5263	0.5276	0.5289	0.5302
3.4	0.5315	0.5328	0.5340	0.5353	0.5366	0.5378	0.5391	0.5403	0.5416	0.5428
3.5	0.5441	0.5453	0.5465	0.5478	0.5490	0.5502	0.5514	0.5527	0.5539	0.5551
3.6	0.5563	0.5575	0.5587	0.5599	0.5611	0.5623	0.5635	0.5647	0.5658	0.5670
3.7	0.5682	0.5694	0.5705	0.5717	0.5729	0.5740	0.5752	0.5763	0.5775	0.5786
3.8	0.5798	0.5809	0.5821	0.5832	0.5843	0.5855	0.5866	0.5877	0.5888	0.5899
3.9	0.5911	0.5922	0.5933	0.5944	0.5955	0.5966	0.5977	0.5988	0.5999	0.6010
4.0	0.6021	0.6031	0.6042	0.6053	0.6064	0.6075	0.6085	0.6096	0.6107	0.6117
4.1	0.6128	0.6138	0.6149	0.6160	0.6170	0.6180	0.6191	0.6201	0.6212	0.6222
4.2	0.6232	0.6243	0.6253	0.6263	0.6274	0.6284	0.6294	0.6304	0.6314	0.6325
4.3	0.6335	0.6345	0.6355	0.6365	0.6375	0.6385	0.6395	0.6405	0.6415	0.6425
4.4	0.6435	0.6444	0.6454	0.6464	0.6474	0.6484	0.6493	0.6503	0.6513	0.6522
4.5	0.6532	0.6542	0.6551	0.6561	0.6571	0.6580	0.6590	0.6599	0.6609	0.6618
4.6	0.6628	0.6637	0.6646	0.6656	0.6665	0.6675	0.6684	0.6693	0.6702	0.6712
4.7	0.6721	0.6730	0.6739	0.6749	0.6758	0.6767	0.6776	0.6785	0.6794	0.6803
4.8	0.6812	0.6821	0.6830	0.6839	0.6848	0.6857	0.6866	0.6875	0.6884	0.6893
4.9	0.6902	0.6911	0.6920	0.6928	0.6937	0.6946	0.6955	0.6964	0.6972	0.6981
5.0	0.6990	0.6998	0.7007	0.7016	0.7024	0.7033	0.7042	0.7050	0.7059	0.7067
5.1	0.7076	0.7084	0.7093	0.7101	0.7110	0.7118	0.7126	0.7135	0.7143	0.7152
5.2	0.7160	0.7168	0.7177	0.7185	0.7193	0.7202	0.7210	0.7218	0.7226	0.7235
5.3	0.7243	0.7251	0.7259	0.7267	0.7275	0.7284	0.7292	0.7300	0.7308	0.7316
5.4	0.7324	0.7332	0.7340	0.7348	0.7356	0.7364	0.7372	0.7380	0.7388	0.7396

Table 3.   Four-place logarithms (base 10).   5.50–9.99

x	0	1	2	3	4	5	6	7	8	9
5.5	0.7404	0.7412	0.7419	0.7427	0.7435	0.7443	0.7451	0.7459	0.7466	0.7474
5.6	0.7482	0.7490	0.7497	0.7505	0.7513	0.7520	0.7528	0.7536	0.7543	0.7551
5.7	0.7559	0.7566	0.7574	0.7582	0.7589	0.7597	0.7604	0.7612	0.7619	0.7627
5.8	0.7634	0.7642	0.7649	0.7657	0.7664	0.7672	0.7679	0.7686	0.7694	0.7701
5.9	0.7709	0.7716	0.7723	0.7731	0.7738	0.7745	0.7752	0.7760	0.7767	0.7774
6.0	0.7782	0.7789	0.7796	0.7803	0.7810	0.7818	0.7825	0.7832	0.7839	0.7846
6.1	0.7853	0.7860	0.7868	0.7875	0.7882	0.7889	0.7896	0.7903	0.7910	0.7917
6.2	0.7924	0.7931	0.7938	0.7945	0.7952	0.7959	0.7966	0.7973	0.7980	0.7987
6.3	0.7993	0.8000	0.8007	0.8014	0.8021	0.8028	0.8035	0.8041	0.8048	0.8055
6.4	0.8062	0.8069	0.8075	0.8082	0.8089	0.8096	0.8102	0.8109	0.8116	0.8122
6.5	0.8129	0.8136	0.8142	0.8149	0.8156	0.8162	0.8169	0.8176	0.8182	0.8189
6.6	0.8195	0.8202	0.8209	0.8215	0.8222	0.8228	0.8235	0.8241	0.8248	0.8254
6.7	0.8261	0.8267	0.8274	0.8280	0.8287	0.8293	0.8299	0.8306	0.8312	0.8319
6.8	0.8325	0.8331	0.8338	0.8344	0.8351	0.8357	0.8363	0.8370	0.8376	0.8382
6.9	0.8388	0.8395	0.8401	0.8407	0.8414	0.8420	0.8426	0.8432	0.8439	0.8445
7.0	0.8451	0.8457	0.8463	0.8470	0.8476	0.8482	0.8488	0.8494	0.8500	0.8506
7.1	0.8513	0.8519	0.8525	0.8531	0.8537	0.8543	0.8549	0.8555	0.8561	0.8567
7.2	0.8573	0.8579	0.8585	0.8591	0.8597	0.8603	0.8609	0.8615	0.8621	0.8627
7.3	0.8633	0.8639	0.8645	0.8651	0.8657	0.8663	0.8669	0.8675	0.8681	0.8686
7.4	0.8692	0.8698	0.8704	0.8710	0.8716	0.8722	0.8727	0.8733	0.8739	0.8745
7.5	0.8751	0.8756	0.8762	0.8768	0.8774	0.8779	0.8785	0.8791	0.8797	0.8802
7.6	0.8808	0.8814	0.8820	0.8825	0.8831	0.8837	0.8842	0.8848	0.8854	0.8859
7.7	0.8865	0.8871	0.8876	0.8882	0.8887	0.8893	0.8899	0.8904	0.8910	0.8915
7.8	0.8921	0.8927	0.8932	0.8938	0.8943	0.8949	0.8954	0.8960	0.8965	0.8971
7.9	0.8976	0.8982	0.8987	0.8993	0.8998	0.9004	0.9009	0.9015	0.9020	0.9025
8.0	0.9031	0.9036	0.9042	0.9047	0.9053	0.9058	0.9063	0.9069	0.9074	0.9079
8.1	0.9085	0.9090	0.9096	0.9101	0.9106	0.9112	0.9117	0.9122	0.9128	0.9133
8.2	0.9138	0.9143	0.9149	0.9154	0.9159	0.9165	0.9170	0.9175	0.9180	0.9186
8.3	0.9191	0.9196	0.9201	0.9206	0.9212	0.9217	0.9222	0.9227	0.9232	0.9238
8.4	0.9243	0.9248	0.9253	0.9258	0.9263	0.9269	0.9274	0.9279	0.9284	0.9289
8.5	0.9294	0.9299	0.9304	0.9309	0.9315	0.9320	0.9325	0.9330	0.9335	0.9340
8.6	0.9345	0.9350	0.9355	0.9360	0.9365	0.9370	0.9375	0.9380	0.9385	0.9390
8.7	0.9395	0.9400	0.9405	0.9410	0.9415	0.9420	0.9425	0.9430	0.9435	0.9440
8.8	0.9445	0.9450	0.9455	0.9460	0.9465	0.9469	0.9474	0.9479	0.9484	0.9489
8.9	0.9494	0.9499	0.9504	0.9509	0.9513	0.9518	0.9523	0.9528	0.9533	0.9538
9.0	0.9542	0.9547	0.9552	0.9557	0.9562	0.9566	0.9571	0.9576	0.9581	0.9586
9.1	0.9590	0.9595	0.9600	0.9605	0.9609	0.9614	0.9619	0.9624	0.9628	0.9633
9.2	0.9638	0.9643	0.9647	0.9652	0.9657	0.9661	0.9666	0.9671	0.9675	0.9680
9.3	0.9685	0.9689	0.9694	0.9699	0.9703	0.9708	0.9713	0.9717	0.9722	0.9727
9.4	0.9731	0.9736	0.9741	0.9745	0.9750	0.9754	0.9759	0.9763	0.9768	0.9773
9.5	0.9777	0.9782	0.9786	0.9791	0.9795	0.9800	0.9805	0.9809	0.9814	0.9818
9.6	0.9823	0.9827	0.9832	0.9836	0.9841	0.9845	0.9850	0.9854	0.9859	0.9863
9.7	0.9868	0.9872	0.9877	0.9881	0.9886	0.9890	0.9894	0.9899	0.9903	0.9908
9.8	0.9912	0.9917	0.9921	0.9926	0.9930	0.9934	0.9939	0.9943	0.9948	0.9952
9.9	0.9956	0.9961	0.9965	0.9969	0.9974	0.9978	0.9983	0.9987	0.9991	0.9996

Table 4. Square roots. 1.00–5.49

x	0	1	2	3	4	5	6	7	8	9
1.0	1.000	1.005	1.010	1.015	1.020	1.025	1.030	1.034	1.039	1.044
1.1	1.049	1.054	1.058	1.063	1.068	1.072	1.077	1.082	1.086	1.091
1.2	1.095	1.100	1.105	1.109	1.114	1.118	1.122	1.127	1.131	1.136
1.3	1.140	1.145	1.149	1.153	1.158	1.162	1.166	1.170	1.175	1.179
1.4	1.183	1.187	1.192	1.196	1.200	1.204	1.208	1.212	1.217	1.221
1.5	1.225	1.229	1.233	1.237	1.241	1.245	1.249	1.253	1.257	1.261
1.6	1.265	1.269	1.273	1.277	1.281	1.285	1.288	1.292	1.296	1.300
1.7	1.304	1.308	1.311	1.315	1.319	1.323	1.327	1.330	1.334	1.338
1.8	1.342	1.345	1.349	1.353	1.356	1.360	1.364	1.367	1.371	1.375
1.9	1.378	1.382	1.386	1.389	1.393	1.396	1.400	1.404	1.407	1.411
2.0	1.414	1.418	1.421	1.425	1.428	1.432	1.435	1.439	1.442	1.446
2.1	1.449	1.453	1.456	1.459	1.463	1.466	1.470	1.473	1.476	1.480
2.2	1.483	1.487	1.490	1.493	1.497	1.500	1.503	1.507	1.510	1.513
2.3	1.517	1.520	1.523	1.526	1.530	1.533	1.536	1.539	1.543	1.546
2.4	1.549	1.552	1.556	1.559	1.562	1.565	1.568	1.572	1.575	1.578
2.5	1.581	1.584	1.587	1.591	1.594	1.597	1.600	1.603	1.606	1.609
2.6	1.612	1.616	1.619	1.622	1.625	1.628	1.631	1.634	1.637	1.640
2.7	1.643	1.646	1.649	1.652	1.655	1.658	1.661	1.664	1.667	1.670
2.8	1.673	1.676	1.679	1.682	1.685	1.688	1.691	1.694	1.697	1.700
2.9	1.703	1.706	1.709	1.712	1.715	1.718	1.720	1.723	1.726	1.729
3.0	1.732	1.735	1.738	1.741	1.744	1.746	1.749	1.752	1.755	1.758
3.1	1.761	1.764	1.766	1.769	1.772	1.775	1.778	1.780	1.783	1.786
3.2	1.789	1.792	1.794	1.797	1.800	1.803	1.806	1.808	1.811	1.814
3.3	1.817	1.819	1.822	1.825	1.828	1.830	1.833	1.836	1.838	1.841
3.4	1.844	1.847	1.849	1.852	1.855	1.857	1.860	1.863	1.865	1.868
3.5	1.871	1.873	1.876	1.879	1.881	1.884	1.887	1.889	1.892	1.895
3.6	1.897	1.900	1.903	1.905	1.908	1.910	1.913	1.916	1.918	1.921
3.7	1.924	1.926	1.929	1.931	1.934	1.936	1.939	1.942	1.944	1.947
3.8	1.949	1.952	1.954	1.957	1.960	1.962	1.965	1.967	1.970	1.972
3.9	1.975	1.977	1.980	1.982	1.985	1.987	1.990	1.992	1.995	1.997
4.0	2.000	2.002	2.005	2.007	2.010	2.012	2.015	2.017	2.020	2.022
4.1	2.025	2.027	2.030	2.032	2.035	2.037	2.040	2.042	2.045	2.047
4.2	2.049	2.052	2.054	2.057	2.059	2.062	2.064	2.066	2.069	2.071
4.3	2.074	2.076	2.078	2.081	2.083	2.086	2.088	2.090	2.093	2.095
4.4	2.098	2.100	2.102	2.105	2.107	2.110	2.112	2.114	2.117	2.119
4.5	2.121	2.124	2.126	2.128	2.131	2.133	2.135	2.138	2.140	2.142
4.6	2.145	2.147	2.149	2.152	2.154	2.156	2.159	2.161	2.163	2.166
4.7	2.168	2.170	2.173	2.175	2.177	2.179	2.182	2.184	2.186	2.189
4.8	2.191	2.193	2.195	2.198	2.200	2.202	2.205	2.207	2.209	2.211
4.9	2.214	2.216	2.218	2.220	2.223	2.225	2.227	2.229	2.232	2.234
5.0	2.236	2.238	2.241	2.243	2.245	2.247	2.249	2.252	2.254	2.256
5.1	2.258	2.261	2.263	2.265	2.267	2.269	2.272	2.274	2.276	2.278
5.2	2.280	2.283	2.285	2.287	2.289	2.291	2.293	2.296	2.298	2.300
5.3	2.302	2.304	2.307	2.309	2.311	2.313	2.315	2.317	2.319	2.322
5.4	2.324	2.326	2.328	2.330	2.332	2.335	2.337	2.339	2.341	2.343

Table 4. Square roots. 5.50–9.99

x	0	1	2	3	4	5	6	7	8	9
5.5	2.345	2.347	2.349	2.352	2.354	2.356	2.358	2.360	2.362	2.364
5.6	2.366	2.369	2.371	2.373	2.375	2.377	2.379	2.381	2.383	2.385
5.7	2.387	2.390	2.392	2.394	2.396	2.398	2.400	2.402	2.404	2.406
5.8	2.408	2.410	2.412	2.415	2.417	2.419	2.421	2.423	2.425	2.427
5.9	2.429	2.431	2.433	2.435	2.437	2.439	2.441	2.443	2.445	2.447
6.0	2.449	2.452	2.454	2.456	2.458	2.460	2.462	2.464	2.466	2.468
6.1	2.470	2.472	2.474	2.476	2.478	2.480	2.482	2.484	2.486	2.488
6.2	2.490	2.492	2.494	2.496	2.498	2.500	2.502	2.504	2.506	2.508
6.3	2.510	2.512	2.514	2.516	2.518	2.520	2.522	2.524	2.526	2.528
6.4	2.530	2.532	2.534	2.536	2.538	2.540	2.542	2.544	2.546	2.548
6.5	2.550	2.551	2.553	2.555	2.557	2.559	2.561	2.563	2.565	2.567
6.6	2.569	2.571	2.573	2.575	2.577	2.579	2.581	2.583	2.585	2.587
6.7	2.588	2.590	2.592	2.594	2.596	2.598	2.600	2.602	2.604	2.606
6.8	2.608	2.610	2.612	2.613	2.615	2.617	2.619	2.621	2.623	2.625
6.9	2.627	2.629	2.631	2.632	2.634	2.636	2.638	2.640	2.642	2.644
7.0	2.646	2.648	2.650	2.651	2.653	2.655	2.657	2.659	2.661	2.663
7.1	2.665	2.666	2.668	2.670	2.672	2.674	2.676	2.678	2.680	2.681
7.2	2.683	2.685	2.687	2.689	2.691	2.693	2.694	2.696	2.698	2.700
7.3	2.702	2.704	2.706	2.707	2.709	2.711	2.713	2.715	2.717	2.718
7.4	2.720	2.722	2.724	2.726	2.728	2.729	2.731	2.733	2.735	2.737
7.5	2.739	2.740	2.742	2.744	2.746	2.748	2.750	2.751	2.753	2.755
7.6	2.757	2.759	2.760	2.762	2.764	2.766	2.768	2.769	2.771	2.773
7.7	2.775	2.777	2.778	2.780	2.782	2.784	2.786	2.787	2.789	2.791
7.8	2.793	2.795	2.796	2.798	2.800	2.802	2.804	2.805	2.807	2.809
7.9	2.811	2.812	2.814	2.816	2.818	2.820	2.821	2.823	2.825	2.827
8.0	2.828	2.830	2.832	2.834	2.835	2.837	2.839	2.841	2.843	2.844
8.1	2.846	2.848	2.850	2.851	2.853	2.855	2.857	2.858	2.860	2.862
8.2	2.864	2.865	2.867	2.869	2.871	2.872	2.874	2.876	2.877	2.879
8.3	2.881	2.883	2.884	2.886	2.888	2.890	2.891	2.893	2.895	2.897
8.4	2.898	2.900	2.902	2.903	2.905	2.907	2.909	2.910	2.912	2.914
8.5	2.915	2.917	2.919	2.921	2.922	2.924	2.926	2.927	2.929	2.931
8.6	2.933	2.934	2.936	2.938	2.939	2.941	2.943	2.944	2.946	2.948
8.7	2.950	2.951	2.953	2.955	2.956	2.958	2.960	2.961	2.963	2.965
8.8	2.966	2.968	2.970	2.972	2.973	2.975	2.977	2.978	2.980	2.982
8.9	2.983	2.985	2.987	2.988	2.990	2.992	2.993	2.995	2.997	2.998
9.0	3.000	3.002	3.003	3.005	3.007	3.008	3.010	3.012	3.013	3.015
9.1	3.017	3.018	3.020	3.022	3.023	3.025	3.027	3.028	3.030	3.032
9.2	3.033	3.035	3.036	3.038	3.040	3.041	3.043	3.045	3.046	3.048
9.3	3.050	3.051	3.053	3.055	3.056	3.058	3.059	3.061	3.063	3.064
9.4	3.066	3.068	3.069	3.071	3.072	3.074	3.076	3.077	3.079	3.081
9.5	3.082	3.084	3.085	3.087	3.089	3.090	3.092	3.094	3.095	3.097
9.6	3.098	3.100	3.102	3.103	3.105	3.106	3.108	3.110	3.111	3.113
9.7	3.114	3.116	3.118	3.119	3.121	3.122	3.124	3.126	3.127	3.129
9.8	3.130	3.132	3.134	3.135	3.137	3.138	3.140	3.142	3.143	3.145
9.9	3.146	3.148	3.150	3.151	3.153	3.154	3.156	3.158	3.159	3.161

Table 4.   Square roots.   10.0–54.9

x	0	1	2	3	4	5	6	7	8	9
10	3.162	3.178	3.194	3.209	3.225	3.240	3.256	3.271	3.286	3.302
11	3.317	3.332	3.347	3.362	3.376	3.391	3.406	3.421	3.435	3.450
12	3.464	3.479	3.493	3.507	3.521	3.536	3.550	3.564	3.578	3.592
13	3.606	3.619	3.633	3.647	3.661	3.674	3.688	3.701	3.715	3.728
14	3.742	3.755	3.768	3.782	3.795	3.808	3.821	3.834	3.847	3.860
15	3.873	3.886	3.899	3.912	3.924	3.937	3.950	3.962	3.975	3.987
16	4.000	4.012	4.025	4.037	4.050	4.062	4.074	4.087	4.099	4.111
17	4.123	4.135	4.147	4.159	4.171	4.183	4.195	4.207	4.219	4.231
18	4.243	4.254	4.266	4.278	4.290	4.301	4.313	4.324	4.336	4.347
19	4.359	4.370	4.382	4.393	4.405	4.416	4.427	4.438	4.450	4.461
20	4.472	4.483	4.494	4.506	4.517	4.528	4.539	4.550	4.561	4.572
21	4.583	4.593	4.604	4.615	4.626	4.637	4.648	4.658	4.669	4.680
22	4.690	4.701	4.712	4.722	4.733	4.743	4.754	4.764	4.775	4.785
23	4.796	4.806	4.871	4.827	4.837	4.848	4.858	4.868	4.879	4.889
24	4.899	4.909	4.919	4.930	4.940	4.950	4.960	4.970	4.980	4.990
25	5.000	5.010	5.020	5.030	5.040	5.050	5.060	5.070	5.079	5.089
26	5.099	5.109	5.119	5.128	5.138	5.148	5.158	5.167	5.177	5.187
27	5.196	5.206	5.215	5.225	5.235	5.244	5.254	5.263	5.273	5.282
28	5.292	5.301	5.310	5.320	5.329	5.339	5.348	5.357	5.367	5.376
29	5.385	5.394	5.404	5.413	5.422	5.431	5.441	5.450	5.459	5.468
30	5.477	5.486	5.495	5.505	5.514	5.523	5.532	5.541	5.550	5.559
31	5.568	5.577	5.586	5.595	5.604	5.612	5.621	5.630	5.639	5.648
32	5.657	5.666	5.675	5.683	5.692	5.701	5.710	5.718	5.727	5.736
33	5.745	5.753	5.762	5.771	5.779	5.788	5.797	5.805	5.814	5.822
34	5.831	5.840	5.848	5.857	5.865	5.874	5.882	5.891	5.899	5.908
35	5.916	5.925	5.933	5.941	5.950	5.958	5.967	5.975	5.983	5.992
36	6.000	6.008	6.017	6.025	6.033	6.042	6.050	6.058	6.066	6.075
37	6.083	6.091	6.099	6.107	6.116	6.124	6.132	6.140	6.148	6.156
38	6.164	6.173	6.181	6.189	6.197	6.205	6.213	6.221	6.229	6.237
39	6.245	6.253	6.261	6.269	6.277	6.285	6.293	6.301	6.309	6.317
40	6.325	6.332	6.340	6.348	6.356	6.364	6.372	6.380	6.387	6.695
41	6.403	6.411	6.419	6.427	6.434	6.442	6.450	6.458	6.465	6.473
42	6.481	6.488	6.496	6.504	6.512	6.519	6.527	6.535	6.542	6.550
43	6.557	6.565	6.573	6.580	6.588	6.595	6.603	6.611	6.618	6.626
44	6.633	6.641	6.648	6.656	6.663	6.671	6.678	6.686	6.693	6.701
45	6.708	6.716	6.723	6.731	6.738	6.745	6.753	6.760	6.768	6.775
46	6.782	6.790	6.797	6.804	6.812	6.819	6.826	6.834	6.841	6.848
47	6.856	6.863	6.870	6.877	6.885	6.892	6.899	6.907	6.914	6.921
48	6.928	6.935	6.943	6.950	6.957	6.964	6.971	6.979	6.986	6.993
49	7.000	7.007	7.014	7.021	7.029	7.036	7.043	7.050	7.057	7.064
50	7.071	7.078	7.085	7.092	7.099	7.106	7.113	7.120	7.127	7.134
51	7.141	7.148	7.155	7.162	7.169	7.176	7.183	7.190	7.197	7.204
52	7.211	7.218	7.225	7.232	7.239	7.246	7.253	7.259	7.266	7.273
53	7.280	7.287	7.294	7.301	7.308	7.314	7.321	7.328	7.335	7.342
54	7.348	7.355	7.362	7.369	7.376	7.382	7.389	7.396	7.403	7.409

Table 4.   Square roots.   55.0–99.9

x	0	1	2	3	4	5	6	7	8	9
55	7.416	7.423	7.430	7.436	7.443	7.450	7.457	7.463	7.470	7.477
56	7.483	7.490	7.497	7.503	7.510	7.517	7.523	7.530	7.537	7.543
57	7.550	7.556	7.563	7.570	7.576	7.583	7.589	7.596	7.603	7.609
58	7.616	7.622	7.629	7.635	7.642	7.649	7.655	7.662	7.668	7.675
59	7.681	7.688	7.694	7.701	7.707	7.714	7.720	7.727	7.733	7.740
60	7.746	7.752	7.759	7.765	7.772	7.778	7.785	7.791	7.797	7.804
61	7.810	7.817	7.823	7.829	7.836	7.842	7.849	7.855	7.861	7.868
62	7.874	7.880	7.887	7.893	7.899	7.906	7.912	7.918	7.925	7.931
63	7.937	7.944	7.950	7.956	7.962	7.969	7.975	7.981	7.987	7.994
64	8.000	8.006	8.012	8.019	8.025	8.031	8.037	8.044	8.050	8.056
65	8.062	8.068	8.075	8.081	8.087	8.093	8.099	8.106	8.112	8.118
66	8.124	8.130	8.136	8.142	8.149	8.155	8.161	8.167	8.173	8.179
67	8.185	8.191	8.198	8.204	8.210	8.216	8.222	8.228	8.234	8.240
68	8.246	8.252	8.258	8.264	8.270	8.276	8.283	8.289	8.295	8.301
69	8.307	8.313	8.319	8.325	8.331	8.337	8.343	8.349	8.355	8.361
70	8.367	8.373	8.379	8.385	8.390	8.396	8.402	8.408	8.414	8.420
71	8.426	8.432	8.438	8.444	8.450	8.456	8.462	8.468	8.473	8.479
72	8.485	8.491	8.497	8.503	8.509	8.515	8.521	8.526	8.532	8.538
73	8.544	8.550	8.556	8.562	8.567	8.573	8.579	8.585	8.591	8.597
74	8.602	8.608	8.614	8.620	8.626	8.631	8.637	8.643	8.649	8.654
75	8.660	8.666	8.672	8.678	8.683	8.689	8.695	8.701	8.706	8.712
76	8.718	8.724	8.729	8.735	8.741	8.746	8.752	8.758	8.764	8.769
77	8.775	8.781	8.786	8.792	8.798	8.803	8.809	8.815	8.820	8.826
78	8.832	8.837	8.843	8.849	8.854	8.860	8.866	8.871	8.877	8.883
79	8.888	8.894	8.899	8.905	8.911	8.916	8.922	8.927	8.933	8.939
80	8.944	8.950	8.955	8.961	8.967	8.972	8.978	8.983	8.989	8.994
81	9.000	9.006	9.011	9.017	9.022	9.028	9.033	9.039	9.044	9.050
82	9.055	9.061	9.066	9.072	9.077	9.083	9.088	9.094	9.099	9.105
83	9.110	9.116	9.121	9.127	9.132	9.138	9.143	9.149	9.154	9.160
84	9.165	9.171	9.176	9.182	9.187	9.192	9.198	9.203	9.209	9.214
85	9.220	9.225	9.230	9.236	9.241	9.247	9.252	9.257	9.263	9.268
86	9.274	9.279	9.284	9.290	9.295	9.301	9.306	9.311	9.317	9.322
87	9.327	9.333	9.338	9.343	9.349	9.354	9.359	9.365	9.370	9.375
88	9.381	9.386	9.391	9.397	9.402	9.407	9.413	9.418	9.423	9.429
89	9.434	9.439	9.445	9.450	9.455	9.460	9.466	9.471	9.476	9.482
90	9.487	9.492	9.497	9.503	9.508	9.513	9.518	9.524	9.529	9.534
91	9.539	9.545	9.550	9.555	9.560	9.566	9.571	9.576	9.581	9.586
92	9.592	9.597	9.603	9.607	9.612	9.618	9.623	9.628	9.633	9.638
93	9.644	9.649	9.654	9.659	9.664	9.670	9.675	9.680	9.685	9.690
94	9.695	9.701	9.706	9.711	9.716	9.721	9.726	9.731	9.737	9.742
95	9.747	9.752	9.757	9.762	9.767	9.772	9.778	9.783	9.788	9.793
96	9.798	9.803	9.808	9.813	9.818	9.823	9.829	9.834	9.839	9.844
97	9.849	9.854	9.859	9.864	9.869	9.874	9.879	9.884	9.889	9.894
98	9.899	9.905	9.910	9.915	9.920	9.925	9.930	9.935	9.940	9.945
99	9.950	9.955	9.960	9.965	9.970	9.975	9.980	9.985	9.990	9.995

ANSWERS TO STARRED EXERCISES

SECTION 1.1

3. (a) $\frac{34}{5}$ (c) $\frac{9}{5}$ (e) $\frac{31}{10}$ (g) $\frac{3176}{1000}$

4. (a) $\frac{4}{7}$ (c) $\frac{5}{7}$

5. (a) $\frac{5}{6}$ (c) $\frac{37}{40}$ (e) $\frac{2}{9}$ (g) $\frac{1}{3}$

6. (a) 0.56 (c) $0.\overline{36}$ (e) $0.\overline{5}$ (g) $0.\overline{285714}$

7. (a) $\frac{4}{9}$ (c) $\frac{37}{99}$ (e) $\frac{a_1}{9}$

8. Figure A.1.1.1

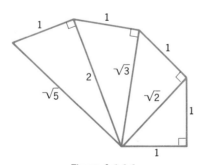

Figure A.1.1.1

SECTION 1.3

1. $4x - 16$

3. $14x - 14y + 28$

4. $\frac{1}{4}bx - b$

6. $6y + 2a$

7. $2y - \frac{2}{3}x$

9. $9a^2 - 12x^2$

10. $3x + 4$ 12. 0 14. $-3abc - \frac{5}{2}$

16. $6x^2 - 6y^2 + 2xy - 2y$ 18. $5a^2x + 7a^2 - 6a$

21. -3 23. -20

SECTION 1.4

1. $5x^7$ 3. $8x^6y^3$ 4. $-6a^7x^4$

6. $x^2 - 6ax + 9a^2$ 7. $-27x^6y^{15}$ 9. $u^3 + uv$

10. $u^3 - uv + 1$ 12. $4x^5 + 4x^3 - 4x^2 - 4$

14. $a^2 - b^2 - c^2 - 2bc$ 16. $36x^4y^3 - 4x^2y^5$

18. $25x^3 - 75x^2 + 75x - 25$ 20. $6x^5 - 16x^4y - xy^3 + 3y^4$

22. $abx^3 - a^3x^2 + b^2x$ 24. $2ax^3 + (2b - a)x^2 + (2c - b)x - c$

26. $rsvp$ 28. (a) 3   (b) 20

30. (a) 3   (b) -7

SECTION 1.5

1. $x^2 - 1$ 3. $x^2 - 6x + 9$

4. $4x^2 + 4x + 1$ 6. $25x^2 - 10x + 1$

7. $16t^2 + 40t + 25$ 9. $16t^2 - 25$

10. $x^2 + 4x + 4$ 12. $4y^2 - 12xy + 9x^2$

13. $8x^3 + 12x^2 + 6x + 1$ 15. $4r^2 + 12rs + 9s^2$

16. $4r^2 - 12rs + 9s^2$ 18. $x^3 + 6ax^2 + 12a^2x + 8a^3$

19. $x^2 - \frac{1}{4}$ 21. $x^3 + \frac{3}{2}x^2 + \frac{3}{4}x + \frac{1}{8}$

22. $x^3 - \frac{3}{2}x^2 + \frac{3}{4}x - \frac{1}{8}$ 24. $\frac{1}{4}x^2 - x + 1$

25. $\frac{1}{8}x^3 + \frac{3}{4}x^2 + \frac{3}{2}x + 1$ 27. $\frac{1}{4}x^2 + \frac{1}{3}x + \frac{1}{9}$

28. $x^6 - 3a^2x^4 + 3a^4x^2 - a^6$ 30. $x^{12} - 3x^8 + 3x^4 - 1$

31. $x^6 + 3a^2x^4 + 3a^4x^2 + a^6$ 33. $x^4 - 4x^3 + 4x^2$

SECTION 1.6

1. $3(x + 2)$ 3. $a(b + 1)$ 4. $b(a + c)$

6. $ab(a + b)$ 7. $x(y + 2)$ 9. $2x(x - 2)$

10. $(x + c)(x - c)$ 12. $4(3 + x)(3 - x)$ 13. $(4x + a)(4x - a)$

15. $(x - 2)^2$ 16. $(x + 3)^2$ 18. $(x - 8)^2$

19. $(x - 2)(x^2 + 2x + 4)$ 21. $(2x + 3)(4x^2 - 6x + 9)$

22. $xy(xy^2 - y + 1)$ 24. $(8x + y)(2x - y)$

25. $3x(5x - 6)$ 27. $[(a + c)x + b + d][(a - c)x + b - d]$

28. $3(x + 11)^2$ 30. $5(x - 3)^2$

31. $2(ax - 2)(a^2x^2 + 2ax + 4)$ 33. $2(2x + a)(4x^2 - 2ax + a^2)$

SECTION 1.7

1. $x^2 + 5x + 4$
3. $x^2 + 3x - 4$
4. $x^2 - 5x + 4$
6. $x^2 - 8x + 15$
7. $x^2 - 2x - 15$
9. $6x^2 - x - 15$
10. $6x^2 + x - 15$
12. $12x^2 + 5x - 2$
13. $12x^2 - 5x - 2$
15. $12x^2 - 11x + 2$
16. $9x^2 - 12x + 4$
18. $6x^2 - 5x - 6$
19. $5x^2 - 7x + 2$
21. $10x^2 + 33x - 7$
22. $5x^4 + 7x^2 + 2$

SECTION 1.8

1. $(x + 3)(x + 1)$
3. $(x + 3)(x - 1)$
4. $(x + 4)(x + 2)$
6. $(x - 5)(x - 3)$
7. $(x - 2)(x + 1)$
9. $(x - 9)(x - 3)$
10. $(x - 5)(x - 7)$
12. $(x - 2)(x - 10)$
13. $(x + 7)(x - 4)$
15. $(x - 12)(x + 2)$
16. $(x + C)(x - A)$
18. $(x - r)(x - 1)$
19. $(2x + 1)(x + 2)$
21. $(2x + 3)(x - 1)$
22. $(3x + 2)(2x + 5)$
24. $(2x + 3)(x + 10)$
25. $(25x + 3)(x + 2)$
27. $(2x - 5)(x - 10)$
28. $(4x - 1)(3x - 2)$

SECTION 1.9

1. $x(x + 1)(x - 1)$
3. $x^2(x - 5)(x - 1)$
5. $3x^3(x - 7)(x - 5)$
7. $(x + a)(x - a)(x^2 + ax + a^2)(x^2 - ax + a^2)$
9. $x(2x - 1)(x + 2)$
11. $(x + 1)(x - 1)^2$
13. $3(x + 2)^2(x - 2)^2$
15. $[x^2 + (x - 1)^2](2x - 1)$
17. $-(x + 2)^2(x - 1)(x - 3)$
19. $(x + 1)(x + 15)(x - 3)$
21. $(x + 3y + 1)(x + 3y - 1)$
23. $-2(6x - 7)(x + 1)$

SECTION 1.10

1. $\dfrac{1 + 2x}{x^2 - 1}$
3. $\dfrac{10x}{2x^2 + 1}$
5. $\dfrac{x - 2}{2}$
7. $\dfrac{x^2 + 4}{x - 1}$
9. $\dfrac{x + 3}{x - 5}$
11. 5
13. $2x^2 + 3x + 2$
15. $2x^2 - 2x - 5$
17. $x^3 - 2x + 3$
19. $4x^2 + 15x - 10$

SECTION 1.11

1. $x - 5 + \dfrac{15}{x + 5}$
3. $2x^2 + 10x + 50 + \dfrac{400}{x - 5}$

5. $x^2 + 4 + \dfrac{9}{x^2 - 2}$

7. $x^2 - 4 - \dfrac{x^2 - 6x - 5}{2x^3 + x + 1}$

9. $(x - 1)(x + 2)(x - 3)(x + 4)$

11. $(x - a)^2(x - b)$

SECTION 1.12

1. (b) $3.1416 = \frac{31416}{10000}$ (c) $(1 + \sqrt{2})(1 - \sqrt{2}) = (1)^2 - (\sqrt{2})^2 = -1$
2. Take for example

$$a = 1 + \sqrt{2} \quad \text{and} \quad b = 1 - \sqrt{2}$$

and note that

$$a + b = 2 \quad \text{and} \quad ab = -1. \quad \square$$

3. $x^4 - y^4$

4. $x^3 - (3 + a)x^2 + (1 + 3a)x - a$

5. $2x^4 + x^3 - 4x^2 + 11x - 4$

6. $x^4 - 2x^2y^2 + y^4$

7. $x^4 - x^3y + xy^3 - y^4$

8. $x^4 - ax^3 - 3a^2x^2 + 5a^3x - 2a^4$

9. $(3x + 10)(3x - 10)$

10. $(x + \sqrt{5})(x - \sqrt{5})$

11. $(3x + 2)(9x^2 - 6x + 4)$

12. $(x + 7)(x + 1)$

13. $(5x + 7)(x + 1)$

14. $(2x - 3)(x + 3)$

15. $2a(3x^2 + 4)$

16. $(a^2x^2 + b^2)(ax + b)(ax - b)$

17. $2x(4x + 5)(2x - 1)$

18. $a^4(bx + 1)(bx - 1)(b^2x^2 + bx + 1)(b^2x^2 - bx + 1)$

19. $(3x - 2)(x + 2)$

20. $(x + a)(2x^2 - 2ax + 2a^2 - 1)$

21. $x - 4$

22. $x^2 - 2ax + a^2$

23. $x + 2$

24. $2(x^2 - ax + a^2)$

25. $1 - 5x$

26. $(c - a)x + (d - b)$

27. $5x^3 + 4x^2 + x + 1$

28. 2

29. $x^2 - 3x + 5$

30. $2x + 1$

31. $x^3 - x + \dfrac{x}{x^2 + 1}$

32. $(3x^2 + 1)(x + 2)(x^2 - 2x + 4)$

33. $(x + 2)(x + 5)(x + 6)(x + 1)$

34. If p is odd, then p is of the form $2n + 1$ with n an integer. Therefore

$$p^2 = (2n + 1)^2 = 4n^2 + 4n + 1 = 2(2n^2 + 2n) + 1.$$

Setting

$$2n^2 + 2n = q$$

we have

$$p^2 = 2q + 1.$$

This shows that p^2 is odd. $\square$

35. If p is not divisible by 3, then p is of the form $3n + 1$ or $3n + 2$. In the first case

$$p^2 = (3n + 1)^2 = 9n^2 + 6n + 1 = 3(3n^2 + 2n) + 1,$$

and in the second

$$p^2 = (3n + 2)^2 = 9n^2 + 12n + 4 = 3(3n^2 + 4n + 1) + 1.$$

In both cases p^2 is of the form $3q + 1$ and thus not divisible by 3. $\square$

SECTION 2.1

1. $\sqrt{50} = 5\sqrt{2} \cong 5(1.414) = 7.070$
3. $\sqrt{180} = 6\sqrt{5} \cong 6(2.236) = 13.416$
5. $\sqrt{0.5} = \sqrt{\frac{1}{2}} = \frac{1}{2}\sqrt{2} \cong \frac{1}{2}(1.414) = 0.707$
7. $\sqrt{0.0019} = \sqrt{\frac{19}{10000}} = \frac{1}{100}\sqrt{19} \cong \frac{1}{100}(4.359) \cong 0.044$
9. $\sqrt{\frac{3}{4}} = \frac{1}{2}\sqrt{3} \cong \frac{1}{2}(1.732) = 0.866$
11. $\sqrt{\frac{24}{49}} = \frac{2}{7}\sqrt{6} \cong \frac{2}{7}(2.449) \cong 0.700$
13. $\frac{1}{2}\sqrt{2}$ 15. $\frac{1}{5}(\sqrt{6} - 1)$ 16. $\sqrt{11} + \sqrt{10}$
18. $\frac{1}{2}(5 + \sqrt{15})$ 19. $\frac{1}{3}(7 + 2\sqrt{10})$ 21. $\dfrac{3 - \sqrt{3x}}{3 - x}$
22. $\sqrt{3}$ 24. $7\sqrt{2} + \sqrt{14}$ 25. $\sqrt{x} + \sqrt{y}$
27. $-\dfrac{\sqrt{x - h} + \sqrt{x + h}}{2h}$
28. $\frac{8}{15}\sqrt{15}$
31. $\sqrt{98} = 7\sqrt{2} \cong 7(1.414) = 9.898$
32. (a) 91.056   (b) 87.352

SECTION 2.2

1. $-2, -3$ 3. $-4, 1$ 5. $7, -2$
7. $-\frac{4}{3}, -1$ 9. $\frac{1}{4}, \frac{2}{3}$ 11. $5, 7$
13. $-3 \pm \sqrt{3}$ 15. $-1 \pm \frac{1}{5}\sqrt{10}$ 17. $-\frac{1}{2}$(square already complete)
19. $-\frac{1}{6} \pm \frac{1}{6}\sqrt{73}$ 21. $-\frac{5}{32} \pm \frac{1}{32}\sqrt{89}$

SECTION 2.3

1. $-6 \pm \sqrt{31}$ 3. $\frac{1}{2}(-11 \pm \sqrt{161})$ 5. $-2, -\frac{3}{2}$
7. $\frac{1}{2}(3 \pm \sqrt{15})$ 9. $-1 \pm \sqrt{11}$ 11. $\frac{1}{2}(5 \pm \sqrt{105})$
13. no real roots   15. $\frac{1}{3}(-2 \pm \sqrt{19})$
18. (a) $x = \dfrac{1 \pm \sqrt{1 - y^3}}{y^2}$ (b) $y = \dfrac{-1 \pm \sqrt{1 + 8x^3}}{2x^2}$

SECTION 2.4

1. $0, 0.333, \frac{1}{3}, 0.334, \dfrac{1}{\sqrt{3}}, 1, 1.142, 1.41, \sqrt{2}$

3. $2a < 2b$ 5. $-2a > -2b$ 6. $-b > -2b$
8. $ab > a^2$ 9. $-2a^2 > -2b^2$ 11. $1 - ab < 1 - a^2$
12. $2a < 2b$ 14. $-2a > 3b$ 15. $-2a > -2b$
17. $ab > b^2$ 18. $\sqrt{a} < \sqrt{b}$ 20. $\sqrt{a+b} < \sqrt{a} + \sqrt{b}$

SECTION 2.5

1. $\{-1, 0, 1, 2\}$ 3. $\{-1, 0, 1, 2, 3, 4\}$ 4. $\{1\}$
6. $\{2, 4\}$ 7. $\varnothing$ 9. $\varnothing$
10. A 12. set of all real numbers
13. $\{x: 3 < x \le 4\}$ 15. A
17. $\{0\}, \{1\}, \{2\}, \{0, 1\}, \{0, 2\}, \{1, 2\}, \{0, 1, 2\}$

SECTION 2.6

1. $(0, 2)$ 3. $(0, 1]$ 5. $[0, 2]$
7. $[0, 2]$ 9. $(-3, \infty)$ 11. $(-\infty, \infty)$
13. $[0, 1]$ 15. $(-\infty, -3) \cup (-3, 0) \cup (0, 3) \cup (3, \infty)$
16. bounded above 18. bounded (both above and below)
20. not bounded at all 22. bounded below
24. not bounded at all 26. bounded
28. bounded below

SECTION 2.7

1. $(-\infty, 1)$ 3. $(-\infty, -3]$ 5. $[-1, \infty)$
7. $(-1, 1)$ 9. $(-\infty, 1) \cup (2, \infty)$ 11. $(-\infty, \infty)$
13. $(-\infty, -\frac{1}{5})$ 15. $(-5 - 2\sqrt{6}, -5 + 2\sqrt{6})$
17. $(-\infty, 0) \cup (0, \infty)$ 19. $(0, 1) \cup (2, \infty)$
21. $(-\infty, 1) \cup (2, 3)$

SECTION 2.8

1. $(2, 5)$ 3. $(4, 5)$ 4. $(-\infty, -\frac{5}{3}) \cup (5, \infty)$
6. $(1, \infty)$ 7. $(-\infty, \frac{5}{3}) \cup (\frac{11}{6}, \infty)$ 9. $(-\infty, \frac{1}{2}) \cup (\frac{2}{3}, \infty)$
10. $(1, 2)$ 12. $(-1 - \sqrt{2}, 0) \cup (\sqrt{2} - 1, \infty)$ 13. $(0, \frac{5}{3}) \cup (\frac{5}{2}, \infty)$
15. $(-\frac{5}{3}, -1) \cup (\frac{5}{2}, \infty)$

SECTION 2.9

1. 4
3. 3, 5
5. ± 1

7. (3, 5)
9. [1, 5]
11. $(-\infty, -\frac{1}{4}] \cup [\frac{7}{4}, \infty)$

13. $[-\frac{7}{2}, \frac{5}{2}]$
15. $(\frac{7}{10}, \frac{9}{10})$
17. $\varnothing$

19. $(-\infty, \infty)$
21. $(-\infty, \frac{3}{2})$
23. $(\frac{3}{2}, \infty)$

SECTION 2.10

1. $-\frac{1}{11}(3 + 2\sqrt{5})$
2. $\frac{1}{3}(2\sqrt{2} - \sqrt{5})$
3. $\frac{1}{4}(1 - \sqrt{2})(1 + \sqrt{5})$

4. $-\frac{2}{15}\sqrt{15}$
5. $\frac{11}{10}\sqrt{5}$
6. $\frac{1}{4}, \frac{4}{3}$

7. $-\sqrt{5}$
8. $-8 \pm \sqrt{67}$
9. no real roots

10. $-\frac{3}{4}$
11. $-\frac{3}{2}, 2$
12. no real roots

13. $-\frac{1}{2}\sqrt{5}, \sqrt{2}$
14. $(-\infty, 5)$
15. $(-\infty, -9]$

16. $(-4, 5)$
17. $(-\infty, -4] \cup [5, \infty)$
18. $[-\frac{3}{2}, \frac{3}{2}]$

19. $\varnothing$
20. $(\frac{2}{5}, \frac{6}{5})$
21. $(-\infty, -2] \cup [-\frac{1}{2}, \infty)$

22. $(\frac{1}{3}, 1)$
23. $(-\sqrt{5}, -\sqrt{3}) \cup (\sqrt{3}, \sqrt{5})$

24. $(-2, -1) \cup (1, \infty)$
25. $(-\infty, -\frac{1}{2}) \cup (1, 3)$

26. (1, 2)
27. $(-\infty, \frac{3}{2} - \frac{1}{2}\sqrt{5}) \cup (1, 2) \cup (\frac{3}{2} + \frac{1}{2}\sqrt{5}, \infty)$

28. (a) false   (b) true   (c) true   (d) false

29. With $c < 0$,

$$\frac{1 + \sqrt{c^2}}{c - 1} = \frac{1 - c}{c - 1} = -1.$$

30. With $a < 0$ and $b > 0$,

$$a^2|b^2| - 2ab|ab| = a^2b^2 - 2ab|a||b|$$
$$= a^2b^2 - 2ab(-a)(b)$$
$$= a^2b^2 + 2a^2b^2$$
$$= 3a^2b^2.$$

31. For all real numbers a and b,

$$|a - b| = |a + (-b)| \leq |a| + |-b| = |a| + |b|. \quad \square$$

32. For all real numbers a and b,

$$0 \leq (a - b)^2 = a^2 - 2ab + b^2$$

so that

$$2ab \leq a^2 + b^2$$

and consequently

$$ab \le \tfrac{1}{2}(a^2 + b^2). \quad \square$$

33.   With $r < 0$ and $a < b$,

$$a = \frac{a + ra}{1 + r} < \frac{a + rb}{1 + r} < \frac{b + rb}{1 + r} = b. \quad \square$$

SECTION 3.1

1. (a) II   (c) IV
2. (a) $(2, 3)$ (c) $(-2, 3)$
3. (b) $(\tfrac{1}{2}, \pi)$
4. $(\tfrac{3}{2}, -4)$
5. (a) $(2, 1)$ (c) $(\tfrac{1}{2}\sqrt{2} + \tfrac{1}{2}, -\tfrac{1}{2}\sqrt{2} - \tfrac{1}{2})$
6. $(-1, 6); (2, 2)$
8. $(-\tfrac{1}{2}\sqrt{3}b, \tfrac{1}{2}b)$ or $(\tfrac{1}{2}\sqrt{3}b, \tfrac{1}{2}b)$
9. (a) midpt of $\overline{PQ} = (-\tfrac{1}{2}, 1)$, midpt of $\overline{QR} = (\tfrac{3}{2}, 1)$, midpt of $\overline{PR} = (1, 0)$
10. (a) $r = -1, s = \tfrac{1}{3}$ (c) $r = -\tfrac{1}{3}, s = \tfrac{1}{3}$

SECTION 3.2

1. (a) $\sqrt{26}$ (c) $\sqrt{x_0^2 + y_0^2}$
2. (a) $a = \tfrac{5}{12}$
3. (a) yes   (c) no
4. (a) $(x - 1)^2 + (y - 2)^2 = 4$ (c) $x^2 + y^2 = 49$
6. (a) $(x - 2)^2 + (y + 1)^2 = 2$
7. (a) radius 2; center $(4, 1)$ (c) radius 10; center $(-\sqrt{2}, -\sqrt{3})$
9. 20
10. (a) 10   (c) 6
11. (a) $\tfrac{1}{2}|a - c||b|$

SECTION 3.3

1. (a) $\tfrac{1}{2}$ (c) 0   (e) $-(y_0/x_0)$ (g) 1
2. (a) l_5 (c) l_3 and l_6
3. (a) at $(8, 0)$
4. (b) at $(0, \tfrac{3}{4})$
5. $a = -\tfrac{33}{4}, b = \tfrac{15}{7}$
6. (a) $y_0 = 1$
7. (b) $x_0 = \tfrac{11}{5}$

SECTION 3.4

1. (a) slope 1, y-intercept 4   (c) no slope, no y-intercept
    (e) slope -1, y-intercept -1 (g) slope $-\tfrac{7}{4}$, y-intercept -1
    (i) slope $-b/a$, y-intercept b (k) slope 0, y-intercept $-\tfrac{5}{2}$

2. (a) $y = 5x + 2$ (c) $y = 5x - 2$.
3. (a) $y = 3$ 4. (b) $x = -3$
5. (a) $2x + y - 1 = 0$ (c) $x - 2y + 2 = 0$

SECTION 3.5

1. (a) $y - 3 = 5(x - 1)$ (c) $y - 3 = -5(x - 1)$
   (e) $y - 3 = 5(x + 1)$ (g) $y + 3 = 5(x + 1)$
2. (a) $x + y - 3 = 0$ (c) $3x + y + 1 = 0$

3. (a) $\dfrac{x}{3} + \dfrac{y}{5} = 1$ (c) $\dfrac{y}{5} - \dfrac{x}{3} = 1$

4. (a) $y - 7 = 0$ (c) $x + y - 9 = 0$ (f) $2x + 3y - 25 = 0$

5. (a) $y = -2x + 4$ (c) $\dfrac{x}{2} + \dfrac{y}{4} = 1$

6. $y - 3 = -\frac{1}{4}(x - 3)$

SECTION 3.6

1. $(2, -2)$ 3. $(-\frac{5}{7}, \frac{13}{7})$ 5. same line
7. parallel lines   9. $(0, -\frac{1}{2})$ 11. $(2, 2), (-2, -2)$
13. $(\frac{2}{5}\sqrt{10}, -\frac{4}{5}\sqrt{10}), (-\frac{2}{5}\sqrt{10}, \frac{4}{5}\sqrt{10})$ 15. $(2\sqrt{2}, 0)$ 17. no intersection
19. $(1 + \sqrt{3}, 1 - \sqrt{3}), (1 - \sqrt{3}, 1 + \sqrt{3})$
21. (a) $4x + 3y - 6 = 0$ (c) $x - 9 = 0$

SECTION 3.7

1. (a) (i) $(7, -7)$ (ii) $(-7, 7)$ (iii) $(-7, -7)$
   (b) (i) $(\sqrt{2}, \sqrt{2})$ (ii) $(-\sqrt{2}, -\sqrt{2})$ (iii) $(-\sqrt{2}, \sqrt{2})$
   (c) (i) $(-1, 0)$ (ii) $(1, 0)$ (iii) $(1, 0)$
   (d) (i) $(-\pi, -\pi)$ (ii) (π, π) (iii) $(\pi, -\pi)$
   (e) (i) $(-3, 4)$ (ii) $(3, -4)$ (iii) $(3, 4)$
   (f) (i) $(0, d)$ (ii) $(0, -d)$ (iii) $(0, d)$
   (g) (i) $(\frac{1}{3}, -0.33)$ (ii) $(-\frac{1}{3}, 0.33)$ (iii) $(-\frac{1}{3}, -0.33)$
   (h) (i) $(a - b, -a - b)$ (ii) $(b - a, a + b)$ (iii) $(b - a, -a - b)$
2. (a) $(\frac{1}{2}, \frac{1}{2})$ (b) O (c) $(2\sqrt{2}, \sqrt{2})$ (d) $(-\frac{3}{4}, \frac{5}{4})$
3. $P_2(2 - a, 4 - b)$ 4. (a) 2   (b) 2   (c) $2\sqrt{2}$ (d) 13
5. $(0, \sqrt{3}), (0, -\sqrt{3})$ 6. (a) $\sqrt{3}$ (b) 1   (c) 0
7. (a) $x - y + 5 = 0$ (b) $2x + y - 5 = 0$ (c) $x - 3y + 15 = 0$
   (d) $x + y - 5 = 0$ (e) $\sqrt{3}x - y + 5 = 0$

8. (a) $x + y - 2a = 0$ (b) $\sqrt{3}x - y = 0$
   (c) $3x + y - 6 = 0$ (d) $10x - 9y - 6 = 0$

9. (a) $x + y - 1 - \sqrt{3} = 0$ (b) $x + \sqrt{3}y - 4 = 0$ (c) $x + \sqrt{3}y - 4 = 0$

10. (a) $1, -1$ (b) $-\dfrac{\sqrt{2}}{2}, \dfrac{\sqrt{2}}{2}$ (c) $\dfrac{\sqrt{3}}{2}, -\dfrac{\sqrt{3}}{2}$ (d) 0

11. (a) $(x + 2)^2 + (y - 2)^2 = 2$ (b) $x^2 + y^2 = a^2 + b^2$

12. (a) center at $(3, 4)$, $r = 5$ (b) center at $(-2, 0)$, $r = 1$

13. $b = 7, A = \frac{25}{2}$ 14. $(-\frac{7}{2}, 3), (\frac{25}{2}, -9)$ 15. $\dfrac{x}{3} + \dfrac{y}{3} = 1$

16. $\dfrac{x}{6 + 2\sqrt{3}} + \dfrac{y}{3 - \sqrt{3}} = 1, \quad \dfrac{x}{6 - 2\sqrt{3}} + \dfrac{y}{3 + \sqrt{3}} = 1$

17. The perpendicular bisector of $\overline{PQ}$ is the y-axis:

$$x = 0.$$

The perpendicular bisector of $\overline{PR}$ is the line

$$y - \frac{1}{2}c = \frac{a - b}{c}\left[x - \frac{1}{2}(a + b)\right].$$

The perpendicular bisector of $\overline{QR}$ is the line

$$y - \frac{1}{2}c = \frac{-a - b}{c}\left[x - \frac{1}{2}(-a + b)\right].$$

The three lines meet at the point

$$\left(0, \frac{b^2 - a^2 + c^2}{2c}\right). \quad \square$$

18. Figure A.3.7.1.

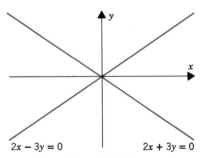

2x − 3y = 0   2x + 3y = 0

Figure A.3.7.1

19. (a) $(-1, 2)$ (b) parallel   (c) $(\frac{13}{2}, \frac{1}{6})$ (d) $(\frac{2}{5}\sqrt{3} - \frac{1}{5}\sqrt{2}, \frac{2}{5}\sqrt{2} + \frac{1}{5}\sqrt{3})$

20. (a) $(7, 5), (7, -3)$ (b) $(4, 6)$ (c) no intersection

(d) $(2 + \frac{1}{2}\sqrt{34}, 3 + \frac{1}{2}\sqrt{34}), (2 - \frac{1}{2}\sqrt{34}, 3 - \frac{1}{2}\sqrt{34})$

(e) $(7, 5)$ (f) $(7, 5), (0, 4)$

21. $(\frac{3}{2}\sqrt{2} - \frac{1}{2}\sqrt{6} + \sqrt{3} - 1, \frac{1}{2}\sqrt{6} - \frac{1}{2}\sqrt{2} - \sqrt{3} + 1)$

SECTION 4.1

1. $f(0) = 1, f(\frac{1}{2}) = \frac{1}{2}, f(1) = 0$ 3. $f(0)$ undefined, $f(\frac{1}{2}) = 2, f(1) = 1$
5. 0 7. -2 9. $f(x - 1) = (x - 1)^2, f(x + 1) = (x + 1)^2$
11. $f(x - 1) = x^2, f(x + 1) = (x + 2)^2$
13. dom $(f) = (-\infty, \infty)$, ran $(f) = [1, \infty)$
15. dom $(F) = (-\infty, \infty)$, ran $(F) = (-\infty, \infty)$
17. dom $(f) = (-\infty, 0) \cup (0, \infty)$, ran $(f) = (0, \infty)$
19. dom $(f) = (-\infty, 1]$, ran $(f) = [0, \infty)$
21. dom $(f) = (-\infty, 1]$, ran $(f) = [-1, \infty)$
23. dom $(h) = (-\infty, 1)$, ran $(h) = (0, \infty)$
25. dom $(g) = [-1, 1]$, ran $(g) = [0, 1]$
27. dom $(f) = (-\infty, \infty)$, ran $(f) = (-\infty, 2]$
29. dom $(f) = (-\infty, \infty)$, ran $(f) = [\frac{11}{4}, \infty)$
31. dom $(f) = (-\infty, \infty)$, ran $(f) = [-1, 0)$
33. dom $(f) = (-\infty, 2] \cup [3, \infty)$, ran $(f) = [0, \infty)$

SECTION 4.2

1. $f(a) = D, f(b) = 0, f(c) = A, f(0) = 0, f(d) = C, f(e) = B$
2. $[a, e]$ 3. $[A, D]$ 4. at b and at 0
5. maximum D taken on at a 6. minimum A taken on at c
7. increases on $[c, d]$; decreases on $[a, c]$ and on $[d, e]$
8. positive on $[a, b)$ and $(0, e]$; negative on $(b, 0)$
10. Figure A.4.2.1   12. Figure A.4.2.1

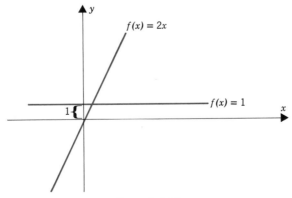

Figure A.4.2.1

14. Figure A.4.2.2
18. Figure A.4.2.3
22. Figure A.4.2.5

16. Figure A.4.2.2
20. Figure A.4.2.4
24. Figure A.4.2.6

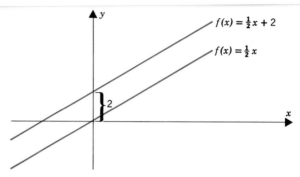

Figure A.4.2.2

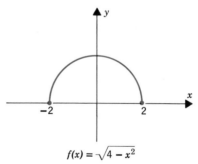

$f(x) = \sqrt{4 - x^2}$

Figure A.4.2.3

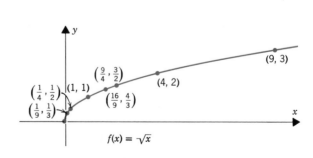

Figure A.4.2.4

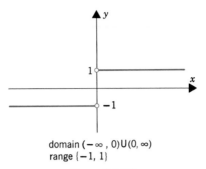

domain $(-\infty, 0) \cup (0, \infty)$
range $\{-1, 1\}$

Figure A.4.2.5

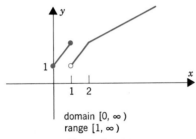

domain $[0, \infty)$
range $[1, \infty)$

Figure A.4.2.6

SECTION 4.3

1. Figure A.4.3.1   3. Figure A.4.3.2

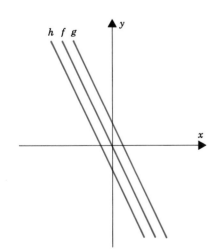

Figure A.4.3.1

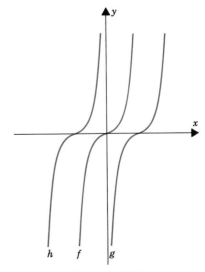

Figure A.4.3.2

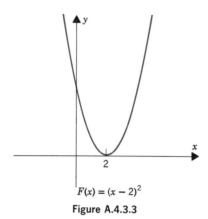

$F(x) = (x - 2)^2$

Figure A.4.3.3

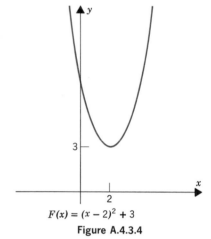

$F(x) = (x - 2)^2 + 3$

Figure A.4.3.4

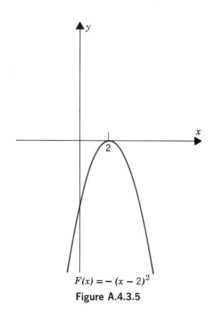

$F(x) = -(x - 2)^2$

Figure A.4.3.5

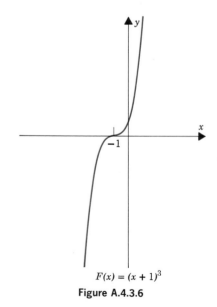

$F(x) = (x + 1)^3$

Figure A.4.3.6

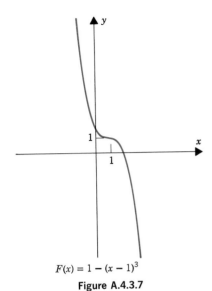

$$F(x) = 1 - (x - 1)^3$$

Figure A.4.3.7

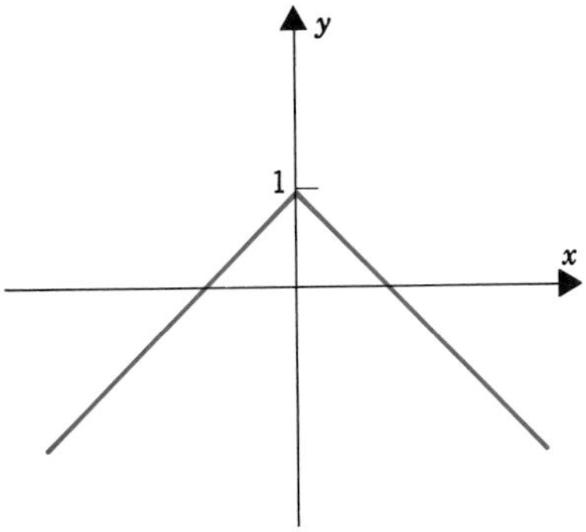

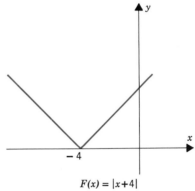

$$F(x) = 1 - |x|$$

Figure A.4.3.8

$$F(x) = |x+4|$$

Figure A.4.3.9

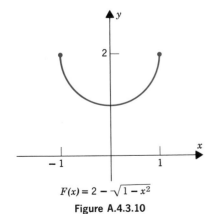

$$F(x) = 2 - \sqrt{1 - x^2}$$

Figure A.4.3.10

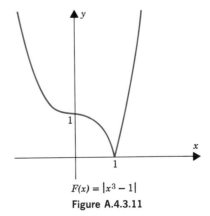

$$F(x) = |x^3 - 1|$$

Figure A.4.3.11

SECTION 4.4

1. even

3. neither odd nor even

5. even

7. odd

9. bounded below

11. not bounded at all

13. bounded above

15. bounded (both above and below)

18. Figure A.4.4.1

23. Figure A.4.4.2

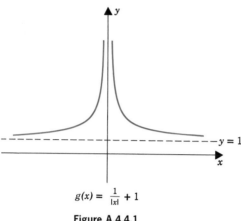

$$g(x) = \frac{1}{|x|} + 1$$

Figure A.4.4.1

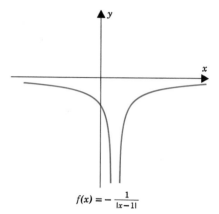

$$f(x) = -\frac{1}{|x-1|}$$

Figure A.4.4.2

SECTION 4.5

1. (a) $(f + g)(x) = 2x^2$

   (b) $(f - g)(x) = -2a^2$

   (c) $(fg)(x) = x^4 - a^4$

   (d) $\frac{f}{g}(x) = \frac{x^2 - a^2}{x^2 + a^2}$

3. (a) $(6f + 3g)(x) = 9\sqrt{x} \quad (x > 0)$

   (b) $(fg)(x) = x - \frac{2}{x} + 1 \quad (x > 0)$

   (c) $\frac{f}{g}(x) = \frac{x - 1}{x + 2} \quad (x > 0)$

5. (a) $(f + g)(x) = \begin{cases} 1 - x, & x \leq 1 \\ 2x - 1, & 1 < x < 2 \\ 2x - 2, & x \geq 2 \end{cases}$

   (b) $(f - g)(x) = \begin{cases} 1 - x, & x \leq 1 \\ 2x - 1, & 1 < x < 2 \\ 2x, & x \geq 2 \end{cases}$

   (c) $(fg)(x) = \begin{cases} 0, & x < 2 \\ 1 - 2x, & x \geq 2 \end{cases}$

7. Figure A.4.5.1   9. Figure A.4.5.2
11. Figure A.4.5.3   13. Figure A.4.5.4
15. even   17. odd

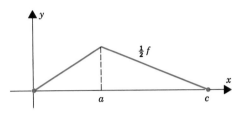

Figure A.4.5.1   Figure A.4.5.2

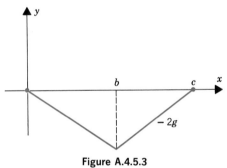

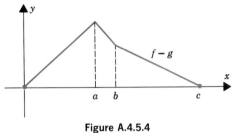

Figure A.4.5.3

Figure A.4.5.4

SECTION 4.6

1. $2x^2 + 5$ 3. $\sqrt{x^2 + 5}$ 5. x with $x \neq 0$

7. $\dfrac{1}{x} - 1$ 9. $\dfrac{1}{x^2 + 1}$ 11. $4x^2 - 4$

13. $(x^4 - 1)^2$ 15. $2x^2 + 1$

17. (a) $ax + b$ (c) $\dfrac{1}{x_3}$ 18. (b) a^2x^2

SECTION 4.7

1. 25   3. $\frac{1}{32}$ 5. 144
7. 18   9. 8   11. $\frac{3}{4}$
13. 46   15. $4a^2b^3$ 17. ab
19. $\dfrac{1}{xy}$ 21. $a - 1$ 23. $\dfrac{a^2}{5a - 3}$

25. $\dfrac{uv}{v - u}$

SECTION 4.8

1. $f^{-1}(x) = \frac{1}{5}(x - 3)$
5. not one-to-one
9. $f^{-1}(x) = [\frac{1}{3}(x - 1)]^{1/3}$
13. $f^{-1}(x) = (x - 2)^{1/3} - 1$
17. $f^{-1}(x) = \frac{1}{3}(2 - x^{1/3})$
21. not one-to-one
25. $f = (f^{-1})^{-1}$

3. $f^{-1}(x) = \frac{1}{4}(x + 7)$
7. $f^{-1}(x) = (x - 1)^{1/5}$
11. $f^{-1}(x) = 1 - x^{1/3}$
15. $f^{-1}(x) = x^{5/3}$
19. $f^{-1}(x) = 1/x$
23. $f^{-1}(x) = (1/x - 1)^{1/3}$
27. Figure A.4.8.1

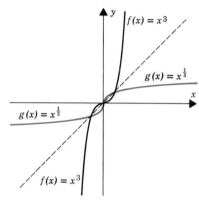

Figure A.4.8.1

SECTION 4.9

1. (a) yes (b) no (c) no (d) yes
2. (a) $(3P - 2Q)(x) = 3x^5 - 4x^2 - 3x + 14$
  $(PQ)(x) = 2x^7 - x^5 - 2x^3 + 8x^2 + x - 4$
3. (a) 1 (c) 8 (e) 5
4. (a) yes (d) yes
5. (a) $P(x) = (x - 1)(x^2 + 5x + 4) = x^3 + 4x^2 - x - 4$
  (c) $P(x) = (x + 2)(x + 1)x^2 = x^4 + 3x^3 + 2x^2$
  (e) $P(x) = x(x^2 - 3)(x + \frac{3}{2}) = x^4 + \frac{3}{2}x^3 - 3x^2 - \frac{9}{2}x$
6. (a) 28 (c) −79

SECTION 4.10

1. (a) 1, 2, 3 are all simple zeros
  (c) $-\frac{5}{2}$ has multiplicity 4
  (e) −1 and 1 have multiplicity 2
  (g) −2 is a simple zero; −1 and 1 have multiplicity 2; 0 has multiplicity 4

2. (a) 1 has multiplicity 3
    (c) 1 is a simple zero, 0 has multiplicity 3

SECTION 4.11

1. $27^{\sqrt{2}}$ 3. $\frac{1}{3}$ 4. $\frac{1}{3}$ 6. 3   7. -1
9. -3 10. 6   12. $-\frac{2}{5}$ 13. $-\frac{8}{5}$ 15. $-\frac{3}{4}$
16. $\log_{10} 15 = \log_{10} 3 + \log_{10} 5 \cong 0.4771 + 0.6990 = 1.1761$
18. $\log_{10} 0.016 = \log_{10} [(10^{-3})(4^2)] = -3 + 2 \log_{10} 4$
$$\cong -3 + 2(0.6021)$$
$$= -1.7958$$
19. $\log_{10} \sqrt[3]{25} = \log_{10} 5^{2/3} = \frac{2}{3} \log_{10} 5 \cong \frac{2}{3}(0.6990) = 0.4660$
21. $\log_{10} 14.4 = \log_{10} [(10^{-1})(3^2)(4^2)] = -1 + 2 \log_{10} 3 + 2 \log_{10} 4$
$$\cong -1 + 2(0.4771) + 2(0.6021)$$
$$= 1.1584$$
22. $\log_{10} 0.9 = \log_{10} \frac{9}{10} = \log_{10} 9 - \log_{10} 10 = 0.9542 - 1 \cong 0.0458$
24. $\log_{10} \sqrt[4]{36} = \frac{1}{4} \log_{10} 6^2 = \frac{1}{2} \log_{10} 6 \cong \frac{1}{2}(0.7782) = 0.3891$
25. $\log_{10} \sqrt{110} = \frac{1}{2} \log_{10} 110 = \frac{1}{2}(2 + \log_{10} 1.1) \cong \frac{1}{2}(2 + 0.0414) = 1.0207$
     $\sqrt{110} \cong 10.5$
27. $\log_{10} t = \frac{1}{5}\log_{10} 1620 = \frac{1}{5}(3 + \log_{10} 1.62) \cong \frac{1}{5}(3 + 0.2095) = 0.6419$
     $t \cong 4.38$
29. $\log_{10} t = 2 \log_{10} 63.9 + 3 \log_{10} 7.2 - 3 \log_{10} 11.5$
$$\cong 2(1 + 0.8055) + 3(0.8573) - 3(1 + 0.0607)$$
$$= 2(1.8055) + 3(0.8573) - 4(1.0607)$$
$$\cong 1.9401 = 1 + 0.9401$$
     $t \cong 87.1$
31. Set
$$\log_B x = \alpha, \qquad \log_b x = \beta, \qquad \log_b B = \gamma$$
     so that
$$x = B^\alpha, \qquad x = b^\beta, \qquad B = b^\gamma. \quad \square$$

32. $\log_5 10 = \dfrac{\log_{10} 10}{\log_{10} 5} \cong \dfrac{1}{0.6990} \cong 1.4306$

34. $\log_5 100 = \dfrac{\log_{10} 100}{\log_{10} 5} \cong \dfrac{2}{0.6990} \cong 2.8612$

36. $x = 100, x = 1$ 38. $x = \sqrt[5]{20}$
40. Figure A.4.11.1
42. By reflection in the y-axis.
43. (a) Figure A.4.11.2

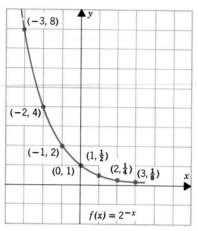

Figure A.4.11.1

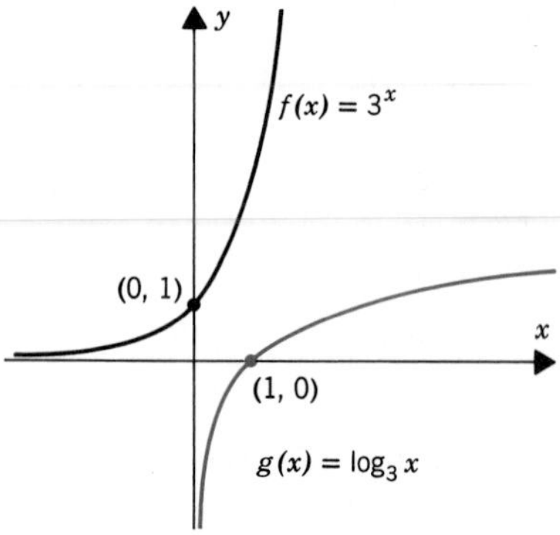

Figure A.4.11.2

SECTION 4.12

1. only (a)   2. (a) and (b)   3. (b) and (d)
4. -1 and 1 are simple zeros
5. 1 is a simple zero; 0 has multiplicity 2
6. -2 and $\frac{5}{2}$ are simple zeros; 0 has multiplicity 3
7. -1 and 1 are simple zeros; 0 and $\pm\sqrt{2}$ have multiplicity 2
8. (a) $f(x - 2) = (x - 4)^2$ (b) $f(x + 2) = x^2$

   (c) $g(x^2) = x^2 + \dfrac{1}{x^2}$ (d) $g\left(\dfrac{1}{x}\right) = \dfrac{1}{x} + x = g(x)$

   (e) $f(g(x)) = \left(x + \dfrac{1}{x} - 2\right)^2$ (f) $f(f(x)) = [(x - 2)^2 - 2]^2$

9. domain $[1, \infty)$; range $[0, \infty)$
10. domain $(-\infty, 0) \cup (0, \infty)$; range $\{-1, 1\}$
11. domain $(-\infty, 0) \cup (0, \infty)$; range $(-\infty, 1)$
12. domain $[0, \infty)$; range $(0, \frac{1}{5}]$
13. $f(f^{-1}(x)) = x$, $1 - [f^{-1}(x)]^3 = x$, $[f^{-1}(x)]^3 = 1 - x$, $f^{-1}(x) = (1 - x)^{1/3}$
14. $f(f^{-1}(x)) = x$, $3[f^{-1}(x)]^{1/5} = x$, $f^{-1}(x) = \left(\dfrac{x}{3}\right)^5$
15. $f(f^{-1}(x)) = x$, $\dfrac{2}{f^{-1}(x)} - 1 = x$, $f^{-1}(x) = \dfrac{2}{1 + x}$
16. $(f \circ g)(-x) = f(g(-x)) = f(g(x)) = (f \circ g)(x)$. □
17. If f and g are bounded on a common domain D, then there are positive numbers M and N such that

$$-M \le f(x) \le M \qquad \text{for all } x \in D$$

and
$$-N \leq g(x) \leq N \qquad \text{for all } x \in D.$$

It follows that
$$-(M + N) \leq f(x) + g(x) \leq M + N \qquad \text{for all } x \in D$$

and
$$-MN \leq f(x)g(x) \leq MN \qquad \text{for all } x \in D. \quad \square$$

18. $\dfrac{f(x + h) - f(x)}{h}$

19. Figure A.4.12.1

20. Figure A.4.12.2

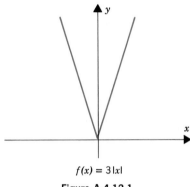

$f(x) = 3|x|$

Figure A.4.12.1

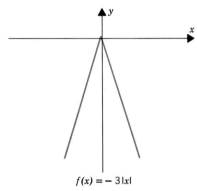

$f(x) = -3|x|$

Figure A.4.12.2

21. Figure A.4.12.3

22. Figure A.4.12.4

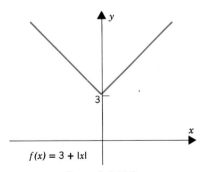

$f(x) = 3 + |x|$

Figure A.4.12.3

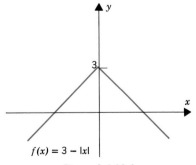

$f(x) = 3 - |x|$

Figure A.4.12.4

23. Figure A.4.12.5
25. Figure A.4.12.7
27. Figure A.4.12.9

24. Figure A.4.12.6
26. Figure A.4.12.8
28. Figure A.4.12.10

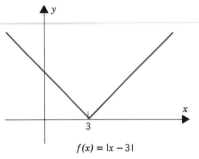

$f(x) = |x - 3|$

Figure A.4.12.5

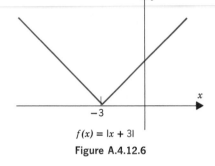

$f(x) = |x + 3|$

Figure A.4.12.6

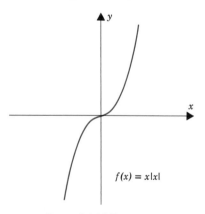

$f(x) = x|x|$

Figure A.4.12.7

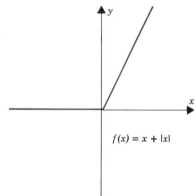

$f(x) = x + |x|$

Figure A.4.12.8

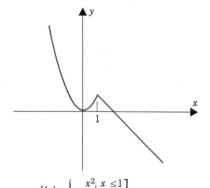

$f(x) = \begin{cases} x^2, & x \leq 1 \\ 2 - x, & x > 1 \end{cases}$

Figure A.4.12.9

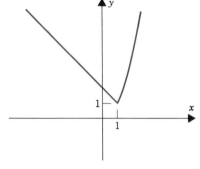

$f(x) = \begin{cases} 2 - x, & x \leq 1 \\ x^2, & x > 1 \end{cases}$

Figure A.4.12.10

29. Figure A.4.12.11
31. Figure A.4.12.13
33. Figure A.4.12.15

30. Figure A.4.12.12
32. Figure A.4.12.14
34. Figure A.4.12.16

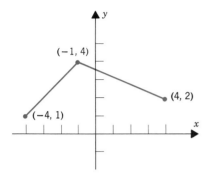

Figure A.4.12.11

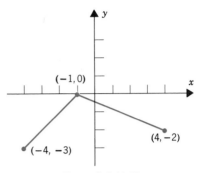

Figure A.4.12.12

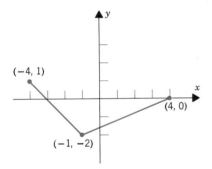

Figure A.4.12.13

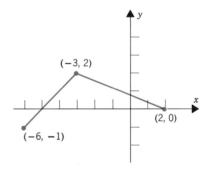

Figure A.4.12.14

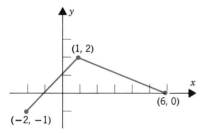

Figure A.4.12.15

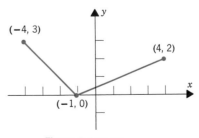

Figure A.4.12.16

35. Figure A.4.12.17   36. Figure A.4.12.18
37. Figure A.4.12.19   38. Figure A.4.12.20

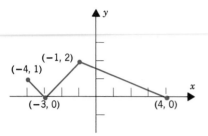

Figure A.4.12.17

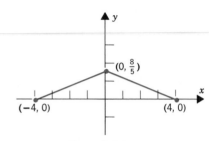

Figure A.4.12.18

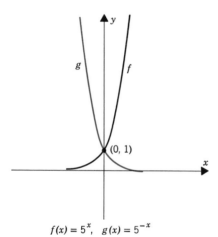

$f(x) = 5^x, \quad g(x) = 5^{-x}$

Figure A.4.12.19

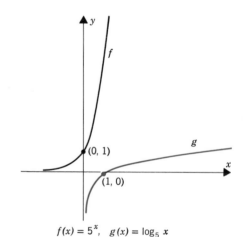

$f(x) = 5^x, \quad g(x) = \log_5 x$

Figure A.4.12.20

SECTION 5.1

1. (a) $\frac{3}{2}\pi$ radians   (c) $\frac{5}{3}\pi$ radians
2. (a) $540°$ (d) $292.5°$
4. length of the arc $= \frac{1}{3}\pi$, area of the sector $= \frac{1}{3}\pi$
6. $r_1 = \sqrt{3}r_2$

SECTION 5.2

1. $\frac{1}{4}\pi \pm 2n\pi$ 3. $\pi \pm 2n\pi$ 5. $-\frac{7}{4}\pi - 2n\pi$
7. $-\pi - 2n\pi$ 9. -1 11. -1

13. -1 15. 0   17. $\frac{1}{2}$

19. $\frac{1}{2}\sqrt{3}$ 21. $\frac{1}{2}\sqrt{2}$ 23. $\frac{1}{2}$

25. 0   27. $\frac{1}{2}\sqrt{3}$

29. All rotations $t \pm 2n\pi$ take the radius vector to the same position as the rotation t.

30. -1 32. $\frac{1}{2}\sqrt{2}$ 34. ± 1

36. $\pm\frac{1}{2}\sqrt{2}$ 38. $\pm\frac{3}{5}$

SECTION 5.3

1. $-\frac{1}{2}\sqrt{3}$ 3. $-\frac{1}{2}\sqrt{3}$ 4. $-\frac{1}{2}$

6. $-\frac{1}{2}\sqrt{3}$ 7. $-\frac{1}{2}\sqrt{2}$ 9. $\frac{1}{4}\sqrt{2}(1-\sqrt{3})$

10. $\frac{1}{4}\sqrt{2}(1+\sqrt{3})$ 12. $\frac{1}{2}\sqrt{2-\sqrt{2}}$ 13. $\frac{1}{2}\sqrt{2-\sqrt{2}}$

15. $\frac{1}{2}\sqrt{2}$ 16. $\frac{1}{2}\sqrt{2}$ 18. $\frac{1}{4}\sqrt{2}(1-\sqrt{3})$

21. (a) $\frac{7}{25}$ (c) $\frac{1}{10}\sqrt{10}$

22. (b) $-\frac{4}{25}\sqrt{6}$ (d) $\frac{1}{10}\sqrt{50-20\sqrt{6}}$

23. In general

$$(*)\qquad \cos^2 \tfrac{1}{2}t + \sin^2 \tfrac{1}{2}t = 1.$$

By the double angle formula

$$(**)\qquad \cos^2 \tfrac{1}{2}t - \sin^2 \tfrac{1}{2}t = \cos t.$$

If we subtract $(**)$ from $(*)$,

$$2 \sin^2 \tfrac{1}{2}t = 1 - \cos t$$

and thus

$$\sin^2 \tfrac{1}{2}t = \tfrac{1}{2}(1 - \cos t). \qquad \square$$

SECTION 5.4

1. $y = 5 \sin (t + \pi)$; amplitude 5, period 2π, y-intercept 0

3. $y = \frac{1}{3} \sin (t + \frac{3}{2}\pi)$; amplitude $\frac{1}{3}$, period 2π, y-intercept $-\frac{1}{3}$

5. $y = 2 \sin (3t + \frac{1}{2}\pi)$; amplitude 2, period $\frac{2}{3}\pi$, y-intercept 2

7. $y = 2 \sin 3t$; amplitude 2, period $\frac{2}{3}\pi$, y-intercept 0

9. π units to the right

11. $\frac{3}{2}\pi$ units to the left

13. $\frac{3}{4}\pi$ units to the right

15. amplitude 1, period π, y-intercept 0; Figure A.5.4.1

16. amplitude $\frac{1}{2}$, period π, y-intercept 0; Figure A.5.4.2

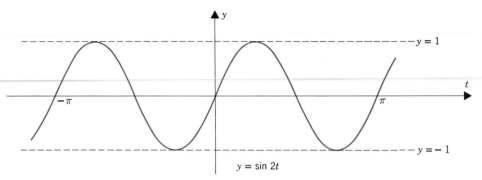

$$y = \sin 2t$$

Figure A.5.4.1

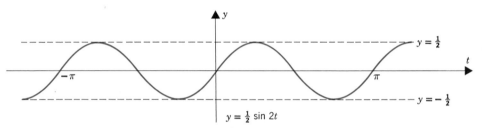

$$y = \tfrac{1}{2} \sin 2t$$

Figure A.5.4.2

19. amplitude 3, period π, y-intercept 0; Figure A.5.4.3

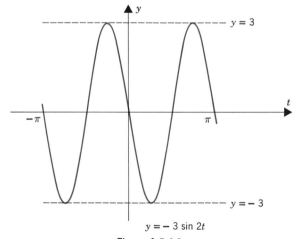

$$y = -3 \sin 2t$$

Figure A.5.4.3

20. amplitude 1, period π, y-intercept 1; Figure A.5.4.4

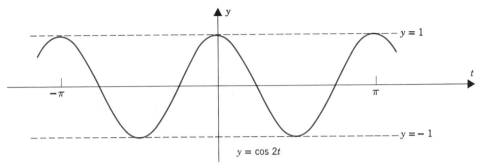

Figure A.5.4.4

23. amplitude 2, period $\frac{8}{3}\pi$, y-intercept 0; Figure A.5.4.5

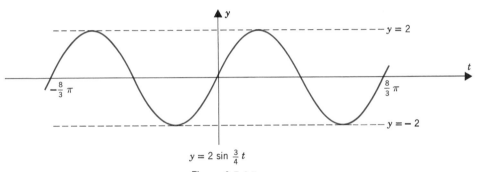

Figure A.5.4.5

24. amplitude 2, period $\frac{8}{3}\pi$, y-intercept 0; the graph is the graph of Figure A.5.4.5 displaced $\frac{4}{3}\pi$ units to the left

SECTION 5.5

1. 1   3. 0   4. $\frac{1}{3}\sqrt{3}$ 6. 1   7. $-\frac{1}{3}\sqrt{3}$

9. $-\sqrt{3}$ 10. $\sqrt{3}$ 12. $\frac{1}{3}\sqrt{3}$ 13. $-\sqrt{3}$ 15. $\frac{1}{3}\sqrt{3}$

16. $\frac{2}{3}\sqrt{3}$ 18. $\sqrt{2}$ 19. -1 21. $-\sqrt{2}$ 22. 2

24. $\sqrt{2}$ 25. 1   27. $-\frac{2}{3}\sqrt{3}$ 28. $\frac{1}{2}\pi$ 30. $\frac{1}{3}\pi$

31. 3π 33. 4

34. $\tan(A + B) = \dfrac{\tan A + \tan B}{1 - \tan A \tan B} \cong \dfrac{2.5 + 1.2}{1 - (2.5)(1.2)} = -1.85$

36. $\tan 2A = \dfrac{2 \tan A}{1 - \tan^2 A} \cong \dfrac{2(2.5)}{1 - (2.5)^2} \cong -0.95$

37. $\tan (2A - B) = \dfrac{\tan 2A - \tan B}{1 + \tan 2A \tan B} \cong \dfrac{-0.95 - 1.2}{1 + (-0.95)(1.2)} \cong 15.36$

39. $\cot 2B = \dfrac{1}{\tan 2B} = \dfrac{1 - \tan^2 B}{2 \tan B} \cong \dfrac{1 - (1.2)^2}{2(1.2)} \cong 0.18$

40.   (a) $\cot (-t) = \dfrac{1}{\tan (-t)} = \dfrac{1}{- \tan t} = -\dfrac{1}{\tan t} = -\cot t.$ $\square$

   (c) $\operatorname{cosec} (-t) = \dfrac{1}{\sin (-t)} = \dfrac{1}{- \sin t} = -\dfrac{1}{\sin t} = -\operatorname{cosec} t.$ $\square$

41.   (b) $\sec (t + 2\pi) = \dfrac{1}{\cos (t + 2\pi)} = \dfrac{1}{\cos t} = \sec t.$ $\square$

42.   Figure A.5.5.1   44.   Figure A.5.5.2

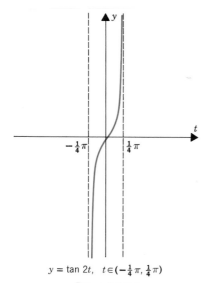

$y = \tan 2t, \quad t \in (-\tfrac{1}{4}\pi, \tfrac{1}{4}\pi)$

Figure A.5.5.1

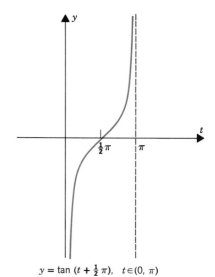

$y = \tan (t + \tfrac{1}{2}\pi), \quad t \in (0, \pi)$

Figure A.5.5.2

47. $\tan 2t = \dfrac{2 \tan t}{1 - \tan^2 t}$

48. $(\tan t + \sec t)(1 - \sin t) = \left(\dfrac{\sin t}{\cos t} + \dfrac{1}{\cos t} \right)(1 - \sin t)$

$$= \dfrac{1 - \sin^2 t}{\cos t} = \dfrac{\cos^2 t}{\cos t} = \cos t. \square$$

50. $(\tan t + \cot t) = \dfrac{\sin t}{\cos t} + \dfrac{\cos t}{\sin t}$

$\qquad = \dfrac{\sin^2 t + \cos^2 t}{\cos t \sin t} = \dfrac{1}{\cos t \sin t} = \sec t \csc t.$ $\square$

52. By Exercise 46

$$1 + \cot^2 t = \csc^2 t.$$

Thus

$$1 = \csc^2 t - \cot^2 t = (\csc t + \cot t)(\csc t - \cot t)$$

and

$$\dfrac{1}{\csc t \cot t} = \csc t - \cot t. \quad \square$$

SECTION 5.6

1. $\frac{1}{3}\pi$ 3. $-\frac{1}{6}\pi$ 5. $\frac{1}{4}\pi$ 7. $\frac{1}{2}\pi$ 9. $\frac{2}{3}\pi$ 11. $\frac{1}{2}$

13. $\frac{1}{2}\sqrt{3}$ 15. $\frac{1}{4}\pi$ 17. $-\frac{1}{4}\pi$ 19. $\frac{1}{4}\pi$ 21. 1.16   23. -0.46

25. 0.28
27. Figure A.5.6.1

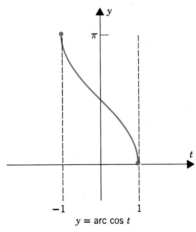

$y = \text{arc } \cos t$

Figure A.5.6.1

28. (b) odd   30. (a) $\sqrt{1 - t^2}$ (b) $\sqrt{1 - t^2}$
32. Figure A.5.6.2
33. Figure A.5.6.3
34. Figure A.5.6.4

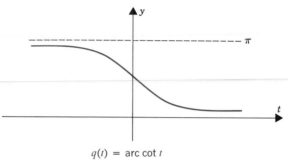

$q(t) = \text{arc cot } t$

Figure A.5.6.2

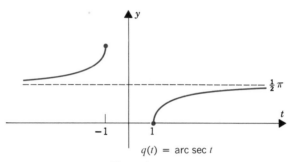

$q(t) = \text{arc sec } t$

Figure A.5.6.3

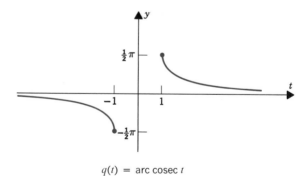

$q(t) = \text{arc cosec } t$

Figure A.5.6.4

SECTION 5.7

1. $a = c \sin A = 2 \sin 50° \cong 1.5320$
    $b = c \cos A = 2 \cos 50° \cong 1.2856$

3. $b = \dfrac{a}{\tan A} = \dfrac{100}{\tan 60°} \cong 57.7367$

$$c = \frac{a}{\sin A} = \frac{100}{\sin 60°} \cong 115.4734$$

5. $\sin A = \dfrac{a}{c} = \dfrac{1.44}{12} = 0.1200, \ A \cong 7°$

  $b = c \cos A \cong 12 \cos 7° \cong 11.9100$

7. $b = \dfrac{a}{\tan A} = \dfrac{4}{\tan 19°} \cong 11.6178$

  $c = \dfrac{a}{\sin A} = \dfrac{4}{\sin 19°} \cong 12.2850$

9. $d = 1000 \tan 65° \cong 2145$ (about 2145 ft)

11. $N = 2 \cos 10° \cong 1.9696$ (about 1.97 miles north)
  $E = 2 \sin 10° \cong 0.3472$ (about 0.35 miles east)

13. $d = \dfrac{7}{\tan 20°} \cong 19.2308$ (about 19.23 ft away)

15. moved about 6.77 ft closer to the wall; about 9.06 ft high

SECTION 5.8

1. $C = 60°, \ a = \dfrac{6 \sin 40°}{\sin 60°} \cong 4.4536, \ b = \dfrac{6 \sin 80°}{\sin 60°} \cong 6.8231$

3. $49 = 64 + 81 - 2(8)(9) \cos A, \ \cos A \cong 0.6666, \ A \cong 48°$
  $64 = 49 + 81 - 2(7)(9) \cos B, \ \cos B \cong 0.5238, \ B \cong 58°$
  $C \cong 180° - 48° - 58° = 74°$

5. $B = 108°, \ a = \dfrac{24 \sin 16°}{\sin 108°} \cong 6.9545, \ c = \dfrac{24 \sin 56°}{\sin 108°} \cong 20.9189$

7. $A = 70°, \ B = 70°$ (isosceles triangle), $c = \dfrac{10 \sin 40°}{\sin 70°} \cong 6.8405$

9. $b^2 = 4 + 16 - 2(2)(4) \cos 12° \cong 4.3504, \ b \cong 2.086$
  $\overset{\textstyle\curvearrowleft}{\text{Table 4}}$

  $\sin A = \dfrac{4 \sin 12°}{2.086} \cong 0.3987, \ A \cong 156.5°$

  $C \cong 180° - 12° - 156.5° = 11.5°$

11. $1.44 = 1.69 + 4 - 2(1.3)(2) \cos A, \ \cos A \cong 0.8173, \ A \cong 35°$
  $1.69 = 1.44 + 4 - 2(1.2)(2) \cos B, \ \cos B \cong 0.7813, \ B \cong 39°$
  $C \cong 180° - 35° - 39° = 106°$

13. $C = 40°, \ a = \dfrac{5 \sin 79°}{\sin 40°} \cong 7.6353, \ b = \dfrac{5 \sin 61°}{\sin 40°} \cong 6.803$

15. $\sin(180° - \theta) = \sin 180° \cos \theta - \cos 180° \sin \theta = (0) \cos \theta - (-1) \sin \theta = \sin \theta$
  $\cos(180° - \theta) = \cos 180° \cos \theta + \sin 180° \sin \theta = (-1) \cos \theta + (0) \sin \theta = -\cos \theta$

16. $d^2 = 25 + 4900 - 2(5)(70) \cos 45° \cong 4430.03$, $d \cong 66.56$
   about 66.56 miles

18. $h \cong 2.7788$ (about 2.78)   20. (c) $\frac{3}{4}\sqrt{15}$

SECTION 5.9

1. $45°$ 3. $90°$ 5. about $127°$
7. $y = \frac{1}{3}\sqrt{3}x + 2$ 9. $y = x - 1$ 11. $y = -\frac{1}{3}\sqrt{3}x + \sqrt{3}$
13. about $22°$ 15. about $17°$ 17. $90°$, $45°$, $45°$
19. about $17°$, $131°$, $32°$

SECTION 5.10

1. (a) $\frac{3}{4}\pi$ (b) $\frac{5}{6}\pi$ (c) $\frac{3}{2}\pi$ (d) $\frac{11}{6}\pi$
2. (a) $30°$ (b) $33.75°$ (c) about $114.6°$ (d) $112.50°$
3. arc length 5π, area 10π
4. $\theta = \frac{1}{2}\sqrt{2}\pi$, area $= \sqrt{2}\pi$
5. $\pm\sqrt{15}$
6. (a) $-\frac{1}{2}\sqrt{2}$ (b) $-\frac{1}{2}\sqrt{2}$ (c) $\frac{1}{2}\sqrt{3}$ (d) $\frac{1}{3}\sqrt{3}$
   (e) $-\sqrt{2}$ (f) 1 (g) $-\sqrt{3}$ (h) 2
7. (a) $\frac{1}{2}\sqrt{3}$ (b) π (c) $\frac{1}{2}\sqrt{3}$
   (d) $\frac{1}{3}\pi$ (e) -1 (f) $\frac{3}{5}$
   (g) $-\frac{1}{3}\pi$ (h) $\frac{4}{5}$ (i) $\frac{4}{3}$
8. (a) $-\frac{17}{25}$ (b) $-\frac{4}{25}\sqrt{21}$ (c) $-\frac{1}{10}\sqrt{70}$ (d) $\frac{1}{10}\sqrt{30}$
9. (a) amplitude $\frac{1}{3}$, period π, y-intercept $\frac{1}{3}$; Figure A.5.10.1

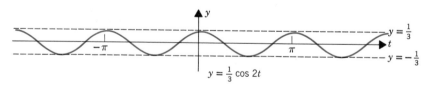

$$y = \frac{1}{3} \cos 2t$$

Figure A.5.10.1

(b) amplitude 2, period π, y-intercept -2; Figure A.5.10.2
(c) amplitude $\frac{1}{2}$, period $\frac{1}{2}\pi$, y-intercept $\frac{1}{2}$; Figure A.5.10.3
(d) amplitude $\frac{1}{4}$, period 4π, y-intercept $\frac{1}{8}\sqrt{2}$; Figure A.5.10.4

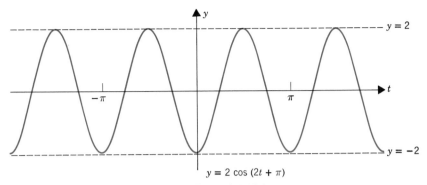

$y = 2 \cos (2t + \pi)$

Figure A.5.10.2

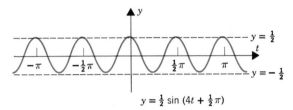

$y = \frac{1}{2} \sin (4t + \frac{1}{2}\pi)$

Figure A.5.10.3

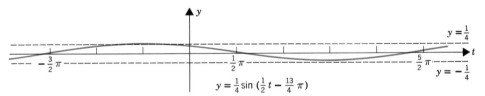

$y = \frac{1}{4} \sin (\frac{1}{2}t - \frac{13}{4}\pi)$

Figure A.5.10.4

10. $\sin A = 0.6000, A \cong 37°$
     $B \cong 90° - 37° = 53°$
     $b \cong 5 \cos 37° \cong 3.9930$

11. $A = 90° - 80° = 10°$
     $a = 4 \sin 10° \cong 0.6944$
     $b = 4 \cos 10° \cong 3.9392$

12. $\cos A = \frac{10}{15} \cong 0.6667, A \cong 48°$
     $B \cong 90° - 48° = 42°$

     $a \cong 15 \sin 48° \cong 11.1465$

13. $B = 90° - 63° = 27°$
     $b = \dfrac{8.91}{\tan 63°} \cong 4.5390$

     $c = \dfrac{8.91}{\sin 63°} \cong 10$

14. $B = 90° - 32° = 58°$
     $a = \tan 32° \cong 0.6249$

     $c = \dfrac{1}{\cos 32°} \cong 1.1792$

15. $\tan A = \frac{15}{25} = 0.6000, A \cong 31°$
     $B \cong 90° - 31° = 59°$

     $c = \dfrac{15}{\sin 31°} \cong 29.1262$

16. $A = 180° - 80° - 70° = 30°$ 17. $c^2 = 4 + 1 - 2(2)(1) \cos 60° \cong 3$

$b = \dfrac{5 \sin 80°}{\sin 30°} \cong 9.8480$ $c = \sqrt{3} \cong 1.732$

$c = \dfrac{5 \sin 70°}{\sin 30°} \cong 9.3970$ $A = 90°,\ B = 30°$

18. $C = 180° - 30° - 75° = 75°$ (isosceles triangle)

$b = 10$

$a = \dfrac{10 \sin 30°}{\sin 75°} \cong 5.1765$

19. $b^2 = 25 + 1 - 2(5)(1) \cos 63° \cong 21.46,\ b \cong 4.63$
$25 \cong 21.46 + 1 - 2(4.63)(1) \cos A,\ \cos A \cong -0.2743,\ A \cong 106°$
$C \cong 180° - 106° - 63° = 11°$

20. $100 = 121 + 144 - 2(11)(12) \cos A,\ \cos A \cong 0.6250,\ A \cong 51°$
$121 = 100 + 144 - 2(10)(12) \cos B,\ \cos B \cong 0.5125,\ B \cong 59°$
$C \cong 180° - 51° - 59° = 70°$

21. $B = 180° - 40° - 60° = 80°$

$a = \dfrac{12.31 \sin 40°}{\sin 80°} \cong 8.0350$

$c = \dfrac{12.31 \sin 60°}{\sin 80°} \cong 10.8250$

22. $a^2 = 9 + 4 - 2(3)(2) \cos 40° \cong 3.81,\ a \cong 1.95$
$9 \cong 3.81 + 4 - 2(1.95)(2) \cos B,\ \cos B \cong -0.1526,\ B \cong 99°$
$C \cong 180° - 40° - 99° = 41°$

23. $4 = 16 + 25 - 2(4)(5) \cos A,\ \cos A = 0.9250,\ A \cong 22°$
$16 = 4 + 25 - 2(2)(5) \cos B,\ \cos B = 0.6500,\ B \cong 49°$
$C \cong 180° - 22° - 49° = 109°$

24. $S = 4 \sin 6° \cong 0.4180$ (about 0.42 miles south)
$E = 4 \cos 6° \cong 3.9780$ (about 3.98 miles east)

25. (a) about 6.25 ft   (b) about 5.72 ft   26. $\frac{15}{2}\sqrt{3} \cong 12.99$

27. about 0.65 miles high   28. $40°,\ 53°,\ 87°$

29. about 0.98 miles high   30.   about 49° wider

31. In view of (5.6.9) we can show that

$$\tfrac{1}{2}\pi - \text{arc sin } t = \text{arc cos } t$$

by showing that

$$\cos\left(\tfrac{1}{2}\pi - \text{arc sin } t\right) = t \quad\text{and}\quad \tfrac{1}{2}\pi - \text{arc sin } t \in [0, \pi].$$

These assertions are verified below:
(i) $\cos\left(\tfrac{1}{2}\pi - \text{arc sin } t\right) = \sin(\text{arc sin } t) = t$
$\qquad\qquad\qquad \cos\left(\tfrac{1}{2}\pi - u\right) = \sin u$

(ii) $-\frac{1}{2}\pi \le \arcsin t \le \frac{1}{2}\pi,$

   $-\frac{1}{2}\pi \le -\arcsin t \le \frac{1}{2}\pi,$

   $0 \le \frac{1}{2}\pi - \arcsin t \le \pi.$ ☐

SECTION 6.1

1. Choose, for instance, the plane that passes through the vertex of the cone and is perpendicular to its axis.
3. Choose any plane that contains the axis of the cone.

SECTION 6.2

1. (a) $y^2 = 8x$; Figure A.6.2.1   (c) $x^2 = 8y$; Figure A.6.2.2
    (e) $y^2 = 2x$; Figure A.6.2.3   (g) $x^2 = 2y$; Figure A.6.2.4

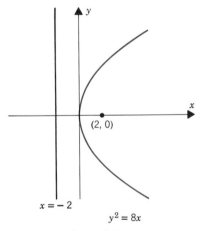

$y^2 = 8x$

Figure A.6.2.1

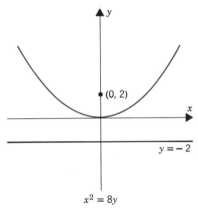

$x^2 = 8y$

Figure A.6.2.2

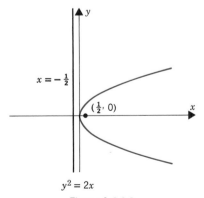

$y^2 = 2x$

Figure A.6.2.3

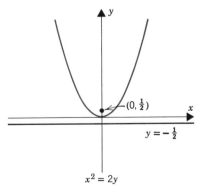

$x^2 = 2y$

Figure A.6.2.4

2. (a) focus $(\frac{1}{2}, 0)$, directrix $x = -\frac{1}{2}$, axis $y = 0$; Figure A.6.2.3

(c) focus $(0, -\frac{3}{4})$, directrix $y = \frac{3}{4}$, axis $x = 0$; Figure A.6.2.5

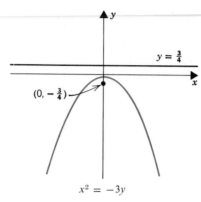

$x^2 = -3y$

Figure A.6.2.5

3. (a) $(0, 0)$, $(\frac{1}{4}, 1)$

4. (a) $(\sqrt{3}, 1)$, $(-\sqrt{3}, 1)$

5. $(0, 0)$, $(-(50)^{1/3}, \frac{1}{5}(50)^{2/3})$

SECTION 6.3

1. (a) $(x - 3)^2 + y^2 = 4$ (d) $(x + 3)^2 + (y - 5)^2 = 4$

2. (b) $(x + 5)^2 + (y - 10)^2 = 4$ (d) $(x - 10)^2 + (y - 5)^2 = 4$

3. (a) $(y - 6)^2 = -6(x + 4)$ (c) $(y - 1)^2 = 6(x - 2)$

4. (a) vertex $(1, 0)$, focus $(\frac{3}{2}, 0)$, axis $y = 0$, directrix $x = \frac{1}{2}$; Figure A.6.3.1.

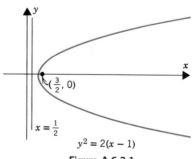

$y^2 = 2(x - 1)$

Figure A.6.3.1

(c) vertex $(-2, \frac{3}{2})$, focus $(-2, \frac{7}{2})$, axis $x = -2$, directrix $y = -\frac{1}{2}$; Figure A.6.3.2

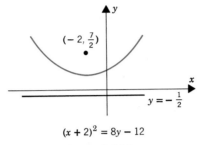

$$(x + 2)^2 = 8y - 12$$

Figure A.6.3.2

(e) vertex $(-\frac{1}{2}, \frac{3}{4})$, focus $(-\frac{1}{2}, 1)$, axis $x = -\frac{1}{2}$, directrix $y = \frac{1}{2}$; Figure A.6.3.3

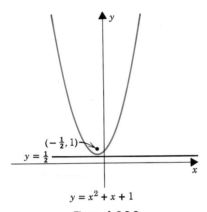

$$y = x^2 + x + 1$$

Figure A.6.3.3

5.   (b) $(x - 1)^2 = 4y$; Figure A.6.3.4

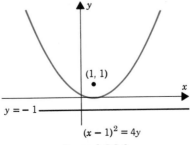

$$(x - 1)^2 = 4y$$

Figure A.6.3.4

(d) $(y + 2)^2 = -6(x - \frac{7}{2})$; Figure A.6.3.5

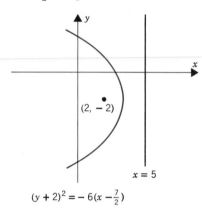

$$(y + 2)^2 = -6(x - \tfrac{7}{2})$$

Figure A.6.3.5

6. $(1, 1)$, $(1, -1)$ 8. $(0, -5)$, $(3, 4)$, $(-3, 4)$

9. vertex $\left(-\dfrac{B}{2A}, \dfrac{4AC - B^2}{4A}\right)$, focus $\left(-\dfrac{B}{2A}, \dfrac{4AC - B^2 + 1}{4A}\right)$,

    directrix $y = \dfrac{4AC - B^2 - 1}{4A}$

12. $144x = -5y^2 + 10y - 149$

SECTION 6.4

1. (a) $(0, 0)$ (b) foci $(\pm \sqrt{5}, 0)$ (c) major axis 6
    (d) minor axis 4; Figure A.6.4.1

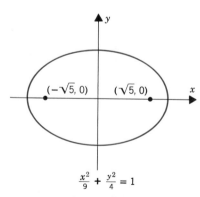

$$\frac{x^2}{9} + \frac{y^2}{4} = 1$$

Figure A.6.4.1

3. (a) $(0, 0)$ (b) foci $(0, \pm\sqrt{2})$ (c) major axis $2\sqrt{6}$
    (d) minor axis 4; Figure A.6.4.2
5. (a) $(0, 0)$ (b) foci $(\pm 1, 0)$ (c) major axis 4
    (d) minor axis $2\sqrt{3}$; Figure A.6.4.3

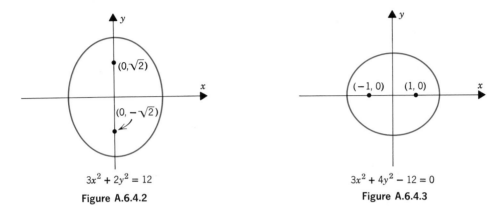

$$3x^2 + 2y^2 = 12$$
Figure A.6.4.2

$$3x^2 + 4y^2 - 12 = 0$$
Figure A.6.4.3

7. (a) $(1, 0)$ (b) foci $(1, \pm 4\sqrt{3})$ (c) major axis 16
    (d) minor axis 8; Figure A.6.4.4

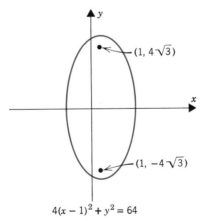

$$4(x - 1)^2 + y^2 = 64$$
Figure A.6.4.4

9. $\dfrac{x^2}{9} + \dfrac{y^2}{8} = 1$

11. $\dfrac{(x - 6)^2}{25} + \dfrac{(y - 1)^2}{16} = 1$

13. $\dfrac{(x - 1)^2}{21} + \dfrac{(y - 3)^2}{25} = 1$

15. $\dfrac{(x - 3)^2}{25} + \dfrac{(y + 1)^2}{9} = 1$

17. $(\pm 2\sqrt{2}, 2)$

SECTION 6.5

1. $\dfrac{x^2}{9} - \dfrac{y^2}{16} = 1$

3. $\dfrac{y^2}{25} - \dfrac{x^2}{144} = 1$

5. $\dfrac{x^2}{9} - \dfrac{(y-1)^2}{16} = 1$

7. $16y^2 - \dfrac{16}{15}(x+1)^2 = 1$

8. transverse axis 2,
   vertices $(\pm 1, 0)$,
   foci $(\pm \sqrt{2}, 0)$,
   asymptotes $y = \pm x$;
   Figure A.6.5.1

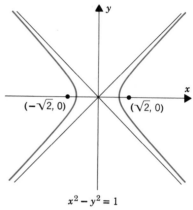

$x^2 - y^2 = 1$

Figure A.6.5.1

10. transverse axis 6,
    vertices $(\pm 3, 0)$,
    foci $(\pm 5, 0)$,
    asymptotes $y = \pm \frac{4}{3}x$;
    Figure A.6.5.2

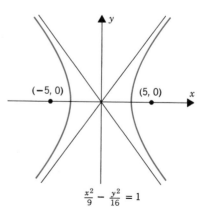

$\dfrac{x^2}{9} - \dfrac{y^2}{16} = 1$

Figure A.6.5.2

12. transverse axis 8,
     vertices $(0, \pm 4)$,
     foci $(0, \pm 5)$,
     asymptotes $y = \pm \frac{4}{3}x$;
     Figure A.6.5.3

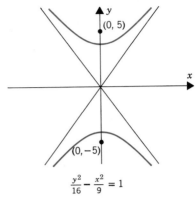

$$\frac{y^2}{16} - \frac{x^2}{9} = 1$$

Figure A.6.5.3

14. transverse axis 6,
     vertices $(4, 3)$ and $(-2, 3)$,
     foci $(6, 3)$ and $(-4, 3)$,
     asymptotes $y = \pm \frac{4}{3}(x - 1) + 3$;
     Figure A.6.5.4

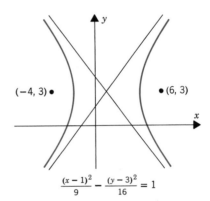

$$\frac{(x-1)^2}{9} - \frac{(y-3)^2}{16} = 1$$

Figure A.6.5.4

16. transverse axis 2,
     vertices $(0, 0)$ and $(-2, 0)$,
     foci $(1, 0)$ and $(-3, 0)$,
     asymptotes $y = \pm \sqrt{3}(x + 1)$;
     Figure A.6.5.5

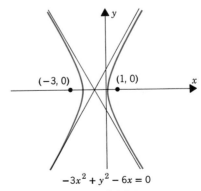

$$-3x^2 + y^2 - 6x = 0$$

Figure A.6.5.5

SECTION 6.6

1. $4y = x^2 - 8x + 12$

2. $24y = -x^2 + 8x + 128$

3. $\dfrac{x^2}{81} + \dfrac{y^2}{4} = 1$

4. $\dfrac{(x-2)^2}{4} + \dfrac{(y-1)^2}{81} = 1$

5. $4x^2 - \frac{4}{3}y^2 = 1$

6. $4y^2 - \frac{4}{3}x^2 = 1$

7. parabola with vertex at $(0, -1)$ and focus at the origin
8. ellipse with foci at $(0, \pm 1)$ and major axis $2\sqrt{3}$
9. hyperbola with foci at $(5, \pm 2\sqrt{5})$ and transverse axis 4
10. hyperbola with foci at $(1 \pm \sqrt{13}, -1)$ and transverse axis 4
11. ellipse with foci at $(-3 \pm \frac{1}{3}\sqrt{6}, 0)$ and major axis 2
12. parabola with vertex at $(5, 2)$ and focus at $(5, 4)$
13. parabola with vertex at $(\frac{3}{2}, -2)$ and focus at $(1, -2)$
14. ellipse with foci at $(1, 1 \pm \sqrt{5})$ and major axis 6
15. ellipse with foci at $(\pm \frac{4}{15}, -2)$ and major axis $\frac{2}{3}$
16. hyperbola with foci at $(-3, 7 \pm 2\sqrt{2})$ and transverse axis $2\sqrt{7}$
17. hyperbola with foci at $(-1 \pm \frac{6}{7}\sqrt{21}, -4)$ and transverse axis $\frac{6}{7}\sqrt{35}$
18. hyperbola with foci at $(-\sqrt{3} \pm \sqrt{5}, -\sqrt{3})$ and transverse axis $2\sqrt{3}$

19. the union of the parabola $x^2 = 4y$ and the ellipse $\dfrac{x^2}{9} + \dfrac{y^2}{4} = 1$

20. the union of the parabola $x^2 = 4y$ and the hyperbola $x^2 - 4y^2 = 1$

SECTION 7.1

9. $(0, 3)$ 11. $(1, 0)$ 13. $(-\frac{3}{2}, \frac{3}{2}\sqrt{3})$ 15. $(-2, 0)$

17. $[1, \frac{1}{2}\pi + 2n\pi], [-1, \frac{3}{2}\pi + 2n\pi]$ 19. $[3, \pi + 2n\pi], [-3, 2n\pi]$

21. $[2\sqrt{2}, \frac{7}{4}\pi + 2n\pi], [-2\sqrt{2}, \frac{3}{4}\pi + 2n\pi]$

23. $[8, \frac{1}{6}\pi + 2n\pi], [-8, \frac{7}{6}\pi + 2n\pi]$ 25. $[2, \frac{2}{3}\pi + 2n\pi], [-2, \frac{5}{3}\pi + 2n\pi]$

SECTION 7.2

1. (a) $[\frac{1}{2}, \frac{11}{6}\pi]$ (b) $[\frac{1}{2}, \frac{5}{6}\pi]$ (c) $[\frac{1}{2}, \frac{7}{6}\pi]$

3. (a) $[2, \frac{2}{3}\pi]$ (b) $[2, \frac{5}{3}\pi]$ (c) $[2, \frac{1}{3}\pi]$

5. symmetry about the x-axis
7. no symmetry about axes or the origin
8. symmetry about both axes and the origin
10. symmetry about both axes and the origin
11. $r^2 \sin 2\theta = 1$ 13. $r = 2$
15. $r = a(1 - \cos\theta)$ 17. the horizontal line $y = 4$
19. the line $y = \sqrt{3}x$ 20. the lines $y = \pm\sqrt{3}x$
22. the circle $x^2 + (y + 2)^2 = 4$

SECTION 7.3

1. circle of radius 2 centered at the pole
3. line through the pole with 30° inclination
5. spiral (Figure A.7.3.1)

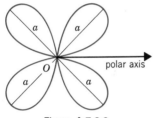

Figure A.7.3.1

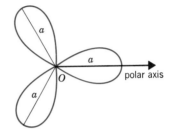

Figure A.7.3.2

7. circle of radius $\frac{1}{2}$ centered at $(\frac{1}{2}, 0)$
9. circle of radius 1 centered at $(1, 0)$ 11. Figure A.7.3.2
13. Figure A.7.3.3   15. Figure A.7.3.4

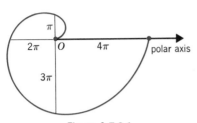

Figure A.7.3.3

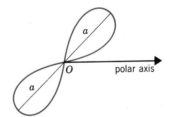

Figure A.7.3.4

SECTION 7.4

1. yes; $[1, \pi] = [-1, 0]$ and the pair $r = -1$, $\theta = 0$ satisfies the equation
3. yes; the pair $r = \frac{1}{2}$, $\theta = \frac{1}{2}\pi$ satisfies the equation
6. $(-1, 0)$, $(1, 0)$ 8. $(\frac{1}{2}[b^2 - 1], b)$
10. $(0, 0)$, $(\frac{1}{4}, \pm\frac{1}{4}\sqrt{3})$

SECTION 7.5

1. Let S be the set of integers for which the statement is true. Since $(1)(2) = 2$ is divisible by 2, $1 \in S$.

    Assume now that $k \in S$. This tells us that $k(k + 1)$ is divisible by 2 and therefore

$$(k + 1)(k + 2) = k(k + 1) + 2(k + 1)$$

is also divisible by 2. This places $k + 1 \in S$.

We have shown that

$$1 \in S \quad \text{and that} \quad k \in S \text{ implies } k + 1 \in S.$$

It follows that S contains all the positive integers. $\square$

4. Let S be the set of integers for which the statement is true. Since $1 = 1^2$, $1 \in S$. Assume now that $k \in S$. This tells us that

$$1 + 3 + 5 + \cdots + (2k - 1) = k^2.$$

It follows that

$$\begin{aligned}
1 + 3 + 5 + \cdots + (2k - 1) + [2(k + 1) - 1] &= k^2 + [2(k + 1) - 1] \\
&= k^2 + 2k + 1 \\
&= (k + 1)^2.
\end{aligned}$$

This places $k + 1 \in S$.

We have shown that

$$1 \in S \quad \text{and that} \quad k \in S \text{ implies } k + 1 \in S.$$

It follows that S contains all the positive integers. $\square$

7. Let S be the set of integers for which the statement is true. Since

$$3^{2(1)+1} + 2^{1+2} = 27 + 8 = 35$$

is divisible by 7, $1 \in S$.

Assume now that $k \in S$. This tells us that

$$3^{2k+1} + 2^{k+2} \text{ is divisible by 7.}$$

It follows that

$$\begin{aligned}
3^{2(k+1)+1} + 2^{(k+1)+2} &= 3^2 \cdot 3^{2k+1} + 2 \cdot 2^{k+2} \\
&= 9 \cdot 3^{2k+1} + 2 \cdot 2^{k+2} \\
&= 7 \cdot 3^{2k+1} + 2(3^{2k+1} + 2^{k+2})
\end{aligned}$$

is also divisible by 7. This places $k + 1 \in S$.

We have shown that

$$1 \in S \quad \text{and that} \quad k \in S \text{ implies } k + 1 \in S.$$

It follows that S contains all the positive integers. $\square$

SECTION 7.6

1. glb 0, lub 2
3. glb 0, no lub
5. glb -2, lub 2
7. glb $\frac{1}{2}$, lub 1
9. glb 0, lub $\frac{1}{2}$
11. glb 1, lub 10
13. glb 0.6, lub $\frac{2}{3}$
15. glb 0, lub $\frac{3}{4}$
18. (a) $0 = \text{glb } S$, $0 \leq (\frac{1}{11})^3 < 0 + 0.001$

SECTION 7.7

1. $x = 3, y = 1$

3. $x = 2, y = 3$

4. $x = \frac{1}{2}, y = 0$

6. $x = \frac{33}{31}, y = -\frac{17}{31}$

7. $x = 1, y = 0, z = 1$

9. $x = \frac{8}{7}, y = \frac{37}{7}, z = \frac{102}{21}$

SECTION 7.8

1. (a)–(h) Figure A.7.8.1.

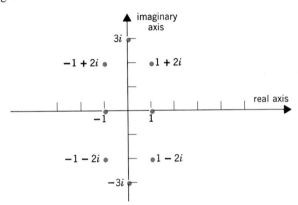

Figure A.7.8.1

2. $8 - i$

4. 1

6. $17 - 7i$

8. $18 - 38i$

10. $-1 + 5i$

12. $-i$

14. $-2 + 2i$

16. i

18. $x = \pm 3i$

20. $-\frac{1}{3} \pm \frac{1}{3}\sqrt{2}\,i$

22. $-\frac{1}{2} \pm \frac{1}{2}i$

26. See Figure A.7.8.2.

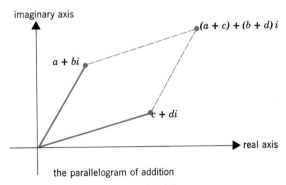

the parallelogram of addition

Figure A.7.8.2

29. $\frac{1}{2} - \frac{1}{2}i$

31. $\frac{8}{5} + \frac{1}{5}i$

32. $-\frac{5}{34} + \frac{3}{34}i$

34. $-\frac{11}{41} - \frac{58}{41}i$

INDEX